CAMBRIDGE LIBRARY COLLECTION

Books of enduring scholarly value

Darwin

Two hundred years after his birth and 150 years after the publication of 'On the Origin of Species', Charles Darwin and his theories are still the focus of worldwide attention. This series offers not only works by Darwin, but also the writings of his mentors in Cambridge and elsewhere, and a survey of the impassioned scientific, philosophical and theological debates sparked by his 'dangerous idea'.

The Philosophy of Zoology

John Fleming (1785–1857) was a minister of the Church of Scotland, but in his time at the University of Edinburgh he had also studied geology and zoology. In the tradition of the country parson who was also a talented and knowledgeable naturalist, he published his first works on the geology of the Shetland Islands while serving there as a minister. His subsequent works led to his being offered the chair of natural philosophy at the University of Aberdeen, and subsequently at the newly created chair of natural history at the Free Church College in Edinburgh. The two-volume Philosophy of Zoology was published in 1822, and the young Charles Darwin is recorded as borrowing it from the library of Edinburgh University in 1825/6. His intention in the book was to 'collect the truths of Zoology within a small compass, and to render them more intelligible, by a systematical arrangement'.

Cambridge University Press has long been a pioneer in the reissuing of out-of-print titles from its own backlist, producing digital reprints of books that are still sought after by scholars and students but could not be reprinted economically using traditional technology. The Cambridge Library Collection extends this activity to a wider range of books which are still of importance to researchers and professionals, either for the source material they contain, or as landmarks in the history of their academic discipline.

Drawing from the world-renowned collections in the Cambridge University Library, and guided by the advice of experts in each subject area, Cambridge University Press is using state-of-the-art scanning machines in its own Printing House to capture the content of each book selected for inclusion. The files are processed to give a consistently clear, crisp image, and the books finished to the high quality standard for which the Press is recognised around the world. The latest print-on-demand technology ensures that the books will remain available indefinitely, and that orders for single or multiple copies can quickly be supplied.

The Cambridge Library Collection will bring back to life books of enduring scholarly value (including out-of-copyright works originally issued by other publishers) across a wide range of disciplines in the humanities and social sciences and in science and technology.

The Philosophy of Zoology

Or a General View of the Structure, Functions, and Classification of Animals

VOLUME 1

JOHN FLEMING

CAMBRIDGE UNIVERSITY PRESS

Cambridge, New York, Melbourne, Madrid, Cape Town, Singapore,
São Paolo, Delhi, Dubai, Tokyo

Published in the United States of America by Cambridge University Press, New York

www.cambridge.org
Information on this title: www.cambridge.org/9781108001656

This edition first published 1822
This digitally printed version 2009

ISBN 978-1-108-00165-6 Paperback

THE
PHILOSOPHY
OF
ZOOLOGY.

THE PHILOSOPHY OF ZOOLOGY;

OR

A GENERAL VIEW OF THE STRUCTURE, FUNCTIONS, AND CLASSIFICATION OF ANIMALS.

By JOHN FLEMING, D. D.

MINISTER OF FLISK, FIFESHIRE,

FELLOW OF THE ROYAL SOCIETY OF EDINBURGH, OF THE WERNERIAN NATURAL HISTORY SOCIETY, &c.

IN TWO VOLUMES.

WITH ENGRAVINGS.

VOL. I.

EDINBURGH:

PRINTED FOR ARCHIBALD CONSTABLE & CO. EDINBURGH:

AND HURST, ROBINSON & CO. LONDON.

1822.

PREFACE.

In preparing this work for the public, the writer was chiefly influenced by a desire to collect the truths of Zoology within a small compass, and to render them more intelligible, by a systematical arrangement. He is not aware that there exists any work in the English language, in which the subject, in its different bearings, has been illustrated in a philosophical manner, or to which a student of Zoology could be referred, as a suitable introduction to the science. There are not wanting, it is true, many disquisitions of great value, on particular departments of the physiology and classification of Animals; for who can enumerate the names of TYSON, LISTER, WILLOUGHBY, RAY, ELLIS, HUNTER, PENNANT, MONRO, and MONTAGU, among the dead, and HOME, KIRBY, and LEACH, among the living zoologists of Britain, without regarding them as extensive benefactors of the science. But the writings of these naturalists, and others which have been noticed in the body of the work, are not only rare, but expensive; so that the task of investigating the facts which have been established, or the theories which have been proposed, can scarcely, in ordinary circumstances, be entered upon. The want, indeed, of such an introduction to the study of the Animal Kingdom, as should serve as an index to the doctrines on which the classification is founded, has frequently been the subject of regret, and may probably be considered as the origin of that indifference to the science which is but too apparent in this country. Botany and Mineralogy have been illustrated by a variety of introductory works, full of enlarged and philosophical views, and professorships have been instituted to accelerate the progress of these sciences: but Zoology has experienced no such fostering care. It has been abandoned to its fate, and suffered to languish under the pernicious influence of peculiar external circumstances.

Among those circumstances which have directly retarded the progress of Zoology in Britain, there is one which has been conspicuously hurtful,—the influence of the dogmas of the Linnean School. There have not been wanting naturalists in this country, who have regarded the twelfth edition of the "Systema Naturæ" as the standard of all excellence in this branch of Natural History, and who have considered the classes, orders, and genera therein established, as sufficient to embrace all the species on the globe. Every attempt to employ characters different from those made use of by LINNÆUS, has been stigmatized as presumptuous innovation; the establishment of a new genus has been condemned as an unnecessary burden imposed on the memory; the new species have been crowded into the established categories, though destitute of the prescribed claims of admission; and all that is valuable in the history of an animal, has been considered capable of being expressed in its trivial name and specific character. Though such has been the practice of the devoted admirers of LINNÆUS, it is not conformable to those principles which regulated the conduct of that enlightened naturalist himself. He examined with the greatest freedom the opinions of his predecessors, and did not suffer the methods which they had employed to regulate the construction of his own divisions. He exhibited the most convincing proofs of the necessity of frequent changes in the arrangement, to keep pace with the progress of science. Within the space of thirty years, his system passed through twelve editions, the greater number of which were revised by himself: these in succession, by the numerous alterations made in the characters, number and distribution of the genera, evinced the depreciated value of those which preceded, and predicted the temporary excellence of all that should follow. The blind adherence of British naturalists to the systematical arrangement of animals which LINNÆUS recommended, which led them to neglect the important services of LISTER and RAY, and reject the methods which these illustrious observers had proposed, and their hostility to every reformation, appear the more remarkable, when it is considered that his Mineralogical System was arrested in its progress by the feeble barriers which WOODWARD and DACOSTA had raised up. Perhaps a part of this influence may be traced to the purchase of the Linnean cabinet by its present illustrious possessor Sir JAMES EDWARD SMITH, and the interest in favour of its former owner which this circumstance could not fail to excite in this country. At all events, there is reason to rejoice that this influence, once so powerful, is on the decline; and to hope that the activity of the present cultivators of the science will atone for the last forty years of zoological listlessness.

In order to form a correct estimate of the zoological merits of LINNÆUS, the "Systema Naturæ" must be regarded as the index, merely, of the names of the different animals, not as the exposition of their history; and the "Amœnitates Academicæ" must be studied, as containing numerous examples of those efforts which can alone add dignity to this department of knowledge.

It will be fortunate for the interests of science, if, in rejecting what is obsolete in the system of LINNÆUS, zoologists do not, at the same time, undervalue that precision in method at which he aimed. This observation appears the more necessary, as there is now much declamation about the worthlessness of Artificial Systems, and the excellence of Natural Methods. But this excellence is more apparent than real. Many of those natural groups which are so much praised are ill defined, and it is even acknowledged by their admirers, that precise limits cannot be assigned to them. Hence it frequently happens, that the definition of the group is applicable to a few genera only, which are considered as its type, and does not embrace other genera which are regarded as belonging to it, but beginning to assume the characters of some of the other neighbouring groups. There is here the use of a method, where there is no precision, and a boasting that the plan of Nature is followed, when that plan is confessedly incomprehensible. Indeed, it often happens, that the admired natural method of one zoologist differs from the censured artificial method of another, merely in the circumstance that different systems of organs have been made choice of as the basis of the respective classifications. Unless zoologists, in the formation of their primary groups, endeavour to determine those characters which all the members possess in common, admitting only such marks into the definition, and practise the same method with all the subordinate divisions, the progress of the science will be unsteady, the student will be startled at its contradictions, and the revolutions in nomenclature become as frequent as the cultivators of the science are numerous.

The ridicule too often thrown out against some of the departments of Zoology, by persons who pretend to considerable intellectual acquirements, may have been prejudicial in its effects, by preventing many from entering upon the study, and by restraining the efforts of others. The Animal Kingdom is considered by many persons as furnishing a delightful field of rational enquiry, but they feel disposed to bestow all their praise on certain subjects of that kingdom which happen to be favourites, and they are ready to stigmatize the remainder as comparatively worthless. They would applaud the student inclined to investigate the instincts of the elephant, but would censure him, as engaging in degrading pursuits, were he detected in examining the habits of a spider, or the structure of a worm. Do such persons consider the wisdom of the plan of Providence as discoverable without an acquaintance with the relation of the particular parts, or a partial view as sufficient to enable them to comprehend the whole? To such judges of Nature, a sneer would perhaps be the most suitable reply; but, as they are numerous, it may be worth while to attempt to teach them sounder views. Much of their error may be traced to the importance which they attach to *size*, without perhaps reflecting, that, if this is to be considered as the best test of the dignity of an animal, the horse must be regarded as more excellent than his rider. Besides, were animals useful to man in proportion to their size, would the worm-like leech ever have been resorted to as a blood-letter, or the insect coccus employed as a dye? On no occasion can the *argumentum ad verecundiam* be employed with more effect, than when addressed to the despisers of the humbler tribes of animated beings. By whose power were the meanest creatures formed? By whose wisdom were all their organs arranged to fit them for the place they occupy? By whose will do they live? By whose bounty is their life sustained? Know, that HE, who, in the beginning, created the Heaven and the Earth, said, " Let the earth bring forth the living " creature after his kind, cattle, and creeping thing, " and beast of the earth after his kind: and it was " so." Is it, then, to be considered as a degrading employment for man to examine those creatures which were formed by GOD?

In collecting the materials of the present work, and preparing it for publication, the author experienced unavoidable difficulties, arising from his remote situation. It was sometimes not in his power, from his time being occupied by other important duties, to consult rare and costly works on Natural History; and the slow progress of his task, which has occupied him for more than three years, has prevented him from quoting some authors, whose publications are already known to the public. The Chapters, for example, on the Organs of Perception, and the Faculties of the Mind, were prepared for the press several months before the publication of the "Physiology of the Mind" by the late Dr BROWN. Had this not been the case, the author would have availed himself of several acute remarks of that discriminating philosopher. Should the reader detect the slightest coincidence of opinion in the two publications, it can only be attributed to the analytical operations having been performed on similar subjects. Another work has appeared more recently, which the author regrets was not before him in the whole course of his enquiries. He refers to Dr BARCLAY'S Treatise on Life and Organization. It should be perused with care by every student of Anatomy and Natural History, as an effectual preservative against the doctrines of Materialism, and deserves a place as well in the library of the Divine as in that of the Physiologist.

In the distribution of the subjects of the following work, it was considered more useful to classify the different organs of animals, and determine the functions which are executed, previous to the systematical arrangement of the species, than to unfold the peculiarities of the various organs as they occur in succession in the different classes. By this method, the student is made acquainted with the varieties of organization and function, and proceeds to the details of methodical distribution, with his mind prepared by general views for conducting the particular investigations. In the distribution of the various groups, the author, after the plan of the "Regne Animal" of M. Cuvier, the most valuable of modern systematical arrangements, begins with the perfect animals, and terminates with those which exhibit the most simple organisation. Suppose an opposite plan to be pursued, many difficulties must present themselves in the course of the arrangement. As the observer ascends in the scale, new organs develope themselves. These are at first so obscure, that he can neither unfold their structure nor guess at their functions, without being guided by his knowledge of the organs of the higher tribes: hence it happens, that the boasted analytical method becomes, in fact, synthetical; and the very terms which are employed to express the characters intimate a knowledge of the peculiarities of the higher divisions.

It was originally intended to have added to each genus in the Vertebral and Molluscous Tribes at least, a list of all the species which have been discovered as natives of the British Isles. But soon perceiving the impossibility of doing this, without exceeding greatly the prescribed limits, the writer relinquished the plan. He has, however, resolved to supply the defect in a separate publication, (for which he has been collecting materials during many years); in which work, he will confine himself to the determination of the specific characters and descriptions of British Animals. The propriety of attempting such a publication, must be acknowledged by every one acquainted with the present chaotic state of the British Fauna.

The Plates which have been added to the present work, consist of figures relating exclusively to British animals. They are not gaudy, but they are correct delineations from nature, for which he is indebted to the pencil of his wife.

With regard to the style, it may be proper to mention, that brevity and perspicuity have been chiefly aimed at. That there are several errors in the composition is readily acknowledged. Those which seem to injure the sense, have been taken notice of in the list of *corrigenda*.

Before concluding, the author embraces this opportunity of expressing his gratitude for the kind assistance he has experienced in prosecuting the work. He must particularly mention the name of Professor JAMESON, who has added so much to the reputation of the University of Edinburgh, as an eminent teacher of Natural History, and to whose exertions the public is indebted for that splendid collection of ANIMALS which adorns the Edinburgh Museum. The work, indeed, was begun in consequence of his recommendation, and he has contributed to its progress by many kind offices.

To PATRICK NEILL, Esq. the author has been under particular obligations; for many valuable hints and sound criticisms, dictated equally by the ties of friendship and attachment to the science

After all his exertions, the writer is aware that much more might be done, to give this treatise stronger claims to public favour, and to render it better deserving of the title which, for the sake of discrimination, he has, perhaps presumptuously, ventured to adopt; and he would only recommend it to the student of Zoology, until a more complete work shall appear, which would be perused by no one with greater eagerness than himself.

The analytical Table of Contents exhibits so fully the method which has been followed, and the subjects which have been treated of, as to supersede the necessity of an Alphabetical Index.

CONTENTS

TABLE

OF

CONTENTS.

VOLUME FIRST.

Page.

CHAP. VIII.

Muscular System.

CHAP. IX.

Nervous System.

CHAP. X.

Organs of Perception.

1. Sense of Touch.

CHAP. XI.

Faculties of the Mind.

CHAP. XIV.

Peculiar Secretions.

CHAP. XV.

Reproductive System.

VOLUME SECOND.

PART I.

ON THE CONDITION OF ANIMALS.

CHAP. I.

Duration of Animals.

CHAP. II.

Distribution of Animals.

CHAP. III.

PART II.

THE METHOD OF INVESTIGATING THE CHARACTERS OF ANIMALS.

CHAP. I.

CHAP. II.

PART III.

ON NOMENCLATURE.

PART IV.

CLASSIFICATION OF ANIMALS.

Divided into Vertebral and Invertebral.

I.

VERTEBRATA.

BIRDS.

I. Fissipedes.

TERRESTRIAL BIRDS.

II.

INVERTEBRATA.

I.

GANGLIATA.

MOLLUSCA.

Division I.—Mollusca Cephala.

Section I. NATANTIA.

Class I.—Cephalopoda.

Order I.—NAUTILIADÆ.

Class II.—Branchifera.

Order I. Branchiæ external.

Order II. Branchiæ Internal.

II.

ANNULOSA.

Order II.—Lepidoptera.

Order III.—Diptera.

Order IV.—Aptera.

II.—MYRIAPODA.

Order I.—Chilognatha.

Order II.—Syngnatha.

Order II.—Habitation internal.

Entozoa.

II.

RADIATA.

Class I.—Echinodermata.

Order I.—Intestine open at both extremities.

Order II.—Intestine open at the mouth only.

Class II.--Acalepha.

Class III. Zoophyta.

Order I.—Carnosa.

Order II.

Order III.

Order IV.

Class IV. Infusoria.

EXPLANATION OF THE PLATES.

(*The Binder is desired to place all the Plates at the end of Vol. I.*)

PLATE I.

PLATE II.

PLATE III.

PLATE IV.

PLATE V.

CORRIGENDA.

VOL. I.

Page 53. line 13. *from the top, after* superior *add* to
—— 125. — 8. ——— *for* facility *read* faculty
—— 238. — 9. ——— consistent —— inconsistent
—— 250. — 21. ——— property —— propriety
—— 263. — 18. ——— been —— become
—— 402. — 9. ———*before* evolution *insert* internal

VOL. II.

Page 121. line 14. *from the top, for* the *read* their
—— 120. — 25. ——— fifth —— fiftieth
—— 201. — 9. ——— division —— denizen
—— 268. *after line* 13. *add*, 1. Body furnished with feet.
—— 579. — 1. *before* Strepsiptera *insert* ORDER VII.

THE

PHILOSOPHY

OF

ZOOLOGY.

THE contemplation of the works of GOD, as exhibited in the material world, forms one of the most suitable, gratifying, and useful employments of his rational creatures. These works present themselves to our notice under different aspects, and require the employment of various methods for their examination. The results of these investigations constitute the different branches of *Natural Science*.

The examination of the forms, magnitudes and motions of the heavenly bodies, is the peculiar business of the *Astronomer*. In the prosecution of his object, he employs measurement and calculation, for the purpose of discovering the laws by which the celestial phenomena are regulated.

The beings which constitute this terraqueous globe, are subject to various changes, in consequence of their mutual actions on one another. These the Experimental Philosopher professes to investigate. When the actions here referred to are accompanied with obvious motions, but do not produce a permanent change in the constitution of those bodies subject to their influence, they constitute those phenomena which the science of *Natural Philosophy* professes to examine and explain. The instruments of investigation employed in this department, like those of the astronomer, are measurement and calculation, aided by experiment. When the actions which take place among bodies, produce a permanent change in their constitution, unaccompanied with motions which admit of measurement, they are considered as belonging to the Science of *Chemistry*. Experiment is the only instrument of investigation which can be employed with safety in this department.

When the beings which constitute this terraqueous globe, are considered as related to one another, exhibiting particular forms, and adapted by their structure to the situations in which they are placed, they are regarded as the subjects of the *Natural Historian*. The establishment of a *System of Nature*, to which all his labours are directed, can only be completed when all the creatures of the globe shall have become known, and their mutual connections ascertained. Observation is the peculiar instrument of research which he employs.

All these sciences mutually explain and illustrate one another. The doctrines of the one are often employed with success, to solve the difficulties which occur in the other; and the instruments of the one may be substituted for those of the other, in particular circumstances, with the happiest results.

CHAP. I.

ON THE DIVISION OF NATURAL OBJECTS, AND THE PECULIAR CHARACTERS OF INORGANIC BODIES.

THE objects which present themselves to the notice of the Natural Historian, on the surface of this globe, exhibit innumerable varieties of form, structure, action and position. But, however diversified in appearance, they readily admit of distribution into various groups, each including numerous species, capable of farther arrangement into subordinate tribes. The most extensive of these groups, are two in number,—the one called the ORGANIZED,—the other the INORGANIC Kingdom. The limits which separate these two divisions, are so well defined, that the distinction has been universally received.

Philosophers and poets, in all ages, have been anxious to point out a certain gradation of perfection in earthly objects,—a CHAIN OF BEING, the links of which consist of all created beings, passing by insensible degrees from the simplest to the most complicated, and constituting one harmonious whole, unbroken and dependent. Crystallization, they say, is the highest link of the inanimate part of the chain, and connects the Mineral with the Vegetable Kingdom. The lichen which encrusts the stone, is but one step higher in the scale of being than the stone itself. The mushrooms and corals, form a bond of union between the Vegetable and Animal Kingdoms; and the vast interval by which Man is separated from his Maker, is occupied by different orders of superior intelligences.

All this appears at first sight plausible, and in some respects in conformity with those arrangements of Nature which we witness taking place in the subordinate divisions of animated beings. But when we examine attentively the characters which distinguish inorganic from organized beings, and animals from vegetables, we perceive, at once, that there are intervening chasms by which different parts of the chain are separated; nor can the most acute observer detect on either side the remaining portions entire, although he may be able to collect a few fragments and disjointed links.

I. *Independence of the different parts of Inorganic Beings.*—The different parts of an inorganic body enjoy an independent existence, while the parts of a body belonging to the organized kingdom, depend on their relative situation for the continuance of their structure and properties. The value of this distinctive mark will appear more obvious by the following examples. If we remove, from a bed of basalt, one of the jointed columns of which it consists, neither the bed nor the column suffer by the disjunction; and the latter retains the same form and structure as before the separation. If the joints be divided from one another, each joint will continue to preserve its character, although no longer in connection with those of which the column originally consisted. If the joint be split into a number of pieces, each fragment will be found to preserve its form and structure as permanently as the concretion from which it was detached. How widely different are the appearances exhibited by an organized body, when subjected to similar treatment!

When we pull off a branch from a tree, the stem itself is injured, while the detached portion speedily exhibits a change in every sensible quality. The leaves wither and drop off, the pliant twig stiffens, and the fresh bark assumes the appearance of a shrivelled crust. Changes equally obvious present themselves when a limb is separated from the body of an animal. Putrefaction soon reduces part of it to earth, and disengages the remainder in air.

In all these circumstances we find, that the instant the parts of an organized being are separated, a destructive process commences, to which there is nothing analogous in the mineral kingdom. The bodies which formerly attracted one another, and in their combinations exhibited to us the finest forms, and executed the most complicated movements, now appear to repel one another, and hasten to have the bond of union dissolved. This character, therefore, which we have now stated, independent of any other, is sufficient to point out the magnitude of that interval which separates the inorganic from the organized kingdom, and divides the living from the dead.

II. *Permanence of Inorganic Bodies.*—If we take a saline mixture, and induce crystallisation, symmetrical bodies are obtained, which are considered as the most perfect models of inorganic existence. These crystals, of whatever size, would continue to exhibit the same form and structure, unless acted upon by some external force of a chemical or mechanical kind. Within, every particle is in its proper place, nor does there exist any power to alter, increase or diminish. But the case is widely different with organized bodies. They acquire definite forms and structures, which are capable of resisting for a time, the ordinary laws by which the changes of inorganic matter are regulated. Internally, however, there is no rest. From the period that the existence of the plant or animal commences, to the day of dissolution, there is no stationary point. Increase and decay succeed by turns. Youth follows infancy, and maturity precedes age. It is thus with the mushroom and the oak,—with the mite and the elephant,—life and death being common to all of them.

III. *Integument distinctive of Organized Bodies.*—The substance of a stone or a crystal, is the same at the surface and the centre. Minerals possess no covering to defend them from external injuries, and preserve their form. When they increase, it is by the addition of matter to their surface; when they decrease, it is by the abstraction of the exterior particles. But organized bodies are enveloped in a covering which differs in structure from the parts within, which defends those parts from the action of external agents, and which is susceptible of extension and contraction. The increase of size is produced by the addition of particles to the interior, and an enlargement of this integument. A diminution of size takes place from the removal of particles from the interior, and a consequent contraction of the covering.

Were it necessary, many characters besides those which have been enumerated, might be exhibited, as distinguishing the inorganic from the organized kingdom. The individuals belonging to the former do not require nourishment and a suitable temperature, neither do they possess a circulating system. Animals and vegetables, on the other hand, stand in need of a supply of food and air, and a suitable temperature, for the continuance of their existence, and are nourished by particles prepared in appropriate organs, and conveyed by peculiar vessels. Inorganic bodies can neither boast of youth nor age, parent nor child; while organized bodies have the power of reproduction as well as the tendency to decay.

Such are the characters by which inorganic and organized bodies may be distinguished, as constituting the two great and primary classes of natural objects. Let us now direct our attention to the structure of the organized kingdom, for the purpose of ascertaining those subordinate divisions of which it is susceptible.

CHAP. II.

ON THE PECULIAR CHARACTERS OF ORGANIZED BODIES.

THE infinite variety of species which constitute the organized kingdom, possess many common properties, independent of the remarkable differences which they exhibit in their structure and appearance. By attending to these common properties, we shall be able to discover some of the qualities of that principle to which they owe their character, and by which their arrangements are regulated.

All organized bodies consist of Solids and Fluids. The former exhibit the appearance of fibres or laminæ, of which cells and tubes are constructed, destined to contain the fluids. These solids and fluids are very differently arranged in the different classes of organized being, although similar in individuals of the same species.

I. *The Characters of the Vital Principle.*—When we examine a plant or an animal as near to the origin of its existence as possible, we witness its embryo or *germ*, small indeed, but possessing a power capable of developing in succession the destined phenomena of existence. By means of this *power*, the germ is able to attract towards it particles of inanimate matter, and bestow on them an arrangement widely different from that which the laws of chemistry and mechanics would have assigned them. The same power not only attracts these particles and preserves them in their new situation, but is continually engaged in removing those which, by their presence, might prevent or otherwise derange its operations.

1. Limited in its power.—But there is a limitation of this power in the exercise of its functions, to the production of a body of a certain magnitude, form, structure, composition and duration.

A. *Magnitude.*—In each particular species, it is restricted in its efforts to the production of a being of a determinate size. When this size has been attained, sometimes by a slow, sometimes by a more rapid growth, the body remains for a time, as it were, stationary as to bulk. The absorption and ejection being equal and opposite, counterbalance each other. There is a proportion likewise preserved between all the parts,—between the roots and the stem,—the limbs and the trunk.

B. *Form.*—In each species, this power is restricted to a determinate form. Hence it is, that the external shape being the same in all individuals of the same species, it becomes an easy matter to recognize them. The seed of a fir-tree never expands into the shape of the ash, nor does the germ of the sheep evolve the appearance of the ox. In the various stages of their existence, certain organized beings are destined to undergo a variety of changes in size and shape; but these, however complicated or numerous, all pave the way for the assumption of the destined forms of the individual. In consequence of the uniformity of these operations, we can predict with certainty, that from the small egg of the butterfly shall burst forth the destructive caterpillar; that this, in its turn, shall appear a dormant pupa, and, in due time, assume the elegant form and variegated colours of the sportive imago.

C. *Structure.*—This same power, which is restricted as to the form and size of the being it is destined to construct, is likewise regulated with regard to *structure*. Thus the germ of the palm-tree is destined to produce a stem, which shall increase by the addition of matter on its central aspect, and the fibres or nerves of whose leaves shall be arranged nearly in straight lines. The germ of the oak, on the other hand, is destined to construct a trunk which shall increase in size by the addition of layers to its circumference, and the nerves of whose leaves shall exhibit a reticular arrangement. The germs of animals, in like manner, are regulated by similar laws. Individuals of the same species are constructed according to the same plan, and are furnished with the same members and organs, although they differ more or less from those of every other species. In short, in each species there is a power capable of producing all the modes of that species, and incapable of producing those of any other.

D. *Composition.*—The power which organized bodies thus possess of attracting towards them the particles of inorganic matter, is not exercised indiscriminately. There is a principle of selection, which displays itself in the production of different substances from the same materials. The plants which grow on the same soil, which are nourished by the same water, and invigorated by the same temperature, select from that soil the particles suited to their nourishment. But all do not select the same particles. The wheat, barley and oats, draw towards them the particles adapted to construct their respective systems.

The horse, the sheep, and the cow, though all feeding in the same field, and consuming the same herbage, select different particles from the same mass, and appropriate these to the production of their peculiar organs.

Even the different members of which an organized body consists, possess this power. From the same circulating fluid, bone, muscle, cartilage, and fat are produced by the selective agency of the particular organs. In consequence of this power, each species, and the separate parts of each species, have an individuality of composition by which they are characterized. Thus, within the bark of the oak, we confidently expect to find wood of a more obviously fibrous structure, and of greater strength, than underneath the bark of the birch-tree: and the flesh of the sheep has always a different flavour from that of an ox or a horse. Even the particles which are secreted from organized bodies, differ according to the species. Thus we find the perfume of the rose different from that of the thyme, and the smell of the herring, from that of the smelt.

E. *Duration.*—The same power which we are now considering, is restricted in its action to a determinate *duration.* It collects the different particles suited to the composition of the individual, with unceasing industry, arranges them with amazing regularity, and, in spite of numerous obstacles, reaches the measure of the standard by which it is regulated. For a time it appears stationary, as to size, structure, and composition. By degrees, however, the functions of this power are exercised with less energy; the fluids decrease in quantity, and the solids become more rigid; the prelude to the total cessation of all its influence over the inorganic matter with which it is surrounded, and the very organs which it has constructed.

The term of duration is very different in different species. While many mushrooms and insects are but the beings "of a summer's day," the stately oak and the voracious pike outlive centuries. In general, when the previous growth of an organized body has been slow, the period of decay is protracted in proportion; and, when maturity has been quickly attained, decay as rapidly succeeds.

2. Possessed of irritability.—There is in this power which we are now considering, a disposition to be acted upon by different external objects, and to exhibit the influence which these exercise by contractile or expansive movements. This faculty is termed *Irritability*. It appears to reside in the fibrous part of organized bodies. Plants exhibit this power in a very remarkable manner, upon exposure to the light of the sun, bending their stems, and turning their leaves in various directions, according to the intensity and incidence of the rays. The pinnated leaves of certain plants exhibit the same power, when touched by any extraneous body, the various leaflets collapsing in rapid succession. This is well displayed in *Mimosa pudica* and *sensitiva*, *Smithia sensitiva*, and some others. The *Hedysarum gyrans*, or Moving Plant, as it has been termed, exhibits a motion in its leaves of a still more remarkable kind, requiring a very warm still atmosphere for its production. The leaf is ternate, and the lateral leaflets approach and recede from one another, in a manner irregular as to time and co-operation. But the example of vegetable irritability which is most accessible in this country, is exhibited by the Barberry-bush. Sir James Edward Smith, who first observed the phenomenon, thus expresses himself: "In this (flower) the six stamens, spreading moderately, are sheltered under the concave tips of the petals, till some extraneous body, as the feet or trunk of an insect, in search of honey,

touches the inner part of each filament, near the bottom. The irritability of that part is such, that the filament immediately contracts there, and consequently strikes its anther, full of pollen, against the stigma. Any other part of the filament may be touched without this effect, provided no concussion be given to the whole. After a while, the filament retires gradually, and may again be stimulated; and when each petal, with its annexed filament, is fallen to the ground, the latter, on being touched, shews as much sensibility as ever *."

In another British plant, the *Cistus helianthemum*, which is very common on dry rocky ground, the filaments, when touched, execute a motion the reverse of that of the Barberry. They retire from the style, and lie down in a spreading form upon the petals. This curious example of vegetable irritability was first pointed out by Dr HOPE, Professor of Chemistry at Edinburgh †.

In animals, this power is displayed by all muscles, in different degrees of intensity, producing directly, the various Involuntary motions; and indirectly, those which are termed Voluntary. In the former of these, an action takes place upon the application of a stimulus, independent of the will; while, in the latter, the action takes place in the muscle, in consequence of volition exercised through the medium of the nerves. When the nerve of a muscle is cut across, and direct communication with the brain thus interrupted, the will ceases to exercise its controul over the motions of the fibres; but involuntary motion, or the peculiar irritability of the muscle, continues in force.

By zoologists in general, the muscular fibre is regarded as deriving this power from nervous energy; since, as they

---

* SMITH's Introduction to Botany, p. 325.

† English Botany, vol. xix. No. 1321.

suppose, every fleshy fibre is accompanied by a nervous filament. Hence they conclude, that the peculiar motions of a muscle, after the section of its nerve, depend upon the portion of nervous pulp which it contains. By HALLER, and a few others, irritability is considered as a quality of the muscular fibre itself; and they adduce in support of their opinion, the well known facts, that parts not muscular are not irritable, and that no proportion exists between the degree of irritability and the number of nerves in any part.

If the degree of irritability of any part be not in proportion to the number or the size of the nerves with which it is supplied, neither is it in the ratio of the number of its muscular fibres. Different muscles, in the same animal, possess dissimilar degrees of irritability; and the same muscles, in different species, likewise vary in the intensity of their action under similar exciting causes. It is impossible to form a conception of irritability, in which the muscular fibre shall not constitute the essential part. It is the only portion of an organized body capable of exhibiting the signs of its action; and, by consequence, the only part in which we have any ground to believe that it exists. The nerves, where present, may excite to action the muscular fibre, by acting as stimulants; but the same actions of contraction and expansion are performed in animals in which no nerves can be traced. In those animals, which belong to the least perfect classes, the nervous pulp is supposed to be diffused throughout the different parts, communicating to the muscular fibre its susceptibility of irritation. There is reason, indeed, from analogy, to acknowledge the propriety of the conjecture respecting this dissemination of the source of nervous energy in the cases to which we refer; otherwise, how could the will of those animals execute its purposes in the production of spontaneous motion. But when the muscular fibre appears to be equally susceptible of irritation, when the nerves are no longer visible to our senses, as when they are aggregated into perceptible filaments, we are disposed to consider the quality as resident in the fibre itself. And this opinion is confirmed, when the attention is directed to the irritability of the vegetable tribes. In this department of the organized kingdom, the existence of nerves has not been demonstrated, nor even rendered probable; yet the expansions and contractions which vegetable fibres exhibit upon the application of stimulants, indicate an irritability the same in kind with that of animals, although its effects are not so rapidly displayed, at the same time that they point out the place of its residence.

M. Lamark, in the introduction to his valuable work, " Histoire Naturelle des Animaux sans Vertebres," Paris, 8vo 1815, vol. i., refers some of the movements which are here considered as indicating the existence of irritability in plants, to the influence of the mechanical or chemical powers, and others, to what he terms *vital orgasm.* All these different actions, however, occur in connection with the vital principle, and their entire dependence on the laws of inorganic matter is a gratuitous assumption. With regard to this *orgasm* of plants, it is identical with the irritability of animals, and seems to have been employed by this author as a convenient ambiguous term, to aid him in his peculiar views of systematical arrangement.

3. Possessed of instinct.—In the exercise of those various powers which produce and modify the size, shape, structure, composition and duration of organized bodies, there is a more immediate reference to the formation of the individual, as it exists unconnected with surrounding objects. The power of irritability, on the other hand, forms a medium of communication not only with the different

parts of the organized body, but with external objects, and gives warning of the approach and retreat of salutary or noxious particles. All its movements, however, are of a passive nature. But the power which is now to be considered, though nearly related to irritability, is more varied in its movements, as it excites us to act upon external objects. It is active, not passive.

As organized bodies, when they begin to enjoy an independent existence, require food for their support, we might conceive it possible for a young plant or animal, by means of the irritability of the vessels, and those determinate powers which have been enumerated, to arrive at maturity, in the absence of counteracting agents, were the different substances necessary for their nourishment, always placed in contact with the vessels fitted for their reception. But the supply of nourishment, even where most abundant, is situated at some distance. Besides, during the whole term of life, obstacles and dangers interrupt and destroy the exercise of the various functions. From the form, size and vigour of the living bodies which we see around us, we may therefore infer the existence of a power which regulates the movements which are necessary to obtain a supply of food, to remove or counteract opposing obstacles, and to fly from impending danger, or repair the injuries which it may occasion. This active agent, which appears to be an inseparable companion of organized existence, we venture to denominate *Instinct*, and shall now endeavour to point out a few instances of its operations.

A. *Regulates the Supply of Food.*—When the seed of a plant is deposited in the soil, under circumstances favourable to its germination, the corcle expands, and, from the nourishment furnished by the cotyledon, evolves the radicle and the stem. The former of these descends into the earth, not according to the laws of gravitation, for it displaces the soil, which is specifically heavier than itself,—while the latter bursts through its earthy covering, and rises to a determinate height in the atmosphere, not in consequence of its specific levity, for its density is greater than that of the circumambient air.

As the radicle is incapable of procuring nourishment for the growth of the future plant, in immediate contact with its surface, it sends out from its substance numerous fibres, which are multiplied in all directions to search after food, and convey it to the proper organs. The number and direction of these fibres, and the distance to which they extend, are regulated by the wants of the plant, and the supply in the soil. The stem in some plants rises unassisted in a vertical direction; in others, it requires the support of a foreign body, or remains prostrate on the ground; and the leaves assume every variety of aspect and position.

Various explanations, intended to account for the descent of the root and ascent of the stem, widely different from the one to which they are here referred, have been offered by different naturalists. One supposes the root to be attracted downwards by the earth, and the stem to be attracted upwards by the sun,—another, that the root is stimulated by the earth, while the stem is stimulated by the air, so that each lengthens in the direction of the greatest excitation,—and a third ascribes the descent of the root to the weight of its watery juices, and the ascent of the stem to the buoyancy of its gaseous secretions. But these hypotheses, while they fail in explaining the phenomena, many of which have not been contemplated, indicate in their authors a strong desire to employ the principles of chemistry or mechanics in the solution of the difficulties attending the examination of organized bodies, and a decided and unphilosophical aversion to admit the existence of the vital principle.

Animals, at their birth, present nearly the same phenomena as plants. When the germ of the egg of a fowl, for example, is excited to action, it begins to arrange and construct the different organs necessary for its future existence, from the materials with which it is surrounded. It then bursts forth from the shell. Nourishment is not now in immediate contact with its absorbent vessels; so that it must employ its feet, bill, tongue and gullet to collect food for the stomach. It accordingly begins to pick the grains, which a mother's tenderness has provided, and executes the various movements of seizing, bruising and swallowing, with all the indifference and dexterity of a confirmed habit.

A quadruped, at its birth, discovers, with almost unerring certainty, the fountain of life, and adjusts its mouth, so as to pump out the milk, without being acquainted with the properties of a vacuum, or the pressure of the superincumbent air. In the explanation of these phenomena, it will be vain to call to our aid the mysterious powers of gravitation and affinity; for instinct, in these cases, exercises a controul over them, and directs their influence. It is equally vain to talk of the pain arising from hunger, and the consequent excitement to exertion, in order to satisfy its cravings: for the exertions themselves are predetermined; instruments are provided to assist; and the end is accomplished by the smallest quantity of labour. This instinctive principle may be considered, therefore, as directing the first movements of every organized body, after its kind, to procure the nourishment necessary for its future growth.

But it is obvious, from the situation in which living beings are placed on the surface of this globe, that many obstacles will obstruct these exertions. Each species will have to contend with difficulties peculiar to its form, and sphere of action. Were there no power, therefore, inherent in an organized body, enabling it to avoid and remove these, life would often be interrupted at its commencement, and few individuals would reach the period of maturity. But the same faculty which directs the first movements of the organized body to obtain a supply of food, likewise assists in removing or counteracting opposing difficulties.

B. *Obviates Difficulties.*—When the roots of a plant, spreading in search of nourishment, meet with interruption in their course, they do not cease to grow, but either attempt to penetrate the opposing body, or to avoid it, by changing their course. Thus, I have repeatedly seen the creeping root of the *Triticum repens*, or Couch-grass, which had pierced through a potato that had obstructed its course: and every one knows, that the roots of a tree will pass under a stone, wall or ditch, and rise again on the opposite side, and proceed in their original direction.

The Shipworm, *Teredo navalis*, an animal which perforates wood, in order to form for itself a habitation, and which is well known as most destructive to the timbers of ships and harbours, in general, lengthens its cell by boring in the direction of the fibres of the plank. Should it happen to meet with the shell of another animal of the same kind, or a knot in the wood, which it is unable to penetrate, it changes its course, so as to avoid the obstacle, either by a slight curvature, or by reversing its original direction.

When two trees of the same kind are planted, the one in a sheltered, and the other in an exposed situation, we witness the display of this faculty, now under consideration, in a very remarkable degree. The former pushes forth its roots in all directions, more especially where there is the greatest supply of nourishment, and the highest temperature; while the latter, which, were it to act in the same manner, would be speedily overturned, multiplies its roots in the direction of the strongest blasts, and these, acting like the stays of a ship's mast, preserve the trunk in its vertical position.

In like manner, the Acorn-shells *(Balani)*, when growing on the firm, even surface of a plank or shell, present a regularity in the margin, and a narrowness of base, widely different from the irregular spreading bottom of the same species, when seated on the friable and uneven surface of a rock. In the former case, the shells are frequently of a cylindrical form; while, in the latter, they are uniformly approaching to conical. In a variety of other cases, which might be brought forward from the Vegetable and Animal Kingdom, where the obstacles are as various as the beings which encounter them, we perceive a unity of purpose, and a striking resemblance among the means employed for its attainment.

This instinctive principle, while it regulates the means to be employed in obtaining food and avoiding obstacles, likewise repairs the injuries to which organized bodies are liable.

C. *Repairs Injuries.*—Plants are exposed to a greater variety of accidents than animals. Destitute of the locomotive faculty, they neither can fight for victory, nor retreat for safety. In some cases, it is true, they close their flowers, or hang down their heads, to defend the unripe pollen from the heat, the wind, or the dew. But though they do not in general possess the means of avoiding injuries, they are abundantly provided with the power to repair them. Thus, when the branch of a tree is torn from the stem, the bark accumulates around the wound, and, in time, covers the offensive scar. The plants with which we form our hedges, still live, though sadly mutilated by the shears, and push forth afresh those branches and leaves necessary to the continuance of their being. The utility, indeed, of many of our culinary vegetables, such as parsley, cives and cresses, depends on the displays of this repairing power. But the indications of its existence which have always appeared to me the most extraordinary, are exhibited by trees. In a young silver-fir, *(Pinus picea)*, for example, I have seen the central vertical shoot cut off, so that the tree must have ceased to increase in height, unless there had been some repairing power, as the lateral branches invariably expand in a horizontal direction. But one of these lateral branches has begun to change its position; and, by bending itself upwards, at last assumed a vertical direction, and became the leading shoot and trunk of that plant, of which it was formerly a subordinate branch. It has sometimes happened, that more branches than one changed towards the vertical direction, and thus rivalled each other in their attempts to repair the loss of the original stem.

Animals, as we have already stated, can, in many instances, protect themselves from accidents, by resistance or retreat. But, when wounded, this renovating power is often exerted in an astonishing degree, in repairing lacerated muscles, cementing broken bones, closing ruptured vessels and supplying the loss of extravasated juices. In the lower orders of animals, the loss of amputated parts is speedily supplied by the production of new organs, as takes place with the tails of serpents, and the claws of lobsters; or the detached parts, when placed in favourable circumstances, evolve themselves into separate and independent individuals, as is the case with the common fresh water polypus.

It would have been easy to have multiplied examples of the display of this instinctive power in repairing injuries, were those which have been produced not sufficient to demonstrate its existence, and mark the characters by which it is distinguished.

4. *Possessed of a Procreative Power.*—The power which we have been considering, as displayed in the arrangement

of the parts of an organized being, and in regulating the motions which are necessary to provide food, remove obstacles, or repair injuries, is likewise exerted, in a determinate manner, in each species propagating individuals of a similar kind. Without this resource, the race would perish with the dissolution of the primitive stock. It is true, that the circumstances under which organized bodies are produced, are widely different in the Vegetable and Animal Kingdoms, and in the subdivisions of these great tribes. But there is at least one point of resemblance. The living power is first employed in the formation of the individual; and, during the developement of the several organs, the procreative instinct is dormant. But when the organized body approaches maturity, there is, as it were, an accumulation of vital energy, which gives birth to a new individual.

These different operations of living beings, which we have thus briefly enumerated, can never be regarded as the effect of their peculiar organization. The organs are formed according to a uniform and determined plan, from inorganic matter, collected from various sources, and arranged, according to the species, on different models. By what power, then, does this organization take place? And what is the nature of that principle which regulates animated beings?

We have already taken notice of the laws by which it is regulated in the constitution, duration, and continuance of organized bodies. These make us acquainted with the existence of a principle, different from any of those which the mechanical or chemical philosophers have investigated with so much success. This power has been denominated *The Living or Vital Principle*; and the phenomena which it exhibits, are included in our idea of *Life* or Vitality.

This vital principle, then, so far as appears to our senses, can only reside in organized bodies. The connection is temporary, and may be dissolved by various circumstances and it is capable of being divided or multiplied by the process of generation. In the various kinds of organized bodies, it exhibits its different qualities, according to the structure of the fabric which it animates; hence we must conclude, either that the same principle is modified by the substances which it pervades, or that there are many different kinds. If we adopt the practice of the chemist, when examining the elementary bodies, and their combinations, and consider those substances distinct in kind, whose affinities and appearances are different, we must conclude, that there are different kinds of vital principles, distinguished by different affinities for the various kinds of inorganic matter, and forming, with these, combinations, whose peculiar characters are obvious and precise. In all organized bodies, we witness an origin, progress and termination; but, in each species, these circumstances exhibit some peculiarity which is not observable in any other species. We are, therefore, led to conclude, that there are as many different kinds of vital principles, as there are species in nature. Hence all the knowledge which we possess regarding the structure and functions of different kinds of organized beings, is, in fact, a knowledge of the attributes of the different vital principles from which these have originated; and our systematical arrangements, an attempt towards their classification. Even the vulgar, talk of the life of plants as distinct in principle from that of animals, although they seldom carry their views on this subject so far as to admit of any inferior division.

II. *Conditions necessary for the Existence of the Vital Principle.*—It is, doubtless, a curious subject of inquiry, to ascertain those conditions which are necessary to the display of this principle of life, or the circumstances under which it is generated or excited to action. Our knowledge of organized bodies, however, is still too imperfect to enable us to accomplish this task. It scarcely warrants us to hope that the object is attainable. We know, however, that the following circumstances are invariably present, whenever the vital principle begins to exhibit its movements.

1. *A Parent.*—Previous to the independent existence of a plant or animal, it is necessary that they shall have formed a part of some other body similar to themselves. This condition is ascertained in so satisfactory a manner, in the case of the more perfect organized beings, as to preclude the possibility of a doubt. There are, however, many plants and animals, with whose manner of growth and mode of propagation, we are so imperfectly acquainted, that we must rely upon the evidence of analogy for the conclusions which we form with respect to their origin. But as all the living bodies whose history has been studied with care, have evidently participated in the existence of other living bodies, before they exercised the functions of life for themselves, the presumption, that the same kind of generation takes place in every organized body, offers all the claims of a legitimate deduction. Life, then, may be considered as proceeding from life, as transmitted from one individual to another, and as dating its actual origin from the period when the voice of Omnipotence uttered the words, "Let the Earth bring forth."

The ancients, whose opinions respecting the nature of generation were necessarily obscure, from the want of proper instruments and methods of observation, considered living bodies as produced in two different ways. In the first, exemplified in the case of the more perfect animals and vegetables, they considered organized bodies as proceeding from other organized bodies, by a process which is termed *Univocal* or Regular Generation. In the second, they supposed that the putrefaction of different bodies, aided by the influence of the sun, generated life in the less perfect organized beings, such as mushrooms and worms. This has been termed *Equivocal* or Spontaneous Generation, and appears to have been devised by the Egyptians, to account for the hosts of frogs and flies which appeared on the banks of the Nile, on the ebbing of its periodical inundations. It was adopted by ARISTOTLE, and still continues to be supported by a few naturalists.

We have already stated, that the origin of life by univocal generation, is demonstrated by direct proof and powerful analogies. Nor is it at all difficult to give an explanation of those appearances on which the whole fabric of the theory of equivocal generation rests.

In the case of plants, some have supposed that the growth of the fungi, the mushroom among dung, and the other parasitical plants which appear on putrid flesh and fruit, might be regarded as examples of the truth of this theory. But the microscope makes us acquainted with seeds of these plants, and experiments prove that these seeds are prolific. The characters by which the different species may be distinguished, though minute, are permanent; and individuals of the same species appear in a variety of situations,—circumstances these, not to be looked for, in beings generated by corruption, or formed from the fortuitous concourse of atoms.

The animalcules which make their appearance in water in which vegetable or animal substances have been infused, seem at first sight to favour this ancient doctrine. But, in these cases, the species have determinate characters, exhibit always the same proportion of parts, and transmit their vitality to their descendants, after the manner of animals of larger growth. Is it probable, therefore, that if these animalcules were produced by the spontaneous aggregation of particles contained in these infusions, that they should be controlled by the same laws which regulate the offspring of univocal generation, as to form, structure and duration; and that the same beings should be able to continue, by transmission, that life which they derived from accident? The most rational explanation which can be given of the appearances of these plants and animals, in such places, is derived from the consideration of the smallness of their seeds or eggs, which may be carried about by the winds, and showered down along with the rains, so as to enter with facility into every situation. They cannot, however, expand, unless in circumstances favourable to their future growth. Thus the soil, in which alone the beautiful little plant called *Monilia glauca* makes its appearance, is the surface of putrid fruit; while the small animal termed *Vibrio aceti*, requires for its growth, vinegar, which has been exposed for some time to the air. This view of the matter is still farther strengthened by the almost indefinite length of time, during which, certain seeds retain their power of germinating; and certain animals, in a torpid state, the faculty of reviviscence.

We are thus led from the phenomena which organized beings present, to reject the theory of equivocal generation, and to admit the important conclusion of HARVEY, *Ex ovo omnia.* But the manner in which these eggs or germs are produced, has furnished matter of curious speculation to the physiologist. By some, it has been supposed, that the germs of all the plants and animals that have been, or ever shall be, in the world, were really all formed within the first of their respective kind,—to be brought forth in a determinate order. This *Theory of Evolution*, as it has been termed, is in a great measure the result of microscopic observations, assisted by preconceived views. There is one circumstance, however, which not only receives no explanation upon the principles of this theory, but stands in opposition, viz. the production of *hybrids*, or the offspring of the union of the males and females of different species or genera. Thus, in the case of the mule, the produce of the ass and the mare, which partakes of the qualities of both parents, it is obvious, that the pre-existent germ of the female was more than stimulated to life by the sexual union,—that its structure was likewise changed, and that all the germs of its future offspring were likewise annihilated, since, mules are seldom fertile. It fails, likewise, to account for the superior fertility of cultivated plants or domesticated animals

Before dismissing this part of our subject, it is necessary to take notice of those facts illustrative of the origin of organized beings, which have been ascertained by the researches of modern geologists. In investigating the structure and composition of the rocks which constitute the crust of the earth, it is observed, that they enclose the remains of animals or vegetables, more or less altered in their texture. Presupposing that those rocks on which all the others rest are the most ancient; and after dividing them according to their age, as determined by their superposition, it has been ascertained, that the organic remains found in the older rocks differ from those which occur in the more recent strata, and that they are all different from the plants and animals which now exist on the surface of the globe. It likewise appears, that the petrifactions contained in the newer strata, bear a nearer resemblance to the existing races, than those which belong to the rocks of an older date. That the remains of those animals which have always been the companions of man, are only to be found in the most recent of the alluvial deposites. In the older rocks, the impressions of the less perfect plants, such as ferns and reeds, are more numerous than those of the dicotyledonous tribes, and the remains of shells and corals abound, while there are few examples of petrified fish. In the more recent strata, the remains of reptiles, birds and quadrupeds, occur, all of them differing from the existing kinds.

Attempts have been made to account for these circumstances by supposing, that the present races of animals and vegetables, are the descendants of those whose remains have been preserved in the rocks, and that the difference of character may have arisen from a change in the physical constitution of the air, or the surface of the earth, producing a corresponding change on the forms of organized beings. The influence of cultivation on vegetables, of domestication on animals, and of climate on man himself, may be considered as strengthening the conjecture. But there are several difficulties which present themselves to those who adopt this opinion. The effect of circumstances on the appearance of living beings, is circumscribed within certain limits, so that no transmutation of species was ever ascertained to take place;—and it is well known, that the fossil species differ as much, nay more, from the recent kinds, as these last do from one another. It remains, likewise, for the abettors of this opinion, to connect the extinct with the living races, by ascertaining the intermediate links or transitions. This task, we fear, will not be executed speedily.

There is yet another view of the matter which suggests itself. If the seeds of some plants, and the eggs of certain animals, be so minute as to be excluded with difficulty from any place to which air and water have access, and if they are capable of retaining, for an indefinite length of time, the vital principle, when circumstances are not favourable to its evolution, the crust of the earth may be considered as a mere receptacle of germs, each of which is ready to expand into vegetable or animal forms, upon the occurrence of those conditions necessary to its growth.

According to this view, the germs of the ferns and palms first expanded their leaves, and afterwards those of the staminiferous vegetables. With regard to animals, it may be supposed that the germs of the zoophytes only, were first disclosed; afterwards those of the testaceous mollusca; and, finally, those of the vertebral animals: That the organized beings of the first periods flourished during the continuance of the circumstances which were suitable to their growth; and that the change which prepared the way for the evolution of those which lived at a subsequent period, contributed to the extinction of the earlier races.

According to this statement, there is little difficulty in accounting for the extinction and revival of the different races of the less perfect animals and vegetables, whose germs appear, even at present, to be regulated according to such circumstances. But it offers no solution of the difficulty attending the preservation of the germs of the more perfect animals, many of which are inseparably connected with the parent, and require the continuance of her life to preserve vitality until the period of evolution. If, then, the present races of quadrupeds did not exist at the time when the mammoth and the other extinct quadrupeds, whose bones CUVIER has described with so much accuracy, were the denizens of our plains, at what period, and under what peculiar physical circumstances, were they called into being? Is the generation of organized beings simultaneous or successive? Have they all been created at once; but, in the progress of time, so modified by the influence of external agents, as now to appear under different forms? Or have they been called into being at different periods, according as the state of the earth became suitable for their reception *. The latter supposition is countenanced by many geological documents.

* See CUVIER's "Recherches sur les Ossemens Fossiles de Quadrupedes," 1812.

2. *Moisture.*—After the germ has been produced by the vital force of the parent, a determinate quantity of moisture appears to be necessary to enable it to exercise the functions of an independent being. In the case of the seeds of plants, and the eggs of all the aquatic animals, the vital spark, though existing, is unable to display its energies when placed beyond the influence of water. Even after the germs have expanded into maturity, the abstraction of moisture frequently produces a cessation of every function. This assertion may be verified, by placing a snail in a dry situation during the summer. Animation becomes suspended, and will continue so for several years, but will return upon moistening the shell with water. Similar effects may be observed with many of the infusory animalcules.

These facts have given rise to the opinion which very generally prevails, that the vital principle resides in the solids, as solids, and that, by means of moisture, it exerts its influence on inorganic matter. Others, on the contrary, considering that the fluids furnish the materials for the construction of the different organs, and are indispensably necessary in the various processes of absorption, secretion, and every other function, consider that the vital principle resides in the fluids. HARVEY and HUNTER assert, that both the blood and the chyle have life; while ALBINUS is disposed to grant it even to the excrement. It is, however, probable, that the living principle resides both in the solids and fluids, as we can form no conception of an organized body that does not consist of vessels and contained fluids; neither have we an idea of life that does not include the motion of these fluids. The vital principle may be retained, though inactive, even after a considerable part of the fluid has been extracted; and, in the case of the snails and animalcules, revive upon the addition of water. But there is not a single experiment which countenances the idea of the vital principle being retained by the solids after absolute desiccation.

The quantity of fluids necessary to the exercise of the functions of organized bodies is very various, according to the species. How different the dry powdery appearance of many of the lichens, compared with the watery tremellæ, or the hard stony substance of the common coral (Corallina polymorpha), from the gelatinous and transparent medusæ. In all cases, the quantity of fluid is greatest during the period of growth, and decreases beyond the term of maturity. Hence the plumpness and flexibility of youth,—the aridity and stiffness of age.

3. *Temperature.*—The two circumstances which we have already pointed out as necessary to the commencement and continuance of the vital functions, are nevertheless insufficient for the purpose, without the aid of a suitable temperature. Unless supplied with heat, the seeds of plants do not germinate; and the eggs of animals are not hatched. Even after the commencement of the living action, animation is suspended or destroyed, when the temperature sinks below a certain degree, which differs according to the species. In some animals, reviviscence and torpidity may be produced by turns, by the communication or abstraction of caloric.

In many cases, where an elevated temperature, or one higher than the surrounding medium is required, as in some plants during the fecundation of the seed, and in warm blooded animals, organs are provided which occasion the evolution of caloric; and, when this fluid ceases to be produced, the functions of life are suspended or destroyed. In general, when the temperature descends to the freezing point, animation is either totally destroyed or suspended; and, below that temperature, circulation ceases, and the fluids congeal. Life is likewise destroyed by a temperature approaching the boiling point. The tendency of the fluids to fly off in vapour, becomes too great for the vital power to restrain.

As the medium in which organized bodies live, is subject to great variations of temperature, it is obvious, that unless Nature had provided some means of defence, many races would soon be extinguished, and others confined in their operations. But, in all the situations in which organized bodies are placed, the means of protection against the vicissitudes of temperature are liberally provided. The locomotive faculty assists their escape from the scene of danger; and, where this is wanting, a covering of a body which conducts heat imperfectly, is bestowed to prevent the diminution of temperature, and various means are every where employed to check its pernicious increase. It is probable, that all organized beings, vegetable as well as animal, have an inherent power of generating cold or heat, according as circumstances require. Some curious experiments were performed in illustration of this subject by HUNTER *.

But the resources of organized bodies, employed to secure a suitable temperature, may be discovered in the circumstances of their physical distribution. Some are found subsisting under the influence of a vertical sun, while others survive the piercing colds of the arctic regions; some prefer the sea-shore, others the summit of the mountains; some live in the water, others on the dry land. To these different circumstances, there is an admirable adaptation of structure and disposition with respect to temperature; so that no part of the earth or the waters can be considered as destitute of life. In these different stations, the different species can perform all the functions of exist-

* Phil. Trans. 1775,-1778.

ence, the condition of the living body being suited to the place of its residence. But it may be asked, Is there no particular degree of heat in which the living principles display their energies with the greatest vigour? To this it may be answered, That the species of organized beings are much more numerous towards the equator, than in any of the other districts of the earth; that their numbers decrease as we approach the poles; and that the same decrease may be observed on leaving the sea-shore, and advancing towards the mountains. These facts seem to indicate a mean temperature of about 84° Fahr. the average mean heat of the equatorial regions, as the most favourable for the increase of living beings; since they decrease in number and variety as we approach the poles, with the progress of the diminution of temperature.

The influence of temperature, in modifying the appearances of organized bodies, is by no means considerable. Among plants, the size appears to be influenced by this cause, the tallest being found in the warm countries, while those of the colder regions are of more humble growth. It is probable, however, that, in this case, the agency of winds in diminishing the size, is more effectual than the decrease of temperature. With regard to animals, size appears but little influenced by temperature. The whales of the arctic seas are the largest animals in the world; while the land quadrupeds of the warm regions are larger than those in temperate climates. But the irregularities which present themselves, forbid the establishment of a general rule.

In every case, then, we witness heat necessary, not only to enable the vital principle to execute its first movements, but the functions of existence in every period. It appears, indeed, to be the principal exciting cause of vital energy, from the commencement to the termination of its actions.

4. *Atmospheric Air.*—If we suppose the existence of a germ ready to expand, and furnished with a sufficient quantity of moisture, and a suitable temperature, still, there is another condition necessary to enable the living power to proceed to execute its functions,—the presence of atmospheric air. The decomposition of this fluid takes place wherever there is living action. The portion of oxygen which it contains disappears, and carbonic acid is substituted in its place. These effects are produced by the seeds of plants, and the eggs of animals. Nor is this decomposition confined to the first stages of life: for we find the stems and leaves of plants, —the skin and lungs of animals, requiring a constant supply of air; and, when this is withheld, decay and death supervene. Judging from the most accurate experiments which have been made on the subject, it appears to be demonstrated, that living bodies produce a superfluous quantity of carbon, and that the presence of the oxygen gas of the atmosphere is required to convert this into carbonic acid, for the purpose of removing it more easily from the system. The azotic gas in the atmosphere does not appear to exert any positive influence on living bodies. It apparently serves only to dilute the oxygenous portion.

5. *Nourishment.*—For the purpose of supplying materials for the increase of the structure, and the repair of the waste which always results from living action, nourishment is necessary for all organized bodies, and is sought after, at stated periods, whenever the system is exercising its functions. It is more or less fluid; and differs both with regard to its composition and quantity, according to the species.

Life, then, quires the germ to have been detached from a parent, supplied with moisture, excited by heat, and furnished with air and food.

III. *Modifications of the Vital Principle.*—Having thus attempted to ascertain the conditions necessary to the commencement of life, and to the exercise of its functions, let us now attend to some of the phenomena presented by organized bodies during the continuance thereof. When all the organs are in such a degree of perfection as to be able, easily and durably, to perform the functions of life, the body is said to be in a state of *Health*. The ancients recognized a Goddess, who was supposed to preside over this condition, whom the Greeks termed Υγιεια, and the Romans *Salus*.

When we consider the situation in which organized bodies are placed on the surface of the globe, the changes which take place in the quantity and nature of their food, the variations in the temperature of the surrounding elements, besides the occurrence of an infinite variety of accidents,—all calculated to retard or derange their motions, we may expect to find many individuals, in which the harmony between the different parts is interrupted, and the functions of particular organs suspended,—producing a condition the opposite of health, termed *Disease*. The uninterrupted influence, indeed, of external objects, acting either directly or indirectly, precludes the possibility of an organized body enjoying perfect health for any length of time. These obstacles, however, are in part removed, by the power which each individual possesses of accommodating itself to circumstances; of varying, within a certain range, its form, structure, and actions; of repairing injuries; and of continuing to live, though in part mutilated. Such displays of the vital power are familiar to the Physician, and constitute the very basis of the healing art.

Besides the destructive influence of extersal objects, we may refer many of the diseases which assail organized bodies, to some imperfection or derangement of the vital principle itself. This is obviously the case, when an organ is either increased or diminished beyond its ordinary size, thereby destroying the proportion which should prevail between all the members; or, when one organ ceases to exercise its own appropriate functions, and performs those for which other organs are constructed. Organized beings, in which such modifications prevail, are termed *Monsters*. That monsters should occur among organized beings, need not surprise us, if we reflect upon the number of the vessels in which the living process is performed; the different properties of the matters acted upon; the endless varieties of combination and decomposition, of absorption and secretion, which take place. It may rather excite our astonishment, that, among living beings, so few are to be found, whose parts do not preserve their true proportions, or whose organs exercise functions which are foreign to their nature.

Monsters may, with propriety, be divided into two kinds. In the first, there is an excess or deficiency of a particular part, by which the relative proportion of the different organs is destroyed. The common garden-carrot may be produced as an example. In the wild state, the root is small, at the same time that it appears fitted to act both as an organ of nourishment and support to the stem with which it is connected. When cultivated, the root attains a much larger size, and the original proportion between it and the stem is greatly changed. The same holds true with regard to the turnip, and nearly all the roots which are cultivated for culinary purposes. The cabbage has become monstrous by the excess of its leaves; the tulip by the size and the shape of its petals; and our most highly esteemed fruits, by the disproportionate magnitude of the pericarp. In the Animal Kingdom such obvious instances of monstrosity of this kind are of rarer occurrence. There is usually a greater harmony among the parts; so that the increase or diminution of one organ is accompanied with corresponding changes in all the others. But, when observed with attention, the different *breeds* of our domestic animals exhibit such monstrosities. The size of the body, when compared with the limbs; the presence or absence of horns; the length or shortness of fur, are all indications of the deranging influence of domestication.

The second kind of monstrosity, and undoubtedly the most remarkable, consists in the substitution of function, which sometimes takes place among the organs. In consequence of this change, parts are produced in situations where they do not occur in the healthy state; while other parts disappear, which are essentially necessary to the harmony of the whole. Thus with regard to plants, Sir JAMES E. SMITH has observed (and we have witnessed the same appearance,) the double-flowering cherry with the pistil changed into a leaf *. But although the form and function of the organ were thus changed, the new production was not foreign to the system, as it resembled exactly the common leaves of the branches. The stamens of the rose are frequently converted into petals; and, in consequence of the change, acquire the agreeable perfume peculiar to the organ into which they have been metamorphosed. On the other hand, when the stamens of the Meadowsweet, (Spiræa ulmaria,) upon the flower becoming double, change into petals, they lose all the fragrance which they would have possessed as stamens, the petals into which they have passed being scentless. In all these instances of monstrosity, we observe, that, although an alteration takes place, both in the structure and function of an organ, it is only in exchange for the character of another organ, peculiar to the system to which it belongs. The pistil of the cherry did

* Introduction to Botany, p. 275.

not assume the leaf of the rose; nor did the stamens of the rose change into the petals of the meadowsweet. In no case is there a mixture of the character of other species. Each preserves, amidst its greatest changes, that individuality of constitution by which its vital principle is distinguished.

In animals, monstrosities of a similar kind prevail. Thus in the human ovarium, concretions of hair, and even teeth have been found; substances which, in ordinary cases, are fabricated in the uterus; and all this in the absence of sexual intercourse *.

In many parts of the human body, encysted tumors have been observed, covered internally with skin, producing hairs. That such organs have assumed the functions of the skin, is obvious from the portions of detached cuticle and hairs found in the cyst, which have been thrown off in succession. In these instances, which occur in other animals as well as man, the hair corresponds, in appearance, with that which grows upon other parts of the body. Hence when encysted tumors of this kind form in sheep, they contain wool. Tumors likewise occur on different parts of the body, producing horny excrescences. A figure of a production of this kind, cut from the head of a woman in Edinburgh, in 1671, and deposited in the Library of the College, is given by Sir ROBERT SIBBALD †. Similar instances have occurred in England, and other countries. Although usually termed *Horns*, they are essentially different from the armature of the ruminating quadrupeds. They have no core upon which they have been formed; consequently they are destitute of a central cavity. They

* Dr BAILIE's "Account of a Change of Structure in the Human Ovarium."—Phil. Trans. 1789, vol. lxxix. p. 71.-78.

† "Scotia Illustrata," tab. xi. fig. 23. part i. p. 60.

are not connected with the skeleton, but produced by the skin, and differ in nothing but shape from the nails of the fingers or toes. Where such excrescences have been observed on sheep, they have resembled the substance of the hoof *.

In all these different instances of monstrosity, in the Animal Kingdom, (and many similar cases might have been enumerated,) in which certain organs perform functions which are unnatural to them, we still find limits assigned even to these aberrations. The new parts, how extraordinary soever they may appear, are still similar to the parts of some other portion of the system. The organs which characterise one species, are never produced on other species as monstrosities; so that there is no mixture or transmutation. The functions of a system may thus be deranged, and the harmony of its parts destroyed; but to such irregularities there are bounds which are never exceeded †.

These occurrences, which take place during the continuance of life, serve greatly to shorten its duration. Neither dwarfs nor giants survive the usual period of existence. The attainment of the senile ευθανασια, so anxiously wished for by all, only takes place when there is an equilibrium between all the organs; when each organ exercises its own peculiar actions; and when the body escapes the destructive influence of external accidents. Previous to the dissolution of the fabric, we witness a rigidity of the vessels; a deficiency of the fluids; and a cessation of motion. The vital principle then deserts the body which it has construct-

---

* See Sir Everard Home's "Observations on Horny Excrescences of the Human Body."—Phil. Trans. 1791, vol. lxxxi. p. 95.-105.

† The reader will find some short but pertinent remarks on this subject, by Dr Barclay, "On the Causes of Organization."—Memoirs of the Wernerian Natural History Society, vol. ii. part. 2. p. 537.-546.

ed, and surrenders it to the influence of the laws of inorganic matter. This last state is termed *Death*.

We have already stated, that the vital principle, in the formation of an organized body, acts in direct opposition to the laws of chemistry or mechanics. With the cessation, therefore, of the influence of the one, and the continued combined operations of the others, we may anticipate very remarkable changes. Perhaps the appearances of death which first present themselves to our notice, proceed from the influence exerted by the laws of mechanics. In obedience to the power of gravitation, the pliant twig hangs down, and the slender stem bends. In animals, the body falls to the ground; the pressure of the upper parts flattens those on which the others rest; the skin stretches out; and the graceful rotundity of life is exchanged for the oblateness of death. The laws of chemistry then appear to operate, in the production of the cadaverous smell, the prelude to putrefaction, when dust returns to dust.

CHAP. III.

THE DISTINGUISHING CHARACTERS OF ANIMALS AND VEGETABLES.

Having endeavoured to ascertain those characters which are common to all organized bodies, it now remains that we mark the limits which separate these into the two great kingdoms of Animals and Vegetables, into which they have been divided by the universal consent of mankind. The undertaking may be regarded as peculiarly difficult, if we judge from the variety of definitions which have been given, and the still unsettled state of opinion among naturalists on the subject. Part of this difficulty, however, may be referred to the employment of ambiguous phrases, and to inattention to the number and character of those properties which are common equally to plants and animals. But the greatest share may be traced to the practice of examining the doubtful objects, without attending to all their relations,—by comparing the less perfect animals with some of the qualities of the most perfect plants, and by allowing our opinions to be influenced by circumstances connected with mere size and form.

When we compare together those animals and plants, which are considered as occupying the highest stations in each kingdom, we perceive that the characters by which they may be distinguished, are obvious and well defined. But when we descend to the animals and plants which occupy the lowest stations, and perceive that they are less complicated in their structure; exercise few functions besides those which are essential to living bodies; and, in consequence, present only obscure points of difference, we may be led into the supposition that, at a certain link of the chain, the two kingdoms coalesce. When, however, we examine all the characters which the imperfect plants and animals exhibit, we shall be able to trace the relations which connect these minute and obscure species, with those in which the characteristics of the kingdom to which they belong are more fully developed. In order to illustrate this subject still farther, and attempting to guard against the errors into which others have been betrayed, we have already enumerated those characters which are essential to the existence of organized bodies; and propose now to consider the marks by which animals may be distinguished from plants.

I. *Animals differ from Plants in Composition.*—The essential elements of organized matter appear to be carbon, oxygen, hydrogen, and azote, together with alkaline and earthy salts. These ingredients are variously combined, according to the species; but, when examined in a general view, they appear to unite according to a different plan in each kingdom.

The solid parts of all plants, termed the woody fibres, contain carbon, oxygen, and hydrogen, with scarcely a trace of azote. Sometimes there is a quantity of silica incorporated with the common integuments. The solid parts of animals consist of lime or magnesia, united with carbonic or phosphoric acids. In those beings, of both kingdoms, which appear to be destitute of solid parts, the points of difference are still numerous. We find the mucilage or gum of soft plants, differing widely from the gelatine or albumen of soft animals; the former being destitute of azote, which enters as a constituent in the latter. In some plants, substances of an animal nature, (or abounding in azote,) have been detected; not, however, constituting a whole plant, but only occurring in certain situations, and always in company with other substances of a decidedly vegetable nature; or consisting only of carbon, hydrogen, and oxygen. In the soft animals, there is no extensive combination of carbon, hydrogen, and oxygen, into which azote does not enter; or, in other words, no substance of a vegetable kind. In consequence of this difference of composition, animal and vegetable matters may be easily distinguished, when burning. The odour of each is so peculiar, that the test may be safely employed by the most inexperienced. Besides, as vegetables abound in oxygen, they have a tendency, after death, to become acid by its new combinations with carbon and hydrogen; whereas, the soft parts of animals, after death, are disposed to become alkaline, the azote entering into new combination with the hydrogen, and forming ammonia.

By the aid of this chemical character, the sponge and coral, the only beings which make any thing like an approach to the vegetable tribes, may be demonstrated to agree with the other members of the animal kingdom, in the gelatine or albumen of their soft parts, and the earthy salts which constitute their coverings or support.

2. *Animals differ from Plants in Structure.*—When we examine the structure of the solid parts of vegetables, we find them consisting of fibres or threads, which lie parallel to one another. In the solid parts of animals, on the other hand, the earthy salts are arranged in plates, forming cells.

Each fibre of a plant constitutes a tube or vessel for the circulation of the sap. As far as observation goes, these tubes are cylindrical throughout. They are aggregated into fagots, which diminish in magnitude as they proceed to the extremities of the plant; but the decrease is owing to the separation of fibres to form smaller fagots, not by the subdivision of the tubes themselves. The vessels of animals differ widely from such an arrangement. They do not constitute the solid parts. They are all of them conical, give off branches, and diminish by subdivision. The solid parts, then, of animals, are cellular;—those of vegetables vascular;—the vessels of plants are parallel and cylindrical; those of animals are irregular and conical. The former are simple; the latter are branched. The organs of support in plants consist of the vessels themselves; in animals, they are chiefly secreted salts.

3. *Animals differ from Plants in their Action.*—The preceding observations on the composition and structure of organized beings, establish the existence of characters sufficiently marked to warrant the conclusion, that animals and vegetables belong to different systems; that there is a model peculiar to each; and that, by attending to the ap-

pearances which they exhibit even after death, it is practicable to ascertain the kingdom to which any individual belongs. When we attend to the display of the living principle in the species of each division, we likewise perceive differences so very remarkable, as to dispel all the doubt and obscurity in which the subject at first appears to be involved. These differences, however, can only be distinctly traced, by a comparison of the most perfect animals with the most perfect plants; and by continuing the contrast, until we approach the lowest station of each kingdom.

When any object comes in contact with the point of my finger, I feel its *presence*, and my finger is said to possess the power of *Sensation*. When the same body comes in contact with my hair, there is no idea of its presence excited; and the hair is said to be destitute of sensation. When I compress or cut across the nerve which communicates between the brain and the finger, the faculty of sensation is suspended or destroyed. The same object may come in contact as before, but there is no feeling excited, intimating its presence. In the hair there is no nerve; and, consequently, division or compression occasions no change. I conclude, therefore, from the evidence of my own experience, that the power of feeling is inseparably connected with the presence and condition of the nerves; and that where there are no nerves, there can be no sensation. I observe, likewise, in the case of animal bodies, that the nerves are only dispersed through those parts which consist of what is termed animal matter, or combinations of carbon, hydrogen, azote and oxygen; and appear to be incapable of communicating the power of sensation to any of the combinations of the earthy salts.

In examining the different branches of the animal kingdom, we observe the presence of a nervous system, more or less developed, according to the station occupied by the species. In the animals of the higher classes, the nervous filaments can be distinctly traced; but in proportion as the absorbing, secreting, and circulating vessels of the system diminish in size, the nerves experience a corresponding decrease; and when these vessels can no longer be detected, on account of the smallness of their diameter, the nervous filaments likewise elude our observation. But in every animal, the most minute, we infer, from analogy, the existence of these vessels; and the belief of the corresponding existence of the nerves, is sanctioned by the same authority. Since all animals may thus be considered as possessing nerves, the faculty of sensation may be regarded as common to every individual of the Animal Kingdom.

Let us now turn our attention to the most perfect plants; to those in which the organs are most numerous, and most complicated in their structure. In these, however, there are no nerves, nor any substance which resembles them in form, structure and function; for the claims of the pith of plants to be considered as analogous, have been abandoned by common consent. If nerves do not exist in the most perfect plants, have we reason to anticipate their discovery in imperfect plants, which occupy a lower scale, whose organization is more simple, and whose living principles possess fewer attributes? Here analogy forbids the expectation; the microscope and the scalpel announce that it is hopeless. We conclude, therefore, that as vegetables are destitute of nerves, they are likewise destitute of the faculty of sensation.

When any object touches my finger, and excites the idea of its presence, it is the nerve which communicates the sensation. But I can remove my finger from the object, and thus cause the sensation to cease, or bring it again in contact, and renew the impression. These actions are expressed by the phrase *Voluntary Motion.* This faculty is confined to certain organs, and certain conditions of these organs. If the nerves of the finger are cut across or compressed, I am unable to communicate to my finger the requisite power. Sensation has ceased, and along with it, voluntary motion. In the one case, the sensation, by the ligature or division of the nerve, is prevented from ascending; in the other, the excitement to voluntary motion is prevented from descending, by the interposition of the same means. I conclude, therefore, that the presence of the nerves is essentially necessary to voluntary motion. Here, however, there is another condition necessary. The organs to which the nerves proceed, must be so constructed as to admit of motion. My fingers possess this requisite qualification in an eminent degree; but when many other parts of my body are touched by any object, even though I feel its presence, I cannot move away the part without a general movement of the whole body. In many cases, I cannot even accomplish my object by any movement of the body, as when the substance is situated in some central part, so constructed as not to obey the will, as is the case with gouty or urinary concretions. Voluntary motion, then, depends on the presence of nerves, and the structure of the parts to which these are distributed.

As all animals may be considered as possessing nerves, and consequently, the essential requisite of voluntary motion; the organs of animals are likewise so constructed, as to admit of its display in the variety of actions which they perform. In a few animals, however, such as the sponge, the displays of this power are scarcely discernible. But as in all such animals there are some soft parts, and as they all possess nerves, we may infer from analogy the existence of this faculty, although its operations are imperceptible to *our senses*.

Though plants possess many organs apparently well adapted to the display of voluntary motion, yet the total absence of any thing like a nervous system, the essential condition of the faculty, and the want of animal matter on which it may exert its energies, forbid us to entertain a belief of its existence.

Besides possessing the faculty of sensation and voluntary motion, I likewise am able to move my limbs in such a manner, as to change the position not of one organ merely, but of my whole body, or to shift from one place to another. This new action is termed *Locomotion*. It requires for its performance not merely the conditions requisite for sensation and voluntary motion ; but likewise an arrangement of organs so constructed, as by their action on the surrounding elements, whether of air, earth or water, the body may be displaced. Quadrupeds, birds, reptiles and fishes, possess such an arrangement of organs, and exhibit the locomotive power in a great degree of perfection. But as we descend in the scale, we find many animals in which such an organization does not exist, and that live on the same spot from the commencement to the termination of their existence. These animals, however, are all *natives of the water* ; and although they be thus stationary themselves, the fluctuations of the element in which they live, produce a variety in the scene, and daily bring new objects in contact with their organs of sensation.

Among the invertebral animals, in which this faculty is not present in every species, there does not appear to be any link of the chain, or any system of organs connected with other functions, which regulate the presence or absence of locomotion. The *Monas*, usually considered as the lowest term of animal life, and in which, neither mouth nor vessels can be perceived, is an animalcule which resides in water, and performs all its locomotive evolutions with considerably rapidity. The *Oyster*, on the other hand, in which a heart, bloodvessels, brain, gills, and stomach, may be easily observed, has one valve of its shell cemented to the rock, and depends on the bounty of the waves for all the objects of its sensation and nourishment.

Plants are destitute of sensation, voluntary motion, and locomotion. They live and die in the same spot. Where the seed first strikes root, there the plant continues, unless transported by some foreign power, and passes through the various stages of growth, maturity and decay. Plants have been found alive, even when detached from the soil; but such plants possess true roots, either as organs of support or nourishment; and only quit their first station by force. This is frequently the case with the floating duckweed, and with the fuci which form the celebrated *Mar do Sargasso*, or sea of sea-weeds, in the great ocean.

4. *Animals differ from Plants in their Nutrition.*—If we attend to the organs of nourishment in the perfect animals, we perceive that there is an alimentary canal, situated towards the centre of the body, and exhibiting in some part of its course an enlargement, termed a *Stomach*. As we descend in the scale of animal life, this tube experiences remarkable changes in form and position, but continues solitary in each individual. When all the other vessels of the body cease to be conspicuous, by reason of the diminutive size of the animal under examination, this organ may often be perceived. Indeed, judging from analogy, we have reason to conclude, that in every individual of the Animal Kingdom, there is a particular cavity destined for the reception of the food.

In the most perfect vegetables, on the other hand, we find the alimentary tubes so numerous, that we are unable to count them; and in no case do we observe any common cavity into which they empty their contents, or any enlargement which bears a resemblance to a stomach. In the inferior tribes of vegetables, the organs of nutrition appear to be constructed according to the same model. In consequence of these peculiar arrangements, the absorbing vessels of nutrition in plants, arise on the surface; while those of animals originate on the inside of the alimentary canal. Dr ALSTON, who regarded this last character as constituting an obvious and essential mark of distinction between plants and animals, fancifully termed a plant an *inverted animal* *.

In all animals, in which the intestinal canal is of sufficient size to be obvious to the senses, we perceive that crude matter is conveyed to it by the mouth; and that after a certain time, the useless part is thrown out as excrement. In plants, all the food which enters their tubes, appears to be in a state of solution; and all the superfluous quantity is dissipated in gaseous or aqueous exhalations.

Some have attempted to insinuate, that plants can live on inorganic matter, while animals can be supported only on that which is or has been organized matter, either of a vegetable or animal nature †. A moment's reflection, however, on the similarity between the elementary substances, of which all organized bodies are composed, will not fail to induce us to assign to them a common origin. All the larger animals feed on the remains of animals and vegetables, while the plants feed on the juices of the soil. But we are utterly ignorant of the particular state of combination, in which the atoms of the nourishing substances may

* "Tirocinium Botanicum Edinburgense," Edin. 8vo. 1753;—a work which may still be perused with advantage by the student of botany.

† SMITH's Introduction to Botany, p. 5.

in, at the time of the absorption by animals or plants. We know, that upon being absorbed, they enter into combinations depending on the living principle; but where is the proof, that animals can form new combinations only from those bodies already in *living union?* The Cheese Mite and the Blue Mould, are both supported by the same food; and the observation applies to many dung-beetles and mushrooms. How many plants and animals appear to subsist on water only?

In the course of the preceding observations, we have attempted to mark the characters by which plants and animals may be distinguished from each other. Instead of giving a definition of these organized beings, in order to draw the line of separation, we have preferred a description in detail. Many of the definitions which have been given, do not embrace all the species of the two classes, but serve to characterize the largest and most perfect merely. They indicate a limited acquaintance with the extensive range of living beings. Such are the definitions of THEOPHRASTUS, JUNGIUS, TOURNEFORT, PONTEDERA, LUDWIG, BOERHAAVE, LINNÆUS and MIRBEL; which do not surpass in point of clearness or acuteness, the discriminations of ARISTOTLE: From this last naturalist nearly all of them appear to have derived their origin.

CHAP. IV.

ON THE POLITY OF NATURE.

HAVING, in the preceding chapter, taken a general view of the constitution of those great classes into which the creatures of this Globe may be divided, we come now to trace the relation in which these classes stand to one another; or to examine what has been termed the *Polity of Nature*,

When we attentively consider the present condition of the inorganic kingdom, it will appear obvious, that the continuance of its existence and arrangement does not depend on the presence of organized beings. Independent of its vegetable covering, of the animals by which it is peopled, and of man himself, this globe could still revolve in its orbit, and act its part in the solar system, though naked, silent and lifeless. Changes would, indeed, take place on its surface, in consequence of the laws of chemistry and mechanics, and independent of the aid of living bodies. The prominent parts would be worn down; the hollows would be filled up; and its outline would assume an inclination every where at right angles with the direction of gravity. These changes have, in part, been accomplished; and have impressed on the different strata the peculiar characters of their structure and superposition.

The organized kingdom, on the other hand, could not exist alone. The beings of which it is composed may be considered as the parasites of this planet, and depend on its present movements for the exercise of their functions. Were the earth to approach nearer the sun, or recede to a greater distance, organized existence would be destroyed, by the mere change of temperature.

The presence of the inorganic kingdom, is necessary to the existence both of plants and animals, as furnishing them with the elements of their composition, and a place of residence. Plants may be considered as more immediately dependent on the inorganic kingdom than animals. Their attachment to the soil is more obvious; and the changes which take place in its condition, exert a more direct influence over them. The *lichens* cover the exposed surface of the rock; and, by the retention of moisture, accelerate its decomposition. The *mosses* next establish themselves in the hollows and crevices; and, by degrees, prepare a soil for the *stameniferous* vegetables. These last attempt to establish themselves in such a manner, as to banish to the still barren districts the first fabricators of their soil. The most perfect plants are, in general, independent of the Animal Kingdom, or able to subsist without their presence. The members of that great family, indeed, are their enemies rather than their friends.

The animal is, in a great measure, dependent on the vegetable kingdom for food and shelter. Some animals live directly on plants, as their only nourishment, others live on the flesh of other animals, but these last are, in general, supported by vegetable food. Hence, we may assert with confidence, that if the vegetable kingdom were to perish, the extinction of life, in the more perfect animals at least, would inevitably follow. Some of the less perfect animals are more independent in their condition. The *infusoria* appear to subsist by decomposing water. They, however, prepare a suitable repast for the *annulose* and *molluscous* tribes; and these, in their turn, contribute to support the vertebral races. In both kingdoms, therefore, the smallest and most obscure species are subservient to the welfare of those which are larger and more perfect.

In viewing the relation of these great classes of beings to one another, we perceive an admirable adaptation of means, to the establishment and continuance of the present order of things. The surface of our globe exhibits a great variety of *situation* for the residence of plants. Part is occupied by land, and part is covered with water. The land varies, in composition and moisture; the water in its contents and motion; and both vary in their temperature. But however different these situations appear to be, there are plants peculiarly adapted for each, in which they flourish with the greatest vigour, and where they are only restrained within fixed limits, by the physical characters of their station.

The condition of the Earth, which thus presents different situations for the species of the Vegetable Kingdom, influences the species of the Animal Kingdom in a similar manner. But animals are not only dependent on the physical character of their *station*, but on the presence of those vegetables on which they subsist, whether directly or indirectly. In the existing arrangements, animals are distributed, with regard to plants, in such a manner, as that a supply of food may be readily obtained; limited, however, so as to prevent the excessive increase of any particular species. In their turn, animals influence the growth of plants, by keeping many species within due bounds, and by assisting the dissemination and nourishment of others. But amidst this variety of action and reaction, and of temporary derangement, circumstances always arise, by which irregularities are checked, losses compensated, and the balance of life preserved.

LINNÆUS, from the contemplation of this subject, concluded, contrary to the generally received opinion, that animals were created on account of plants, not plants on account of animals. The defence of this opinion rests on the consideration, of animals having organs suited to cut and bruise vegetables as food, and by these operations, sometimes contributing to preserve an equal proportion among the species; and on the following reasoning,—that the iron was not made for the hammer, but the hammer for the iron,—the ground not for the plough, but the plough for the ground,—the meadow not for the scythe, but the scythe for the meadow *. The exclusive consideration of the indirect consequences of the actions of animals, has obviously betrayed LINNÆUS into this opinion. That it is erroneous, may be easily demonstrated, by the employment of his own method of reasoning. Plants, we know, are furnished with roots to penetrate the soil for nourishment and support; and fishes have fins adapted for swimming. Now, if the soil was not made for plants, but plants for the soil;

* "Amœnitates Academicæ," vol. vi. p. 22.

if the sea was not made for fish, but fish for the sea, then, instead of considering animals as created on account of plants, we must draw the mortifying conclusion, that both animals and vegetables were created on account of inorganic matter; the living for the sake of the dead.

All that we know with certainty on the subject, amounts to this, That the organized kingdom is dependent on the inorganic; that animals are greatly dependent on vegetables, and that the different tribes in each kingdom have determinate mutual relations. Judging from the mode of action peculiar to the species of each kingdom, we are led to conclude, that vegetables are superior in the scale of being to minerals; that animals are superior plants; and that they constitute a harmonious whole, in which the marks of power, wisdom and goodness, are every where conspicuous.

CHAP. V.

ON THE SUBSTANCES WHICH ENTER INTO THE COMPOSITION OF THE BODIES OF ANIMALS.

ALTHOUGH the attention of many eminent chemists has been directed to the examination of the composition of animal bodies, a great deal remains undetermined, in this difficult department of experimental research. The elementary principles which occur in the Animal Kingdom, have been ascertained with considerable precision; but the binary, ternary, or other compounds which these form, have not been investigated with so much success. As these various ingredients are brought into union in the animal system, by the agency of the vital principle, their state of combination may be expected to differ widely from the ordinary results of elective attraction. When such compounds of organization are submitted to analysis, the influence of the vital principle having ceased, the products obtained, may be regarded in many cases as modifications of the elements of the substance, occasioned by the processes employed, rather than the display of the number or nature of the ingredients as they existed previous to the analytical operations. Errors, we know, are to be guarded against in the analysis of mineral waters, arising from combinations taking place during the process, which did not exist in the compound previous to analysis*. How much more necessary is it, to exercise caution in drawing our conclusions regarding the composition of animal bodies?

In this section I shall enumerate the elementary bodies, which are considered as entering into the composition of the parts of animals, and then consider the combinations which these form, or the substances in which they have been detected.

I. Elementary Substances.

1. *Carbon.*—This substance exists in various states of combination, in the fluids as well as the solids of every animal. It has never been detected in a separate state in any animal substance. In the lungs, however, it appears to occur in the form of charcoal, according to the observations of Dr Pearson †. The lungs, in youth, are light coloured; but they increase in darkness with age, and in old persons they are nearly black. This change of colour is produced, by the deposition of charcoal in the cells of the lungs, and the bronchial glands. Dr Pearson considers

* See Dr Murray's valuable paper, "An Analysis of Sea-water." Edin. Trans. vol. viii. p. 205

† "On the Colouring Matter of the Black Bronchial Glands, and of the Black Spots of the Lungs." Phil. Trans. 1813, p. 159.

this substance as derived from the sooty matter from coal, mixed with the air inspired. If this change to blackness, occurs only in the lungs of the aged inhabitants of cities, the explanation here offered must appear plausible; but if the same change takes place in the lungs of those who live chiefly in the open air, and in the country, we should be disposed to consider the charcoal not as a foreign body, but as a morbid secretion of the organs of respiration.

When animal substances are exposed to a high temperature in closed vessels, the charcoal which is produced, differs considerably from that which is obtained by the same means from vegetables. It is more glossy in appearance, and is incinerated with much greater difficulty.

2. *Hydrogen.*—This gaseous element is universally distributed in the Animal Kingdom. It occurs as a constituent ingredient of all the fluids, and of many of the solids. It is invariably in a state of combination with charcoal; for, as far as we know, it has never been detected in an uncombined or separate state. It has been found in the human intestines, in the form of carburetted hydrogen.

3. *Oxygen.*—This principle is equally widely distributed with the preceding, in the fluids and solids of animals. A constant supply of it from the atmosphere is indispensably necessary to the continuance of animal life. It occurs, not only in combination with other bodies, but probably likewise in a separate state, in the air-bag of fishes, in which it is found varying in quantity, according to the species, and the depth at which the fishes have been caught. It is common, in union with charcoal, forming *carbonic acid.* This acid was first detected by PROUST, in an uncombined state, in urine, and by VOGEL, in blood. The latter chemist "put a quantity of fresh urine into a glass flask, to which was held a bent glass tube, the mouth of which dipped into a vessel containing lime-water. This apparatus

being put under the receiver of an air-pump, the air was slowly exhausted. A great quantity of air bubbles issued from the urine, and the lime-water became milky, indicating the extraction of carbonic acid gas *." By a similar process, it was obtained from blood. When proper precautions were used, DARWIN could not detect any air either in blood, urine, or bile, by the aid of the air-pump. The carbonic acid appeared, however, when these fluids had been exposed to the air, and seemed to be generated in consequence of the removal of atmospheric pressure †. Carbonic acid gas is likewise found in the intestines of quadrupeds; but it chiefly exists in animal bodies, in combination with the alkalies or earths. It is likewise emitted by all animals in the act of respiration.

4 *Azote.*—This gas is very widely distributed as a component part of animal substances. It occurs in almost all the fluids, and in those solid parts which have carbon as a base. The almost universal prevalence of this principle in animal substances, constitutes one of the most certain marks by which they may be distinguished from vegetables. Azote likewise occurs, in an uncombined state, in the air-bag of some fishes. It was first detected as a secretion, by Dr PRIESTLEY in the air-bag of the roach.

5. *Phosphorus.*—This inflammable body exists, in union with oxygen, in the state of phosphoric acid, in many of the solids and fluids of animals. Its existence, however, in an uncombined state, has not been satisfactorily determined, although there appears a tendency to refer the luminousness of several animals to the slow combustion of this substance. Even phosphoric acid can scarcely be said to exist in a separate state, being found in combination with potash, soda, ammonia, lime, or magnesia.

---

* Annals of Philosophy, vii. p. 56.

† Phil. Trans. 1774, p. 345.

6. *Sulphur.*—In combination with other bodies, sulphur exists in considerable abundance in animal substances. It can scarcely be said to occur in a separate state in animals; at least the experiments which may be quoted as encouraging such a supposition, are by no means decisive. United with oxygen, in the form of sulphuric acid, it exists in combination with potash, soda, and lime.

7. *Fluoric Acid.*—BERZELIUS has detected this acid in bones and in urine, in a state of combination with lime *.

8. *Muriatic Acid.*—This acid exists, in combination with an alkali, in a great number of the animal fluids, as with soda and ammonia in urine. I have observed, that several species of the genus Julus, when taken by the hand, emit an odour so much resembling this acid, as to render the opinion probable, that they secrete it, when irritated, in a free state.

9. *Iodine.*—This interesting substance has been obtained by Dr FYFE †, from the residue of the incineration of the common sponge of the shops. The water in which sponge had been infused, and which, consequently, contained its gelatinous ingredient, yielded no traces of it. The portion which remained, after infusion in water, or its albuminous part, afforded, after being burnt to ashes, distinct traces of its existence. We may conclude, therefore, that the iodine is in combination with the albumen or insoluble portion. In the sea-weeds, on the other hand, the portion which is soluble in water contains the iodine; thereby furnishing a very striking point of difference in its mode of combination in the Vegetable and Animal Kingdoms.

10. *Potash.*—This alkali exists in combination with the sulphuric, muriatic, or phosphoric acids; but it is far from

---

* Annals of Philosophy, ii. p. 416.

† "On the Plants from which Iodine can be procured."—Edin. Phil. Journ. i. 256.

abundant in animal fluids. It is much more common in the vegetable kingdom, especially in land plants.

11. *Soda.*—This alkali is present in all the fluids in various states of combination, and is more abundant than the preceding. It gives to many of the secretions the alkaline property of changing vegetable blues into green. It is found in union with the carbonic, phosphoric, sulphuric, and muriatic acids.

12. *Ammonia.*—As the elements of this alkali exist in all the fluids, and many of the solids of animals, it is frequently produced during putrefaction, and the decompositions of analysis. These elements are likewise found united in the system, and the alkali then appears in union with the various acids, as the phosphoric, muriatic, and lactic.

13. *Lime.*—This earth, of which the hard parts of animals, such as bones and shells, are chiefly composed, is of universal occurrence. It is always in a state of combination, and chiefly with the carbonic or phosphoric acids.

14. *Magnesia.*—In the animal system, this earth occurs sparingly. It has been detected in the bones, blood, and some other substances, but always in small quantity, and chiefly in union with phosphoric acid.

15. *Silica.*—This earth occurs still more sparingly than the preceding. It is found in the hair, urine, and urinary calculi.

16. *Iron.*—The existence of iron has hitherto only been detected in the colouring matter of the blood, in bile, and in milk. Its peculiar state of combination in the blood, has given rise to various conjectures; but a satisfactory solution of the question has not as yet been obtained. In milk, it appears to be in the state of phosphate.

17. *Manganese.*—The oxide of this metal has been observed, along with iron, in the ashes of hair, by VAUQUELIN.

These simple substances have been detected by chemists in the solids and fluids of animals; but seldom, as we have said, in a free state. Many of them combine, in such various proportions, that it is extremely difficult to determine their true condition, or fix the characters of their natural and permanent results. Without entering into detail, we shall briefly enumerate these various compounds; the methods of procuring them in a separate state; the characters by which they are distinguished; and their uses in the animal economy.

II. Compounds of Organization.

1. *Gelatine.*—This substance occurs, in nearly a pure state, forming the air-bag of different kinds of fishes, and well known in commerce under the name *isinglass.* When in a dry state, it is colourless and translucent; yields with difficulty to the knife; and, upon being broken, exhibits a resinous lustre, and conchoidal fracture.

When thrown into cold water, it becomes soft and increases in bulk, but does not readily dissolve. Hot water dissolves it in large quantities. The solution is translucent; of an opal colour; and gelatinizes as it becomes cold. In this state it is well known by the name of *Jelly.* Chemists term it *tremulous gelatine.* When in a dry state, gelatine may be kept for any length of time; but, when thus united with water, it very soon putrifies. Specimens procured from different substances, appear to vary considerably, even in the dried state, in the quantity of water which they retain. Those which contain the greatest quantity form the least viscid solutions; while those which contain the least, form solutions which are strong and tenacious. Acids dissolve this substance with facility, and form compounds whose properties have not been examined with much care. Alkalies likewise dissolve it readily, especially when assisted by heat. It is insoluble in alcohol.

Dr Thomson found that a concentrated solution of gelatine, yielded, with nitro-muriate of gold, a copious yellowish-white precipitate, soluble by adding water; and, with nitrate of mercury, a very copious curdy precipitate.

When a solution of tannin is dropt into a solution of gelatine, an union takes place, and an insoluble precipitate, of a whitish colour, falls to the bottom. It is on the union of the tannin of the oak bark with the gelatine of the hides, that the process of tanning leather depends.

When exposed to the destructive distillation, it yields a watery liquor, impregnated with ammonia, a fetid oil, and a bulky charcoal, which, by incineration, gives obvious traces of phosphate of lime, and phosphate of soda. It consists of carbon, hydrogen, oxygen, and azote, in proportions which have not been satisfactorily determined.

Gelatine exists in abundance in the different parts of animals, as bones, muscles, skin, ligaments, membranes, and blood. It is obtained from these substances, by boiling them in warm water; removing the impurities, by skimming, as they rise to the surface; or by subsequent straining and clarifying. It is then boiled to a proper consistency. It is the characteristic ingredient of the softest and most flexible parts of animals.

It is extensively used in the arts, under the names Glue and Size, on account of its adhesive quality, and to give the requisite stiffness to certain articles of manufacture. In domestic economy, it is likewise employed in the form of jelly, and in the formation of various kinds of soup. What is termed *Portable Soup* is merely jelly, which has been dried, having been previously seasoned, according to the taste, with different spices.

2. *Albumen.*—This substance derives its name from the Latin term for the white of an egg, that liquid being con-

sidered as albumen in as pure a state as it has hitherto been obtained.

It is a colourless viscid fluid. When dried at a low temperature, it loses four-fifths of its weight, and assumes the appearance of dried gelatine, possessing, however, a greater degree of transparency.

It is readily soluble in cold water, and the solution changes vegetable blues into green. When the solution is heated to 165° Fahr. the albumen coagulates into a white mass, of considerable consistence, which possesses peculiar properties, to be taken notice of afterwards. When albumen is in a dried or coagulated state, it will continue unchanged for a great length of time; but, when in a liquid state, it soon putrifies. During this process, silver is blackened when brought in contact with it, and the smell likewise indicates the presence of sulphuretted hydrogen. The mineral acids, alcohol and ether, when poured into a solution of albumen, coagulate it, in same manner as heat. The alkalies and earths produce no change. The metallic salts likewise occasion copious precipitates. Dr BOSTOCK found, that a drop of the saturated solution of oxymuriate of mercury, falling into water containing $\frac{1}{2000}$th part of its weight of albumen, produced an evident milkiness.

When a solution of tannin is poured into liquid albumen, a copious yellow precipitate falls down, of the consistency of pitch, and insoluble in water; which is formed by the union of the two substances.

Coagulated Albumen was first examined with care by the celebrated HATCHETT *; and several additions have been made to its history by LINK †.

It is insoluble in water, unless when long exposed to its influence; and, even then, only a small portion is taken up,

* Phil. Trans. 1800.

† Annals of Phil. vii. 456.

in a state in which it appears incapable of combining with tannin.

It is likewise insoluble in alcohol; but, when that liquid is allowed to stand upon it, or to be boiled along with it, it becomes muddy. With sulphuric acid, a brownish-yellow solution is obtained, from which the albumen is precipitated by the addition of water. Nitric and muriatic acids, when assisted by heat, likewise dissolve it; the former changing it into gelatine. Acetic acid has no effect upon it, either hot or cold.

Caustic soda and potash dissolve albumen, when assisted by heat, and form compounds resembling soap. Ammonia produces no effect upon it.

It is considered as consisting of carbon, hydrogen, azote, and oxygen, (the azote in greater quantity than in gelatine,) with traces of soda and sulphur.

Albumen exists in great abundance, both in a coagulated and liquid state, in the different parts of animals. Hair, nails, and horn are composed of it. It appears likewise as a constituent of bone and shells; and there are few of the fluid or soft parts of animals, in which it does not exist in abundance. What has hitherto been termed the *Resin of Bile*, is, according to Berzelius, analogous to albumen.

Albumen is extensively used in the arts. When spread thin on any body, it soon dries, and forms a coating of varnish. Its adhesive power is likewise considerable. When rubbed on leather, it encreases its suppleness. But its chief use is in clarifying liquors. For this purpose, any substance abounding in albumen, as the white of eggs, or the serum of blood, is mixed with the liquid, and the whole heated to near the boiling point. The albumen coagulates, and falls to the bottom, carrying along with it the impurities which were suspended in the fluid, and which rendered it muddy. If the liquor contains alcohol, the application of heat is unnecessary.

3. *Fibrin.*—This substance is usually procured for examination, from blood or muscle. From the former it is obtained, by repeatedly washing the coagulated part which appears after blood drawn from an animal has been allowed to rest, until the water ceases to extract any thing soluble; that which remains is the fibrin. Muscle yields this substance after repeated macerations in water, the mass being frequently subjected to pressure, in order to expel the blood and other fluids from the pores.

Fibrin is of a white colour, and soft and elastic, previous to its being dried. It powerfully resists putrefaction, and may be kept under water for a great length of time, without undergoing any change. It is insoluble in cold water. In boiling water, it curls up, and, after the ebullition has ceased some hours, the water acquires a milky hue. A portion of the fibrin has been dissolved, which produces, by the addition of tannin, a precipitate of white flocculi. Strong alcohol and ether partially decompose it, converting it into an adipocirous matter.

Acids dissolve fibrin directly, or alter its nature so as to render it soluble in boiling water. Acetic acid dissolves it entirely, and the solution has the appearance of jelly. With the alkalies, fibrin readily combines, and, according to BERZELIUS *, the compound bears no resemblance to soap. From the alkaline solution, the fibrin is thrown down in a more or less altered state by the addition of acids.

When exposed to the destructive distillation, it curls up, then melts, gives out water, carbonate of ammonia, acetic and carbonic acids, a fetid oil, and carburetted hydrogen gas, indicating the presence of nearly the same ingredients as enter into the composition of albumen. The charcoal

---

* Annals of Phil. ii. p. 23.

which remains is more copious than that which is left by the two preceding substances. It is incinerated with difficulty, as a glassy coating forms on its surface, and excludes the action of the air. The residue consists of the phosphate of soda and lime, together with carbonate of lime. Fibrin we have seen, exists in the blood, and was formerly called the fibrous part of the blood. It likewise exists in all muscles, forming the essential part or basis of these organs. It exhibits many remarkable varieties, as it appears in the flesh of quadrupeds, birds and fishes.

4. *Extractive.*—When the flesh of an animal is boiled for some time in water, the fibrin and albumen remain, and the solution contains the gelatine and other substances soluble in water. If the solution be evaporated to dryness, and treated with alcohol, the gelatine remains undissolved, and a solution of extractive is obtained. It is the *saponaceous extract of meat* of ROUELLE, and the *osmazome* of THENARD.

When freed from the alcohol by evaporation, the extractive is of a reddish-brown colour, semitransparent, has an acrid taste, and an odour similar to the juice of meat. It attracts moisture from the air, and becomes soft. It is soluble in alcohol and water. It is precipitated by the infusion of nut-galls. When heated, it swells, and emits the odour of burning animal matter. Its charcoal affords carbonate of potash. When distilled, it yields ammonia. Its constituent principles appear to be the same with those of the three preceding substances. According to BERZELIUS, it is always combined with the lactate of soda *.

Extractive exists in the muscles of animals, in the blood and in the brain. It communicates the peculiar flavour of meat to soups. In the opinion of FOURCROY, the brown crust of roasted meat consists of it.

* Annals of Phil. ii. p. 202.

The four substances which we have now enumerated, constitute the soft parts of animals, and enter into the composition of the hard parts and of the fluids. They are readily distinguishable from one another. Extractive alone is soluble in alcohol; gelatine is insoluble in cold, but soluble in hot water; albumen is soluble in cold, and insoluble in hot water; and fibrin is equally insoluble in hot and cold water. They are variously mixed or united; and as they consist of the same elementary principles, chiefly carbon, hydrogen, oxygen and azote, it is probable that they are changed, the one into the other, in many cases, by the living principle,—a transmutation which the chemist has succeeded in accomplishing, and which may soon be of advantage in the arts. The proportion of carbon appears to be least in gelatine, and greatest in fibrin *.

The following substances, which occur in the composition of animals, are not so exclusively distributed, although equally entitled to our attentive consideration.

5. *Mucus.*—This term has been applied to various secreted fluids, whose properties differ considerably from one another. Hence the greatest uncertainty prevails respecting the characters common to all kinds of mucus, and those which distinguish particular varieties or species. The following characters are assigned by BERZELIUS to the mucus of the saliva and of the nose, which we shall quote in his own words, for the purpose of avoiding mistakes.

"The *mucus of the saliva* † is readily procured by mixing saliva with distilled water, from which the mucus

* For farther illustrations respecting the nature of gelatine, albumen and fibrin, the reader is referred to an admirable paper by HATCHETT, "Chemical Experiments on Zoophytes; with some Observations on the component Parts of Membranes." Phil. Trans. 1800, p. 327.

† Annals of Phil. ii. p. 380.

gradually subsides, and it may then be collected on the filter and washed. In this state it is white, and would seem to contain phosphate of lime mixed with it. The mucus is quite insoluble in water *; it becomes transparent and horny in the acetic, sulphuric and muriatic acids, but does not dissolve in them, and the alkalies separate nothing from them. The mucus, therefore, contains no earthy phosphate, though its appearance would lead to suspect this earthy salt. It dissolves in caustic alkali, and is again separated from it by the acids. A small proportion escapes the action of the alkali, but yields to muriatic acid, and is not separable from this acid, by an excess of alkali. The mucus of the saliva is very easily incinerated; and though no phosphate of lime is detected in it by the acids in its natural state, a considerable portion of phosphate appears in the ash after combustion."

The *mucus of the nose* presents the following properties: "Immersed in water, it imbibes so much moisture as to become transparent, excepting a few particles that remain opaque; it may then be separated by the filter from the rest of the water, and may be further dried on blotting paper, till it has again lost nearly the whole of the moisture it had imbibed. Mucus thus dried will again absorb water when immersed in it, and resume its transparency; and this alternate wetting and drying may be repeated an indefinite number of times, but it thus gradually becomes yellowish and more resembling pus. Five parts of recent mucus absorbed by .95 parts of water, produce a glairy mass which will not pour from a vessel. When mucus is boiled with water, it does not become horny, nor does it coagulate; the violent motion of ebullition rends it in pieces, but

* This was considered by Dr Bostock as albumen, and his mucus appears to be the Peculiar Matter of the saliva of Berzelius.

when the boiling is discontinued, it is found collected again at the bottom of the vessel, and nearly as mucous as before. I should observe, however, that this mucus naturally contains a little albumen, which must first be extracted by cold water, to enable the remaining mucus to exhibit the above mentioned appearances. The nasal mucous matter dissolves in diluted sulphuric acid; when the acid is concentrated, the mucus is carbonized. Nitric acid at first coagulates it, a number of yellow spots being dispersed through the coagulum; but by continuing the digestion it softens, and it finally dissolves into a clear yellow liquid, containing none of that yellow substance which I have described under Fibrin.

"Acetous acid hardens mucous matter, but without dissolving it, even in a boiling heat. Caustic alkali, at first, renders mucous matter more viscous, and afterwards dissolves it into a limpid flowing liquid. Tannin coagulates mucus, both when softened by the absorption of water, and when dissolved either in an acid or an alkali."

This substance occurs in a liquid state in the animal economy, as a protecting covering to different organs. It necessarily differs in its qualities, according to the purposes it is destined to serve. In the nose, it defends the organ of smell from the drying influence of the air,—in the bladder, it protects the interior from the contact of the acid urine,—while it preserves the gall-bladder from the action of alkaline bile. It does not contain any suspended particles like the blood, but is homogeneous *. When inspissated, it constitutes, in the opinion of some, the basis of the epidermis, horns, nails, feathers. But the difficulty of obtaining it in a pure state, and the discordant characters assigned to it by different chemists, prevent us from reposing confidence in

* Dr Young, Annals of Phil. ii. p. 117.

the accuracy of the analysis of those substances, of which it is considered as forming an essential ingredient.

6. *Urea.*—This substance is usually procured from human urine by the following process: Evaporate the urine to the consistence of a thick syrup, which concretes into a crystalline mass as it cools. Let this be treated with four times its weight of alcohol, at different times, assisted by a gentle heat, and the greater part will be dissolved. By evaporating this solution to a syrupy consistence, the urea remains, and, as it cools, appears in the form of yellowish-white shining laminæ, crossing each other in different directions. According to Dr PROUT, it may be obtained in a state of still greater purity by the following process: "To the extract of urine evaporated to the consistence of a syrup, nitric acid is to be gradually added until the whole is converted to a crystalline mass, which is to be slightly washed with cold water. The nitric acid is then neutralized by a solution of subcarbonate of potash or soda, and the solution evaporated, in order that the nitrate of potash or soda may crystallise, and be thus separated. The fluid which is left is made into a paste with animal charcoal; cold water separates the urea from this paste in a colourless state; it is to be evaporated to dryness, and the mass digested in alcohol, which takes up the urea alone, leaving the saline bodies or other extraneous matters; and from the alcoholic solution, the urea may be obtained pure and in the crystallised state, although it is often necessary to repeat the crystallisation from the alcohol, two or three times *."

When prepared according to this last method, its crystals assume the form of a four-sided prism, transparent and colourless, with a slightly pearly lustre, sp. gr. 1.35. It has a peculiar but not urinous odour, and undergoes no

* Annals of Phil. xi. p. 352.

change from the atmosphere, except a slight deliquescence in very damp weather. In a strong heat it melts, and is partly decomposed, and partly sublimed without change. It is very soluble in water and alcohol. The fixed alkalies and alkaline earths decompose it; it unites with most of the metallic oxides, and forms crystalline compounds with the nitric and oxalic acids.

Urea is readily decomposed by boiling its solution in water, and supplying new water as it evaporates. Carbonate of ammonia is disengaged,—acetic acid is formed, and some charcoal precipitated. It is considered as containing a greater quantity of azote than any of the animal principles already enumerated. Its ingredients have been variously estimated, owing to the impurity of the urea, and the imperfection of the art of analysing animal principles. Dr Prout and M. Berard give its contents as follows:

	Carbon.	Oxygen.	Hydrogen.	Azote.	Total.
Prout,	19.99	26.66	6.66	46.66	99.97
Berard,	19.40	26.40	10.80	43.40	100.00

Urea exists in the urine of the mammalia, when in a state of health. In the human subject, it is less abundant after a meal, and nearly disappears in the disease called Diabetes, and in affections of the liver.

7. *Sugar*.—This well known substance exists in considerable abundance in milk, and in the urine of persons labouring under Diabetes. In the latter fluid, it is to be considered as a morbid secretion of the kidneys, occupying the natural situation of the urea. In milk, however, it exists as a constituent principle, and may readily be obtained by the following process: Evaporate fresh whey to the consistence of honey, dissolve it in water, clarify with the whites of eggs, and again evaporate to the consis-

tence of syrup. On cooling, white cubical crystals will be obtained. It is less sweet than vegetable sugar, and the empyreumatic oil which it yields by distillation, has a smell resembling benzoic acid. The following analyses by GAY LUSAC, THENARD, and BERZELIUS, exhibit its composition :

	Gay Lusac and Thenard.	Berzelius.
Hydrogen,	7.341	7.167
Carbon,	38.825	39.474
Oxygen,	53.834	53.359 *

According to BERZELIUS, it contains more carbon and less oxygen than common sugar. Dr PROUT, however, obtained from diabetic sugar, and sugar of milk,—results so nearly similar to those afforded by common sugar, when submitted to the same mode of analysis, that he regards them all as essentially the same substances, affected a little in their external characters, by small quantities of some extraneous substance. He states the result as follows : Hydrogen 6.66 + carbon 39.99 + oxygen 53.33 =100.00 †.

Honey, which resembles sugar in many of its properties, can scarcely be regarded as a product of the animal kingdom.

8. *Oils*.—The different bodies found in animals, referable to this division, vary greatly as to colour, consistence, smell, and other characters. They possess, however, in common, the properties of the fixed oils, in being liquid, either naturally or when exposed to a gentle heat, insoluble in water and alcohol, leaving a greasy stain upon paper, and in being highly combustible.

* Annals of Phil. v. p. 266.

† Ib xi. p. 354.

It is difficult to determine whether there is an oily principle, to which all the different kinds of oils owe their primary character, or whether there are as many different principles as there are species of oils. The important experiments of M. CHEVREUL "On Fatty Bodies *," countenance the latter opinion. The products of the saponification of the different kinds of fatty bodies, exhibited peculiar characters by which they could be recognised, varying according to the substances from which they were extracted. These products, it must be added, were the result of decomposition, and their characters, consequently, convey but imperfect information, concerning the natural condition of the principles of animal oils.

Fatty matter occurs in the animal system occupying very different situations. In ruminating animals, it adheres chiefly to the intestines; in seals and whales, it is seated under the skin; while in fishes it is chiefly found in the cells of the liver, and among the muscles. In all cases, it appears to be a product of secretion, and to serve as a store of nourishment, being most abundant, when the animal is furnished with a copious supply of food, and gradually diminishing in quantity as the food becomes scarcer, and disappearing when, from want, a lingering death has been produced.

Animal oils may be distinguished into the following kinds.

1. *Spermaceti.*—This substance constitutes the principal part of the brain of the whale, called *Physeter macrocephalus.* It is freed from the oil which accompanies it, by draining and squeezing, and afterwards by the employment of an alkaline lye which saponifies the remainder. It is

* Annals of Phil. xii. p. 186, from Annales de Chimie.

then washed in water, cut into thin pieces with a wooden knife, and exposed to the air to dry.

It is usually in the form of glossy white flakes, semi-transparent and friable. It melts at 112°, and may be distilled without experiencing great change. About 150 parts of boiling alcohol dissolve 1 of spermaceti, which is again deposited as the solution cools.

It is used in medicine, and in the arts to make candles.

2. *Ambergrease.*—This substance is of a whitish colour, brittle, adheres like wax to the edge of the knife with which it is scraped. Its specific gravity varies from 780 to 926. At 122° it melts, and at 212° it is volatilized. It is insoluble in water. Alcohol, and the volatile oils, dissolve it. With alkalies it forms a soap soluble in water. According to BOUILLON LA GRANGE, it is composed of

52.7 Adipocere,
30.8 Rezin,
5.4 Charcoal.

It is found in the intestines of the spermaceti whale, and in those only which are in a sickly state. It appears to be the excrement altered by long retention in the intestines, and therefore scarcely merits a place among the natural ingredients of the animal system. Upon being voided by the animal, it floats on the surface of the sea, and has been found in various quarters of the globe. It usually contains the beaks of cuttle-fish adhering to it.

It is employed in small quantities by druggists and perfumers.

3. *Fat.*—Fat is usually purified by separating the vessels and membranes which adhere to it, by repeatedly washing with cold water, and afterwards melting it along with boiling water. Still, however, it consists of two substances, suet and oil. CHEVREUL attempted to separate them, by

the action of boiling alcohol, which deposits, on cooling, the suet, and retains the oil. The suet he has termed *stearin*, and the oil *elain*. Braconnot effected their separation by a much more simple process. He pressed the fatty or oily matter, rendered sufficiently solid by cold, between folds of soft paper. The oil is imbibed and the fat remains. To free it from the remainder of the oil, the fat is melted with a little oil of turpentine to dilute the oil, and again pressed in paper. The oil is easily extracted from the paper by boiling water. The suet or stearin, when thus obtained, is dry and brittle, with a shining lustre, and resembles spermaceti. It melts at from 130° to 140° F. It is sparingly soluble in alcohol and ether. All fatty substances consist of different proportions of suet and oil. Butter contains from 40 to 65 *per cent.* of suet; hogs-lard 38 *per cent.*; and marrow about 76 *per cent.*

The fat of ruminating animals is termed *tallow*, and is hard and brittle, while the fat of the hog, called *lard*, is soft and semifluid. Its uses as an article of food,—in the making of candles, hard-soap, and ointments, and to diminish friction, are well known.

4. *Oil.*—The various properties of the different kinds of oils, depend, in a great degree, on the mode of preparation, with the exception of the odour, which arises from the kind of animal from which the oil has been derived. *Spermaceti oil* is considered as the thinnest of the animal oils, and the fittest for burning in lamps. It is obtained from the spermaceti, by draining and pressure. *Train oil* is procured by melting the blubber, or external layer of fat, found underneath the skin of different kinds of whales and seals. From the process employed, it contains, besides the oil, gelatine, albumen, and other animal matters, which render it thick, dark-coloured, and disposed to become rancid. *Fish-oil* is sometimes ex-

tracted from the entire fish, (as the sprat, pilchard, and herring, when they occur in too great quantities to be salted,) by boiling in water, and skimming off the oil as it appears on the surface. In general, however, the oil is obtained from the livers of fish, in which it is lodged in cells. As it cannot be procured completely from livers by mere boiling, they are allowed to become a little putrid, that the oil may be more readily extricated, by the rupturing of the cells. But, along with the oil, various impurities are likewise obtained, such as bile, and gelatinous matter; and as these afterwards putrify, they communicate a disagreeable fetid odour to the oil. This fetid smell is greatly heightened by the charring of the oil in the bottom of the boilers, and the consequent production of the empyreumatic odour.

Various methods have been recommended in order to free train and fish oils from the impurities with which they are contaminated. It is obvious, from the nature of these impurities, that they may be in part be removed by churning the oil along with cold, and afterwards with boiling water. But for the complete purification of oil, other methods, more complicated and expensive, have been resorted to. "If it be agitated with a little sulphuric acid, and then mixed with water, the oil, when allowed to settle, swims on the surface, of a much lighter colour than before; the water continues milky, and a curdy matter is observed swimming between the oil and the water *."

In the year 1761, Robert Dossie, Esq. communicated to the Society for the Encouragement of Arts and Manufactures, various processes for the edulcoration of fish oil, and obtained from the Society a bounty of one hundred pounds. The account of these methods, which are still

* THOMSON's Chemistry, 4th Ed. v. p. 476.

considered the best which have been devised, is published in the twentieth volume of their Transactions, pp. 209,–239 *.

---

* As many readers may not have access to this work, I shall here transcribe the different processes.

The first process, "For purifying fish-oil in a moderate degree, and at a very little expense.

"Take an ounce of chalk in powder, and half an ounce of lime slacked by exposure to the air; put them into a gallon of stinking oil, and, having mixed them well together by stirring, add half a pint of water, and mix that also with them by the same means. When they have stood an hour or two, repeat the stirring, and continue the same treatment at convenient intervals, for two or three days; after which, superadd a pint and a half of water, in which an ounce of salt is dissolved, and mix them as the other ingredients, repeating the stirring, as before, for a day or two. Let the whole then stand at rest, and the water will sink below the oil, and the chalk subside in it to the bottom of the vessel."

The second process, "To purify to a great degree, fish-oil without heat.

"Take a gallon of crude stinking oil, or rather such as has been prepared as above mentioned, and add to it an ounce of powdered chalk; stir them well together, several times, as in the preceding process: and after they have been mixed some hours, or a whole day, add an ounce of pearl-ashes, dissolved in four ounces of water, and repeat the stirring as before. After they have been so treated for some hours, put in a pint of water, in which two ounces of salt are dissolved, and proceed as before: the oil and brine will separate on standing some days, and the oil will be greatly improved both in smell and colour. When a greater purity is required, the quantity of pearl-ashes must be increased, and the time before the addition of the salt and water prolonged."—"If the same operation is repeated several times, diminishing each time the quantity of ingredients one half, the oil may be brought to a very light colour, and rendered equally sweet in smell with the common spermaceti oil."

The third process, "To purify fish-oil with the assistance of heat, where the greatest purity is required.

"Take a gallon of crude stinking oil, and mix with it a quarter of an ounce of powdered chalk, a quarter of an ounce of lime slacked in the air, and half a pint of water; stir them together; and when they have stood some hours, add a pint of water and two ounces of pearl ashes, and place them over a fire that will just keep them simmering till the oil appears of a light amber colour, and has lost all smell

As these impurities are chiefly contracted in the preparation of the oil, from the blubber and livers, it would be of consequence to examine, how far the various processes which have been recommended, might be employed with advantage, by the original manufacturer; or rather to devise means by which the oil might be extracted, without, at the same time, obtaining along with it, those substances which render it too thick for burning in lamps, and disagreeable on account of its smell.

Many other substances secreted by different animals, deserve to be enumerated in this place, such as Castor, Civet, and Musk, but their characters are too imperfectly ascertained, to enable us to go into detail. The facts known concerning them, will be given when treating of the organs in which they are formed.

9. *Acids*.—The acids which we have now to enumerate, consist of various proportions of carbon, hydrogen, oxygen

---

except a hot, greasy, soap-like scent. Then superadd half a pint of water, in which an ounce of salt has been dissolved; and having boiled them half an hour, pour them into a proper vessel, and let them stand till the separation of the oil, water, and lime be made, as in the preceding process."—"If the oil be required yet more pure. treat it after it is separated from the water, &c. according to the second process, with an ounce of chalk, a quarter of an ounce of pearl ashes, and half an ounce of salt."

The fourth process, "Which may be practised alone, instead of process the first, as it will edulcorate and purify fish-oil to a considerable degree, so as to answer most purposes, and for process the third, when the whole is performed.

"Take a gallon of crude stinking oil, and put it to a pint of water poured off from two ounces of lime slacked in the air; let them stand together, and stir them up several times for the first twenty-four hours; then let them stand a day, and the lime water will sink below the oil, which must be carefully separated from them. Take this oil, if not sufficiently purified for your purpose, and treat it as directed in process the third, diminishing the quantity of pearl-ashes to one ounce, and omitting the lime and chalk."—See, likewise, some judicious observations in "SCORESBY's Account of the Arctic Regions," vol. ii. p. 408,—432.

or azote. Some of them are peculiar to the animal kingdom, others exist in equal abundance in plants.

A. *Uric Acid.*—This acid, which has likewise been termed *Lithic*, exists in considerable abundance in urine, and in urinary calculi. It is likewise found in the excrements of birds, reptiles and insects. In all the situations in which it is found, it appears to be a production of the kidneys, or organs exercising an analogous function.

B. *Lactic Acid.*—This acid, which the French chemists endeavoured to prove to be the same with the acetic acid, is ascertained by BERZELIUS to be a peculiar acid *. It is of very general occurrence in the animal fluids, in combination with soda or ammonia. It is found uncombined in muscular flesh.

C. *Amniotic Acid.*—This substance was discovered in the liquor amnii of the cow, by VAUQUELIN and BUNIVA, and obtained in white crystals by evaporating that liquid slowly. It is soluble in heated water, and alcohol, and bears the nearest resemblance in its properties to the two acids which have been enumerated. Dr PROUT did not succeed in finding this substance in the uterus of a cow slaughtered in an early period of her gestation †.

D. *Formic Acid.*—This acid is obtained by infusing ants in water, and drawing it off by distillation. The water in the receiver contains the formic acid. By some, it is regarded as analogous to the acetic acid, and by others, as a mixture of the acetic and malic acids.

The preceding acids are peculiar to the animal kingdom; those now to be enumerated are likewise found in the juices of plants.

E. *Benzoic Acid* exists in considerable abundance, in

---

* Annals of Phil. ii. 201.

† Annals of Phil. v. 416.

the urine of cows, from which it is extracted in France for commercial purposes.

F. *Oxalic Acid.*—This acid has hitherto been found only in certain urinary calculi in union with lime.

G. *Acetic Acid.*—This acid has been detected in milk in union with potash, by BERZELIUS.

H. *Malic Acid.*—This acid has hitherto been found in company with the formic acid, in the liquor obtained from the red ant.

In this short enumeration of the principles which enter into the composition of animals, we perceive that carbon, hydrogen, oxygen and azote, are found in the greatest abundance. By combining in different proportions, they exhibit a great variety of separate substances. The earthy salts are likewise abundant; and when they occur in a separate state, they strengthen the albuminous frame-work, and form the skeleton, giving stability to the body, and acting as levers to the muscles. The alkaline salts occur in the greatest abundance in the secreted fluids.

These different principles, variously modified and mixed, constitute the different parts of the animal frame. These have been divided into two kinds,—*Fluids* and *Solids*.

The fluids consist of those juiçes which are obtained from our food and drink, such as the chyle, and are termed *crude;* of the *blood*, prepared from the crude fluids, and destined to communicate to every part of the body, the nourishment which it requires, and of those fluids which are separated from the blood in the course of circulation, such as the bile, and termed *Secreted Fluids.* These are all contained in appropriate vessels, and are subject to motion and change.

The *Solids* are derived from the fluids, and are usually divided into the soft and hard. The *soft* solids consist chiefly of what is termed animal matter, of combinations of carbon, hydrogen, oxygen, and azote. They consist of fibres, which are usually grouped into faggots; of plates, which, crossing one another in various directions, give rise to a cellular structure, or of a uniform pulpy mass.

The *fibrous texture* may be observed in all the muscles, tendons, and ligaments, and in the bones of many animals, especially before birth. These fibres, however minutely divided, do not appear to be hollow, like those of the Vegetable Kingdom.

The *cellular texture* is universally distributed in the form of membranes, which invest every organ, the bundles of fibres in every muscle; and, by forming tubes, with the addition of the fibrous texture, constitute the containing vessels. This substance gives form to all the different parts, and is that particular portion which is first formed, and which constitutes the frame, on and within which the other materials of the system are deposited. It readily expands by the increase of its contents; and, with equal ease, contracts when the distending cause is removed.

The *pulpy texture* is confined to the brain and nerves, the liver, kidneys, and other secreting organs of the system. Its composition appears to the eye homogeneous, and its form is regulated by its cellular envelope.

These *soft solids* are alone capable of possessing the faculty of sensation. By their aid, the nervous energy is exerted on the different parts of the body; and, through them, the impressions of external objects are received.

The *hard solids* consist either of cartilage, which resembles, in its qualities, coagulated albumen, or of bone, formed by various combinations of earthy salts. They are destitute of sensation, and are chiefly employed in defending the system from injury, giving it the requisite stability, and assisting the muscles in the execution of their movements.

The proportion between the solids and fluids, is not only remarkably different in different species, but in the same species in the various stages of growth.

Having now taken a general view of the substances which enter into the composition of animal bodies, and the various textures or tissues which these form, it is our intention to proceed to consider the frame itself, and to examine the various organs, or system of organs, of which it consists, and the uses to which these are subservient in the animal economy. The order in which we proceed, may appear to be deficient in logical precision; but it is impossible to investigate the organs, in connection with their functions, without adopting a method in some degree arbitrary. The skin, however, obviously demands our first attention.

CHAP. VI.

ON THE CUTANEOUS SYSTEM, ITS STRUCTURE, APPENDICES, AND SECRETIONS.

When treating of the characters by which organized bodies might be distinguished from inorganic matter, it was stated, that the former always possess a skin or common integument. This organ, as it exists in animals, is now to be the subject of our consideration.

1. Structure of the Skin.

Animals present remarkable differences with regard to the size, the shape, and the number of their parts; but they all agree in possessing an exterior covering, or skin, to modify their surface, regulate their form, and protect them from the action of the surrounding elements. In the more perfect animals, this organ consists of the following parts:

the cuticle,—the corpus mucosum,—the corium,—the panniculus carnosus,—and the cellular web.

1. *The Cuticle.*—This is likewise termed Epidermis, or Scarf-skin, and appears to be common to all animals. It constitutes the exterior layer of the skin, and covers not only those parts which are exposed to the influence of the atmosphere, or the waters, but the different openings of the extremities of the vessels, and lines the central surface of these with a coating, varying in its nature according to the functions of the organ which it is destined to protect. It is destitute of bloodvessels, nerves, and fibres, and usually consists of thin transparent membrane, possessing little tenacity. Various pores may be observed on its surface, which are the mouths of the vessels of the inner layers of the skin; and it is variously marked by grooves or ridges, depending on the inequalities of the surface which it covers.

It resists putrefaction for a great length of time; and may easily be obtained in a separate state from the other members, by maceration in water. It likewise separates from the inferior layers, and becomes very obvious in the case of blisters. Alcohol loosens its connection with the inner layers very readily, in many of the inferior animals, such as the earth-worm. So far as it has hitherto been subjected to chemical experiment, it appears to consist of coagulated albumen.

The cuticle exists before birth; and may be observed thickest on those places destined to be exposed to the greatest friction; such as the palms of the hands and the soles of the feet. After birth, it is speedily renewed on parts which have been wounded, or increased in thickness in those places which are frequently subjected to pressure. It is almost constantly undergoing changes. It exfoliates in the form of scales, from our own heads and from the skin of horses; in the form of powder, from the skins of parrots; and entire like a sheath, from serpents and the caterpillars of insects.

The cuticle exhibits very remarkable differences in regard to consistence. In those animals which live on the land, it is more rigid in its texture, and scaly and dry on its surface, than in those which reside in the water. In aquatic animals, it is in general smooth, often pliable; and, in many cases, its texture is so soft and delicate, that it appears like mucus. It assumes, likewise, other appearances, such as scales, nails, shells and plates, which deserve the attentive consideration of the naturalist, as furnishing him with important characters for the arrangement of animals.

2. *Mucous web.*—This has been named in honour of the discoverer *rete Malpighii:* and likewise *corpus mucosum*, or *rete mucosum.* It occurs immediately underneath the cuticle, from which, in general, it may be easily disjoined; but it is often so closely attached to the true skin below, as not to be separated even by maceration in water.

It presents more obvious appearances of organization than the cuticle, and consists of soft vascular tissue. It varies in thickness in different species, and even in different parts of the same individual. It is regenerated more slowly than the cuticle, at least its colour, after an injury, is more pale, inclining to white. In the parts of those animals which have been reproduced after amputation, as the head and horns of snails, the new portions seldom present the same intensity of colour as the old ones. Its chemical constitution is unknown. It is more or less coloured, according to the species or variety; and, indeed, is that portion of the skin in which the colour resides. In man, the other layers of the integuments are transparent or colourless; while this layer is white in the European, black in the Ethiopian, and copper-coloured in the American.

It has been supposed by some, that this layer owes it colour to the action of the solar rays; but as it is often found coloured in those parts secluded from the light, into which it enters along with the cuticle, such as the palate, tongue and ears, we must attempt to trace its origin to some other cause. BLUMENBACH refers the colour to carbon, precipitated on the mucus, and combining with it. The proofs, however, are still wanting, of the presence of this ingredient in the particular state of colour which charcoal exhibits. Were the mucous membrane always found either colourless or black, we might be induced to admit the colouring matter to be charcoal; but when this layer, in different animals, exhibits all the tints of the prismatic spectrum, we are inclined to reject the explanation as hypothetical, and wait the result of more decisive experiments. Dr BEDDOES, and afterwards FOURCROY, ascertained, that oxymuriatic acid deprived the skin of a negro of its black colour; but in a few days it returned with its former intensity. That the colour depends on an animal oil secreted by the true skin, is, perhaps, the most plausible conjecture which can be advanced on the subject.

This layer of the skin occurs not only in the warm-blooded animals, but, perhaps, in all the inferior classes *. CUVIER conjectures, that the shell of molluscous animals and the crust of lobsters, and other crustacea, occupy the place of this layer, as they are in immediate contact with the cuticle. But as these parts are employed as organs of protection and of support to the muscles, purposes to which the mucous web is never applied in its most perfect state, analogy appears in opposition to the conjecture.

3. *Corium.*—This is known by the name of the *cutis*

* It is necessary to add, that the existence of this layer is even denied by some.—Edin. Phil. Journ. i. p. 213.

vera, or true skin. It lies immediately underneath the cuticle and mucous web. It is usually destitute of colour. It consists in some animals, as quadrupeds, of solid fibres, which cross one another in every possible direction, and form a substance capable of considerable extensibility and elasticity. In the cetacea, the fibres are arranged vertically, and exhibit an appearance not unlike a section of the sessile boleti. In fishes, it is more uniform in its texture; and in some cases appears a homogeneous gelatinous membrane. It is more obviously organized, than the two members by which it is covered. Bloodvessels and nerves penetrate its substance, and may be observed forming a very delicate net-work on its surface.

The peripheral surface of this layer, is in close contact with the mucous web, from which it is with difficulty detached. When viewed separately, it presents numerous inequalities, which give to the cuticle the appearance of angular compartments on the back of the human hand, and curved ridges on the palms. When this surface is more minutely examined, it appears covered with numerous tubercles, varying in shape and size according to the species or the functions of the different parts. By macerating the corium for some days in water, the structure of these tubercles may be perceived. Each consists of a pencil of fibres, united at their base. When the central fibres are longest, the tubercle is conical; when of equal lengths, the summit is flat.

This portion of the corium has been characterised by a particular name. It is called *Corpus papillare*, and *villous surface* of the skin. It is most obvious in those parts which possess, in the greatest perfection, the sense of touch. It is generally supposed, that the nervous filaments, connected with that faculty, terminate in these tubercles. They are depressed when not in action, but become erect on the application of any stimulant.

This villous surface is observable on the skin, in almost all quadrupeds. It is found in the soles of the feet only, in birds and reptiles. In fishes and the inferior animals, its existence has not been satisfactorily ascertained.

The corium may be observed in quadrupeds, birds, reptiles and fishes. It is less distinct in the mollusca and crustacea; and in the animals of the lower orders, it has not been detected.

From the experiments which have been made, to ascertain the composition of the corium, it appears to consist chiefly of gelatine. Hence it is that part which is employed in the manufacture of glue. In the thin, soft, flexible hides, the gelatine does not possess the same degree of viscidity, as in those of more rigid texture; and the glue obtained from such, is proportionally weak.

It has already been stated, that gelatine unites with tannin, forming a substance which is insoluble in water. It is to this combination that *leather* owes its peculiar properties. The skin is prepared for being converted into leather, by maceration in water or lime-water, in order to remove the cuticle and hairs. The tannin is obtained from oak-bark, by infusion in water. The cleansed skin, now consisting chiefly of gelatine, upon being placed in this infusion, unites with the tannin. The strength of the leather depends on the strength of the hides; the more rigid these are, the more tough and durable is the leather which they furnish.

4. *Muscular web.*—This is likewise termed *Panniculus carnosus.* It varies greatly in its appearance according to the motions which the skin and its appendices are destined to perform. It consists of a layer of muscles, the extremities of whose fibres are inserted into the corium externally, and adhere to the body internally in various directions. This layer is very obvious, in the hedgehog and the porcu-

pine, to assist in rolling up the body and moving the spines, and in birds, in the erection of their feathers. In man, it can scarcely be said to exist, unless in the upper parts, where cutaneous muscles may be observed, destined for moving the skin of the face, cheeks and head. In the skin of the frog, the only cutaneous muscles which can be observed, are seated under the throat; the skin on the other parts of the body being loose and unconnected with the parts beneath.

The use of this layer of the integument, is to corrugate the skin, and elevate the hairs, feathers or spines with which it is furnished.

5. *Cellular web.*—This forms the innermost layer of the common integuments, and rests immediately on the flesh of the body. It consists of plates crossing one another in different directions, and forming a cellular membrane, varying in its thickness, tenacity and contents, according to the species. In frogs it does not exist. The cells of this membrane are filled with various substances, according to the nature of the animal. In general, they contain fat, as in quadrupeds and birds. In some of these, the layer is interrupted, as in the ruminating animals, while it is continuous in others, as the boar and the whale. In birds, while a part of this web is destined for the reception of fat, other portions are receptacles for air. In the moon-fish, the contained matter resembles albumen in its chemical characters.

Where this layer of subcutaneous fat is of considerable thickness, the sensibility of the skin above is greatly diminished. Hence, it is considered as destined to weaken the impressions of external injuries, and to protect against the effect of the changes of temperature in the surrounding element. But it is subservient to more important purposes in the animal economy. The cells are the magazines, into which the superfluous nourishment is conveyed in the form of fat, to be again absorbed according to the wants of the body. Hence, the cells are nearly empty and collapsed, in animals scantily supplied with food. The air-cells of birds give additional buoyancy to their bodies; and, therefore, assist them in their flight.

II. Appendices.

Having examined the structure of the different layers of the skin, we now proceed to consider the *appendices* with which it is furnished, and by which it is fitted for a variety of purposes in the animal economy.

1. *Hairs.*—With the general appearance of hair, every one is familiarly acquainted; but its peculiar structure and mode of growth, have seldom been attentively examined by the zoologist, although they furnish important characters for the discrimination of species.

Hairs differ remarkably not only in their structure, but likewise in their situation. In some cases, they appear to be merely filamentous prolongations of the cuticle, and subject to all its changes. This is obviously the case with the hair which covers the bodies of many caterpillars, and which separates along with the cuticle, when the animal is said to cast its skin. Such cuticular hairs are likewise found on many shells, such as the *Helix rufescens, hispida* and *spinulosa*, the *Arca lactea*, and *Pectunculus pilosus.*

In true hair, the root is in the form of a bulb, taking its rise in the cellular web. Each bulb consists of two parts, an external, which is vascular, and from which the hair probably derives its nourishment; and an internal, which is membranous, and forms a tube or sheath to the hair, during its passage through the other layers of the skin. From this bulb, and enveloped by this membrane, the hair passes through the corium, mucous web and cuticle. It usually raises up small scales of this last layer, which soon become dry and fall off, but do not form the external covering of the hair, as some have supposed.

The hair itself consists of an external horny covering, and a central vascular part, termed *medulla* or *pith*. This horny covering consists of numerous filaments placed laterally, to which different kinds of hair owe their striated appearance. These filaments appear to be of unequal lengths; those nearest the centre being longest; and, consequently, the hair assumes the form of an elongated cone, with its base seated in the skin. This form gives to the hair that peculiar property, on which the operation of *felting* depends.

When we take hold of a hair by the base with the fingers of one hand, and draw it between two of the fingers of the other, from the root towards the point, it feels smooth to the touch; but if we reverse its position, and draw it between the fingers from the point to the root, we feel its surface rough, and it offers a considerable resistance. The surface of the hair must, therefore, consist of eminences pointing to the distal extremity. In consequence of this structure of the surface, if a hair is seized at the middle between two fingers, and rubbed by them, the root will gradually recede, while the point of the hair will approach the fingers; in other words, the hair will exhibit a progressive motion in the direction of the root, the imbricated surface preventing all motion in the opposite direction.

It is owing to this state of the surface of hairs, that woollen cloth, however soft and pliable, excites a disagreeable sensation of the skin, in those not accustomed to wear it. It likewise irritates sores, by these asperities, and excites inflammation. The surface of linen cloth, on the other hand, feels smooth, because the fibres of which it consists, possess none of those inequalities of surface by which hairs are characterized.

If a quantity of wool be spread upon a table, covered with a linen cloth, and pressed down, in different directions, it is obvious that each hair will begin to move in the direction of its root, as if it had been rubbed between the fingers. The different hairs, thus moving in every direction, become interwoven with each other, and unite into a continuous mass. This is the felt with which hats are made. Curled hairs entwine themselves with one another more closely than those which are straight, though flexible, as they do not, like these, recede from the point of pressure in a straight line ; and hence hatters employ various methods to produce curl in the short fur of rabbits, hares, and moles, which they employ. This is accomplished, chiefly, by applying the solution of certain metallic salts to the fur, by a brush; so that, when the hairs dry, the surface which was moistened, contracts more than the other, and produces the requisite curve.

Hair which has been pulled from the skin, is not so fit for felting as that which has been cut by the shears. In the former case, the bulb, at the base, offers considerable resistance to the motion of the hair. This is the reason why even the short furs are cut off by sharp instruments.

Although straight hairs do not form so close and continuous a felt as curled hairs, from their tendency to procede in a straight line, in the direction of the root; yet this property is of great advantage to the hatter. He spreads over the surface of his coarser cloth, a quantity of fine straight fur; and, by pressure, these fine hairs move inwards in the direction of their roots, and thus form a coating; the base of the hairs being inserted in the cloth, while the extremities are free.

It is owing to the asperities of the surface of hair, that the spinning of wool is so difficult. This is, in a great measure, removed, by besmearing it with oil, by which the inequalities are filled up; or, at least, the asperities become less sensible. When the wool is made into cloth, it is necessary to remove the oil, which is done by the process of fulling. The cloth is placed in a trough, with water and clay, and agitated for some time. The oil is removed by the clay and water; while the agitation, acting like pressure, brings the hairs into closer union, and the cloth is taken out, not only cleansed, but felted. The hairs of every thread entwine themselves with those which are contiguous; so that the cloth may be cut without being subject to ravel. It is to this tendency to felt, that woollen cloth, and stockings, increase in density, and contract in dimensions, by being washed. In many places, woollen stuffs are felted, on a small scale, by placing them in running water, or under cascades; and the Zetlanders expose them to the motions of the tides, in narrow inlets of the sea.

The colour of the peripheral tube exhibits very remarkable differences. By some naturalists, the colour is considered as depending on the fluids contained in the pith; while, according to others, the seat of colour is in the horny covering itself. The truth of the latter opinion, in certain cases, at least, is obvious, by the inspection of the largest hairs or spines of the porcupine, in which the pith is white, while the horny covering is partly coloured. The supporters of the former opinion contend, that the central part alone is vascular, and endowed with vitality, as is demonstrated in the disease termed *Plica Polonica*, in which the hair bleeds, when cut; that grief and anxiety have been known to change the colour of the hair, by influencing, as is supposed, the colouring secretions of the central vessels;

and that the colour changes with age, and, in some animals, with the season, without any apparent change in the horny covering. These arguments we consider as conclusive in favour of the opinion which supposes the colour to depend chiefly on the fluids of the contained vessels, and would lead us to infer, that the structure of the central parts of slender hairs is different from that of the stronger kinds, termed Spines. Indeed it has been ascertained, that the pith of the bristles of the wild boar, form two canals; and, in the whiskers of the seal, one canal may be distinctly perceived.

In general, there is a close connection between the colour of the hair, and that of the mucous web. This is displayed in those animals which are spotted; in which the colour of the skin is generally variegated like that of the hair.

Hairs differ remarkably in form. In general they are round. Frequently on the body, they are thickest in the middle. Sometimes they are flat, or two-edged; and, in the whiskers of seals, they are waved on the margins. In many animals, they are long and straight; while, in others, they are crisped, and are then termed *Wool.* When stiff, they are termed *Bristles;* and, when inflexible, *Spines.* They obtain particular names, according to this situation, as beard, eye-lashes, and whiskers. In general, the motions of the hair depend on the movements of the muscular web. In some cases, however, small muscular fibres may be traced to the bulb of the root, by means of which, particular movements may be executed.

Hairs grow by the roots. In some species, they are renewed annually; and, in all, they are readily reproduced. They are the most truly electrical of animal substances. They are probably nourished by capillary attraction, and the supply regulated by the moisture or dryness of the atmosphere.

Hair is found on all the mammalia, whales not excepted; differing greatly, however, in quantity, according to the species, but always most abundant on the parts most exposed, and on those animals which inhabit cold countries. It is found, along with feathers, on different parts of the body of birds, chiefly, however, on the head and neck. It is absent in the reptiles, fishes, and mollusca; but it may be observed on many annulose animals, and even zoophytes, in which it acts the part of an organ of motion.

Of all the substances consisting of animal matter, hair is the most permanent, resisting putrefaction for a great length of time. According to the experiments of HATCHETT, it consists of gelatine and coagulated albumen. VAUQUELIN detected two kinds of oil; one of which was found to vary in colour, according to the colour of the hair from which it was extracted; black hair yielding a black oil; the other always white, without regard to the original colour of the hair. Hair likewise contains iron, oxide of manganese, phosphate and carbonate of lime, silica, and sulphur.

2. *Feathers.*—As nearly related to hairs in the animal economy, in regard to situation, composition, and purpose, we may here take notice of the plumage of birds.

Feathers may be considered as consisting of the following parts: the quill, the shaft, and the web.

The *quill* takes its rise, like the hair, in the cellular membrane, and passes through a tubular opening in the other layers of the skin, which generally exhibit a duplicature at its base, forming a short sheath. The edges and outer side of the distal part of the quill unite with the proximal end of the shaft, leaving a small opening in the middle of the inner side.

The *shaft* consists of a cuticular layer of matter similar to the substance of the quill; and a central portion, of a white colour; and with a texture like cork. The outer side is in general slightly convex; the inner nearly flat, with a groove in the middle; and the whole tapering to a point at the distal extremity. It is generally single in the large feathers of the wing and tail of birds, but frequently double in the feathers on the body.

The *web* usually occupies both sides of the shaft, and consists of the barbs, which are flattened tapering slips of matter, similar to the shaft, placed transversely, adhering thereto by their base, and lying over each other, like the leaves of a book, pointing outwards. The sides of each barb are furnished with barbules, in the same manner as the shaft is decorated with barbs.

Feathers exhibit very remarkable differences with regard to colour. The shaft is frequently of one colour, while the webs are of another; and different parts of the same web may even be observed exhibiting dissimilar tints. The colour of the feathers of some birds, appears to vary with the season, always becoming of a paler hue during the cold of winter. Particular kinds of nourishing food likewise effect similar changes; but the colours, in this case, become deeper. The feathers of several birds, after death, lose their lustre, and frequently become paler in the colour. Thus the orange-coloured white of the belly of the common merganser soon fades after death. These circumstances point out the vitality of this appendix, however difficult it may be to determine the manner of circulation of any fluid, between the parts of the fully formed feather and the body.

Feathers, like hairs, are not only renewed periodically, but they are readily reproduced, if accidentally destroyed. The manner in which they are formed, is variously stated by different authors, few of whom appear to have examined the subject with any degree of attention. The following account is given as the result of personal observation.

When the feather first pushes itself beyond the surface of the skin, it has the appearance of a thorn. If it be opened in this state, it may be observed to consist of a sheath and medulla.

The *sheath* is in the form of a tube, similar in its composition to the quill; open at the base, where it is inserted in the skin; but closed at the other extremity, where its apex possesses considerable hardness. The use of this tunic is to defend the central parts, while soft and tender.

The *medulla* consists of a gelatinous central cord, containing bloodvessels, and covered with a dark glairy fluid. It extends from the apex of the sheath, to its insertion in the cellular web. It may be considered as the organ which forms the different parts of the feather.

The glairy fluid appears densest towards the apex, and, on the outer side, two white threads may be perceived close to each other, which have been formed in it. These threads gradually increase in size, and finally coalesce, leaving, however, on their inner surface, a small groove, more or less obvious, and which is permanent. In the remaining part of the circle of this fluid, surrounding the central cord, the web of the feathers is formed. Each barb appears with one extremity attached to one of the lateral edges of the shaft, and the other, pointing upwards and inwards, assists in forming a sort of tube of barbs round the central cord. The whole is still restrained by the sheath, which frequently bears the impression, on its inner surface, of the marks of the barbs. Although, at this time, both the shaft and the barbs may be perceived in a state fit for exclusion, towards the extremity, yet, near the base, the glairy fluid scarcely exhibits a trace of the rudiments of the remaining part of the feather.

In proportion as this fluid, at the extremity, changes into the shaft and barbs, the central cord withers, and grows hard. At this time, the sheath bursts open at the point, and the withered medulla appears encircled by the shaft and barbs. As the growth takes place from below, new portions of the feather are pushed forwards with the medulla, in the same position. Between the barbs may likewise be observed numerous threads and scales, which are the dried remains of the fluid in which they were formed. These scales, and the withered medulla, either fall off naturally, or are removed by the animal in the act of *preening*.

When that part of the feather containing the web is formed, the remaining part of the medulla becomes enveloped with the matter of the quill, which is a continuation of the shaft. It is pervious at both extremities, for the entrance and egress of the medulla. Perhaps a distinct idea may be formed of the structure of a feather, by considering the whole as at first like an elongated hollow cone, in which a portion of the base retains its form, and constitutes the quill; while the remainder, towards the apex, has opened on one side, to form the shaft and the web; the medulla becoming exposed in the last part, while it continues covered in the former.

Where more shafts than one arise from the same quill, they appear to be formed on opposite sides of the medulla. This part always appears in the centre; and the grooved sides of the shaft face each other.

The quill does not attain its full growth at the first, like the shaft and the web. It continues to increase in length for a considerable period, during which the corresponding part of the medulla continues soft and vascular. As the animal cannot reach the medulla which is contained in the quill, it remains, and forms that tubular chambered membrane, with which all are familiar, as it appears in common writing quills.

The sheath, of which we have already spoken, likewise grows in length with the new parts of the feather. It is thickest at first, and falls off in the form of scales; but, towards the end of the process, it is thin, and adheres closely to the quill, and is that part which requires to be removed by artificial means, in the preparation of quills for writing.

Feathers appear to consist of nearly the same constituent ingredients as hair. They, perhaps, contain less gelatine.

They occur only on birds, and characterise the class. They furnish the most obvious marks for the discrimination of species; and have obtained various names, according to the parts of the body from which they take their rise. These terms will be particularly enumerated when we come to treat of birds.

3. *Horns*.—Horns take their rise in the same situation as hairs or feathers. They may be regarded as hairs agglutinated, and forming a hollow cone. The fibrous structure of horn may be perceived in many animals at the base, where it unites with the skin. At this part it receives the additions to its growth, the apex of the cone being pushed out in proportion as the increase takes place at the root, and on the inner surface. But horns differ remarkably from hair, in having their central cavity filled by a projection of bone or other solid substance from the body beneath.

The different markings of the horns, particularly the transverse ridges, are indications of the different layers of growth; and, in many cases, the number of these ridges corresponds with the years of life.

The colour of the horn is, in general, distributed through the mass; sometimes, however, it is collected into bands or threads. It seldom experiences much change during the life of the animal.

It is permanent, or does not experience those periodical renovations which we have stated to take place with hair and feathers. The deciduous horns of the stag are different in their nature from true horns, and will be afterwards taken notice of.

The term Horn is usually restricted to the coverings of the projections of the frontal bones of oxen, sheep, and similar quadrupeds; but various appendices of the skin, composed of the same materials, and equally permanent, although seated on other parts of the body, may, with propriety, be included under the same appellation. Among these may be enumerated,

a. Beaks.—The substance which covers the external surface of the maxillary bones of birds, is composed of horn. It varies greatly in texture and markings, according to the species. It differs from the horns of quadrupeds chiefly in this circumstance, that the colouring matter is not so completely fixed, but is probably preserved in vessels which are still in communication with the body. At least it is well known, that the colour of the bill varies with the season, and experiences considerable changes after death. This, however, is most remarkable in those birds, the horny covering of whose bills is thin, so that the colour may depend on the fluids below.

Horny beaks are found on birds and many reptiles; and, in the inferior classes, in all those whose jaws are destined to cut hard substances.

b. Hoofs.—Hoofs resemble horns in their manner of growth, and in containing a central support, formed by the termination of the extreme bones of the feet. They grow from the inner surface and base, and are thus fitted to supply the place of those parts which are worn away by being exposed to friction against hard bodies. Hoofs are peculiar to certain herbivorous quadrupeds.

c. Claws.—These resemble hoofs in structure and situation, deriving their origin from the skin, having a bony centre, and occurring at the extremities of the fingers and toes. But they are of more lengthened form, usually tapering to a point, convex on the upper surface, and concave beneath. Many quadrupeds are furnished with them, and they are usually found on the toes of birds, reptiles, and annulose animals with articulated members.

d. Nails.—Nails differ from horns and claws, in the circumstance of not being tubular, but consisting of a plate, generally convex on the outer surface, and concave beneath. Where they occur, they occupy the same position as the claws, covering the upper surface of the last joints of the fingers and toes. They derive their colour from the membranes which are beneath. They grow by the root; so that the parts which are worn off, are soon supplied. They occur on the toes of quadrupeds, birds, and reptiles, and pass, by insensible gradations, into claws.

e. Spurs.—These occur chiefly on what is termed the leg (tarsus) of gallinaceous birds. They are found, likewise, on the ornithorynchus. Like horns, they are supported in the centre by bone.

Horns, hoofs, and similar parts, bear a close resemblance to one another in chemical composition. When heated, they soften, and may be easily bent or squeezed into particular shapes. They consist of coagulated albumen, with a little gelatine; and, when incinerated, yield a little phosphate of lime.

Their use, in the animal economy, is to protect the soft parts from being injured by pressure against hard bodies. They are in general wanting, where the parts are in no danger of suffering from the influence of such agents. When torn off from the base, they are seldom completely renewed, although very remarkable exertions are frequently made by the system to repair the loss *.

4. *Scales.*—These vary remarkably in their form, structure, mode of adhesion, and situation in different animals. In general they are flat plates, variously marked. In some cases, each scale consists of several plates, the lowest of which are largest; so that the upper surface becomes somewhat imbricated. Some scales adhere by the whole of their central surface; while others resemble the human nail, in having the distal extremity free.

The composition of scales is similar to that of the cuticle, with the addition of some earthy salts. They appear to be inserted in that layer of the skin, and to resemble it in many of their properties. When rubbed off, they are easily renewed, and frequently experience the same periodical renovations as the cuticle. In some cases, the cuticle may be observed extended over them, as in the armadillos. In general, however, it only surrounds their base. They derive their colour from the mucous web on which they are placed.

In quadrupeds, scales occur, covering the whole body like a coat of mail, in the armadillos; or certain parts only, as the tail of the beaver and rat. In birds, the scales are found only on the feet. In reptiles, they occur on every part of the body, and are placed laterally in some; while, in others, they are imbricated like the slates of a house. In fishes, the scales are usually imbricated, with the distal edge free, and the epidermis enveloping their base. In some in-

* Tulpius notices some curious cases of this kind, which fell under his own observation:—"Ungues in digitorum apicibus, semel deperditos, iterum renasci novum non est, sed raro id conspicitur fieri, in secundo aut tertio articulo, prioribus amputatis, in quibus tamen non semel eosdem vidimus non secus progerminare debitamque acquirere formam, ac si in digitorum consisterent apicibus, deponente nunquam solicitudinem suam officiosa natura."—*Observationes Medicæ*, lib. iv. cap. 56.

stances, however, where they are remarkably thick, they are placed laterally, and have a covering of the epidermis spread over them. They may be observed on many insects, exhibiting great varieties of form. What are termed feathers on the wings of butterflies, seem to be a variety of scales. They are found, likewise, on the animals belonging to the class termed Annelides by CUVIER, as on some of the species of the Linnæan genus Aphrodita.

5. *Shells.*—Shells consist of layers of an earthy salt, with interposed membranes of animal matter, resembling coagulated albumen. They grow by the addition of layers of new matter to the edges and internal surface. When broken, the animal can cement the edges, and fill up the crack, or supply the deficiency, when a portion is abstracted.

In general, they occur as a covering to the corium, and are protected by the cuticle, as in the shell of the snail; but, in other cases, they are imbedded in a cell of the corium itself, as appears to be the case with the shield which protects the pulmonary cavity of the slug. In some cases, the corium adheres to the shell by a considerable extent of surface, as in many bivalve shells; while, in the univalves, the body only adheres by one or more muscles inserted in the pillar.

Shells differ remarkably in form, texture, and thickness. In some cases, they are thin semi-transparent and elastic, as the shield of the snail and aplysia; while, in others, as the volutes, they possess the opacity and hardness of marble.

They occur simple, like a scale, tube, or cone variously twisted; or in the form of double cups, simply opposed, or variously articulated. In other cases, where the shells are numerous, they cover the corium in an imbricated form, or are distributed over it in detached scales.

Shells are not exclusively the production of the animals termed *Mollusca*. They occur, likewise, as a covering to several of the *Annelides*, as the *Serpula* and *Dentalium*.

As analogous to shells, as to their use in the animal economy, we mention the membranaceous sheaths in which the bodies of several of the annelides reside, and the tubular coverings, some of which are membranaceous, others calcareous, which protect the zoophytes.

The earthy matter of shells is lime, in union with carbonic acid. Phosphate of lime has likewise been detected, but in small quantity. The colour is secreted from the animal, along with the matter of the shell.

6. *Crusts.*—These are, in general, more brittle in their texture than shell. They exhibit remarkable differences as to thickness and composition. They differ from shells chiefly in containing a considerable portion of phosphate of lime, and in a greater subdivision of parts. In some cases, however, as the crusts of the bodies of insects, the earthy matter is almost absent, and they may be regarded as formed of cuticle alone. Where they contain much earthy matter, as in the crusts of lobsters, the epidermis may be detected as a cover, and the corium beneath may be perceived as a very thin film. In many cases, these crusts are renewed periodically; and, in all, they are readily repaired. Crusts occur in insects, the crustacea, and the echinodermata, or sea-urchins, and star-fish.

In examining these different appendices of the skin, we perceive that they pass, by insensible degrees, into one another, as hair into spines, horns into nails, scales into shells, and crusts into membranes. They have all one common origin, namely, the skin; and, independent of secondary purposes, they all serve for protection.

III.—Secretions of the Skin.

Besides these appendices of the skin, which we have enumerated, it yet remains that we take notice of some of its peculiar secretions, and the purposes which these serve in the animal economy. One class of these secretions performs the office of lubricating the skin; another of regulating the temperature of the body; and a third of carrying off the superfluous carbon.

As the skin forms the external covering of the body, it is therefore exposed to the decomposing influence of the atmosphere or the waters. Were no remedy provided, its texture would soon be destroyed; by desiccation in the one case, and maceration in the other. But, for the prevention of these effects, the skin is liberally supplied with vessels, which secrete upon its surface, fluids, varying in their nature, according to the wasting agents which act upon it. They may, however, be divided into two kinds; the unctuous and the viscous; on the nature of which, we shall offer a few observations.

1. *Unctuous Secretions.* --These are confined to animals which have warm blood and the cells of the cellular web filled with fat, as mammalia and birds.

An oily matter is secreted from the whole skin of mammalia, in a manner not very satisfactorily determined. It coats, likewise, the hairs, and serves to prevent the air from drying these parts too much, or the water from wetting them. It keeps the skin of the whale so soft and smooth as to appear like oiled silk; and the seal, and other aquatic mammalia, emerge from the water with their fur as dry as if they had been executing their movements on land.

In some cases, the glands which secrete an unctuous matter, are numerous in particular places, and their openings become obvious to the eye, as in the human nose, and below the under lip, from which the greasy matter may be squeezed, like little worms. These glands are likewise numerous in the arm-pits; and, in many quadrupeds, they occur in little bags near the anus. In birds they unite, and form a compound gland on the rump, the oily matter of which is squeezed out by the bill, and spread over the feathers. What is termed the *scent* of an animal is probably a volatile oil, secreted along with this unctuous matter from these sebaceous glands.

2. *Viscous Secretions.*—In the animals with cold blood, secretions are produced, by the skin, of substances differing in quality from those of warm-blooded animals; but destined to serve the same purposes, namely, to protect the skin from the action of the surrounding element. The fluid which is secreted, has never been examined with care. In some cases, it appears glairy and adhesive, as in the common slug, resembling albumen; or mucous and slippery, as in the common eel, approaching in its properties to gelatine. As this last animal, however, turns whitish when plunged in boiling water, the secretion likewise contains albumen.

The pores from which this viscous matter is secreted, are frequently visible to the naked eye in the larger animals. They are connected, in fishes, with vessels which traverse the body under the skin, and contain the fluid; and they frequently appear on the surface in regular rows.

Besides these two kinds of secretions poured forth by the skin, which may be named protecting secretions, there is a third, peculiar to quadrupeds, known under the name of Sweat.

3. *Sweat.*—In ordinary cases, the sweat, as it escapes from the skin, is in a state of vapour, and invisible to the eye. The quantity, however, which flies off, is considerable; and has been estimated by LAVOISIER and SEGUIN, to amount, in the human body, to 1 lb. 14 oz. in twenty-four hours. When collected on the interior surface of a glass vessel, it possesses the properties of pure water. When obtained from flannel worn on the skin for several days, it yielded to THENARD acetic acid, and common salt, together with an animal matter resembling gelatine, and traces

of phosphate of lime and oxide of iron. But when procured by this method, it is impure, being mixed with the fat of the unctuous secretions.

When the body is subjected to violent exercise, or exposed to a hot atmosphere, the sweat becomes visible to the eye like drops of dew, at the pores from which it issues. In this state, it is fluid like water, but has a perceptibly saltish taste. BERZELIUS collected a few drops of sweat, as they fell from his face, and evaporated them carefully. "The yellowish residue had all the appearance under the microscope, of the usual mixtures of potash and soda, with lactic acid, lactate of soda, and its accompanying animal matter. It reddened litmus, and dissolved in alcohol; and was, without doubt, of the same nature as the analogous matter found in the other fluids."

From the circumstance, that the quantity of sweat perspired, is in proportion to the temperature, being scarcely perceptible when the surface of the body is disagreeably cold, but copious when heated to excess, it has been concluded, that this secretion is intended to regulate the degree of animal heat, and prevent its accumulation beyond certain limits. The deficiency of perspiration in the dog, is made up by the copious discharge of saliva which takes place when the animal is heated by hard exercise.

When the sweat reaches the surface of the body, it begins to evaporate; and, as fluids, in passing to a state of vapour, absorb a quantity of heat, the sweat abstracts caloric from the body, and reduces its temperature. The cooling effect of evaporation, may be felt by any one, upon wetting a finger and exposing it to the air. It will soon become colder than the others which were left dry.

There is still another secretion which takes place from the skin, differing from those which we have already enumerated, and consisting in the emission of carbon.

4. *Carbon.*—When the human hand is confined for some hours in a vessel filled with atmospheric air, a portion of carbonic acid makes it appearance. When the same hand is confined in a jar with mercury, lime-water, or hydrogen gas, no carbonic acid makes its appearance. The skin does not, therefore, give out the carbonic acid. But in the first experiment, a portion of the oxygen disappears, equal to the quantity contained in the carbonic acid that is produced. We here see the origin of the oxygen of the acid; and as no other body is present, capable of furnishing the other ingredient, the carbon, the conclusion that it is given out by the skin is irresistible. In the lower animals, the same emission of carbon from the skin has been ascertained by the labours of SPALLANZANI, PROVENÇAL and HUMBOLDT. They found, that the skins of fishes produced carbonic acid, as well as the gills; and that when frogs were deprived of their lungs, atmospheric air was still decomposed *.

It is, perhaps, difficult to point out the purposes which this emission of carbon from the skin serves in the animal economy. It appears to be a secondary kind of respiration, which, while it removes from the system the superfluous carbon, probably regulates the distribution of animal heat. But of this we shall speak more at large, when we come to treat of the organs of respiration.

There are several circumstances which prove, that the skin of the human body, in particular states, is capable of exerting an absorbing power. Whether the absorption takes place by peculiar vessels, or by the exhaling vessels having their motions reversed, or whether absorption ever takes

* We refer the reader for a minute and candid account of this cutaneous emission of carbon, to "An Inquiry into the changes induced in Atmospheric Air, by the germination of Seeds, the Vegetation of Plants, and the Respiration of Animals;" by DANIEL ELLIS, 8vo, Edin. part i. 1807, and part ii. 1811;—a work which exhibits accurate information and great caution.

place in the state of health, are questions to which no satisfactory answer has been given.

In the view which has been taken of the skin, its layers, appendices and functions, no characters have presented themselves of any great value in the construction of primary divisions in the classification of animals. The cuticle and the corium, perhaps, exist in all animals. In some, the mucous web is the seat of colour; but when this layer can no longer be detected, we still find the remaining layers of the skin, and their appendices, presenting even a brilliant display of colours, intimating, that its absence has been supplied, by the action of some of the other layers. The muscular web is present in some, and almost absent in other animals, even of the same class; and the same observation holds true with respect to the cellular web.

Hair is not peculiar to the mammalia. It is found on birds and even worms. Feathers, however, are peculiar to birds, as nothing analogous has been observed on the bodies of other animals. Horns, scales, shells, and crusts, are found on animals so widely different in form, structure and habits, as to forbid the employment of the characters which they furnish, in the construction of divisions of a high order. Lately, however, CUVIER has distributed animals into four divisions; Vertebral, Molluscous, Articulated and Radiated. The *external* character of the Articulated division, in which insects, lobsters and worms are included, is founded on the appearances presented by the skins of the animals which it includes. In these, the body is either in whole or in part formed into rings, by means of transverse plates of the skin, or the skin contains scales or crusts, placed like rings on the body, or some of its parts.

If the arrangement of the skin and its appendices, in whole or in part, into rings, was exclusively confined to those animals which are regarded as belonging to the annulose division, the investigations of the student would be greatly facilitated by its construction, as the character is obvious and defined. This, unfortunately, is far from being the case. The scales on the body of the armadillo, and on the tail of a rat, are more obviously annulose than the crust of the crab or the body of the aphrodite.

This external character, therefore, is somewhat deficient in precision, or rather in exclusive properties; and when employed to designate a primary division, may mislead or bewilder the student. This is the more to be regretted, as the internal or essential character of the division requires dissection, and is difficult to detect. It is taken from the appearances presented by the nervous system.

CHAP. VII.

ON THE OSSEOUS SYSTEM, INCLUDING THE COMPOSITION, ARTICULATIONS AND ARRANGEMENT OF BONES.

The organs which we examined in the preceding chapter, are generally considered as destined for protection, and are, therefore, placed on the exterior of the body. The bones occupy a different position. They are seated in the interior; and while they likewise assist in protecting many important organs, they, at the same time, give stability to the frame, support to the muscles, and afford levers for the execution of locomotion.

I.—Composition of Bones.

In considering the nature of bone, it will be of advantage to the reader, to be made acquainted with its composition, in order to understand its mode of growth. The different parts of which bones consist, may be reduced to four; the periosteum, cartilaginous basis, earthy matter, and fat.

1. *Periosteum.*—This bears the same relation to the bone as the skin to the body, serving as a covering for its surface, and a sheath for the different cavities which enter it. It varies in thickness, according to the nature of the bone. Its texture is obviously fibrous; and it possesses bloodvessels. Its sensibility indicates the existence of nerves. Where this organ covers cartilaginous processes, it is called Perichondrium.

2. *Cartilaginous basis.*—This part of bone apparently consists of gelatine and coagulated albumen. The gelatine may be obtained by boiling the bone, previously broken into small fragments, for a considerable time in water. If the liquor be sufficiently concentrated, it will gelatinize when cold. The coagulated albumen may be obtained, by removing the earthy salts by means of weak muriatic acid. What remains is termed *cartilage.* It retains the original figure of the bone; and may be considered as the frame, in the cells of which the earthy matter has been deposited. It contains numerous bloodvessels, which reach it by passing through the periosteum.

3. *Earthy Matter.*—The quantity of earthy matter may be ascertained, by exposing the bones for some time to a red heat. The cartilaginous basis is consumed, and the earthy matter is left behind; or the earthy matter may be obtained in solution, by steeping the bone in diluted muriatic acid. It chiefly consists of lime united with phosphoric acid, forming phosphate of lime. Carbonate of lime occurs, but generally in small quantity. Phosphate of magnesia is likewise found, sometimes to the amount of 3 *per cent.* HATCHETT detected a minute portion of sulphate of lime; and BERZELIUS has confirmed the observation of other experimentalists, as to the occurrence of fluate of lime to the extent of 3 *per cent.*

4. *Fat.*—This part of bones bears a close resemblance to the fixed oils. In some bones, as those of whales, it oc-

curs fluid like oil. In the hollow bones of oxen, it appears more like butter, and is the well known substance termed *marrow.*

When we trace the progress of the growth of bone, from its first appearance in the fœtus, unto its perfect state in the adult animal, we perceive, that the cartilaginous basis and the periosteum exist during every period: they are the only parts which are organized, and are capable of enlargement in every direction. At the commencement of the growth of the bone, the cartilaginous basis is soft, flexible, and gelatinous. By degrees the earthy matter is deposited. It proceeds, as it were, from centres near the surface of the cartilage. In the case of flat bones, the earthy matter is deposited like two plates, one above the other, which gradually unite at their edges, leaving a space between them, in which the earthy matter is more cellular than the exterior crusts. This space is termed *diploë.* In each plate, there are one or more centres of ossification, where the earthy matter is first deposited, and from which it is gradually extended over the whole surface. In the long bones, these centres of ossification are more numerous. The central parts are also more cellular than the surface, and called their *cancelli* or lattice-work. The fat, in general, resides in the cancelli, and in the hollow parts, towards the centre of the bone.

The bones increase in size, not as in shells, scales, or horns, by the addition of layers to the internal surface, but by the expansion of the cartilaginous basis; which, when it becomes saturated with earthy matter, is incapable of farther enlargement. This is the reason why the bones of young animals are soft and flexible, while those of old animals are hard and brittle.

The proportion between the cartilaginous basis, and the earthy matter, differs, not only in every animal according to age, the earthy matter being smallest in youth, but, likewise, according to the nature of the bone itself, and the purposes which it is destined to serve. The teeth contain the largest portion of earthy matter. Remarkable differences are likewise observable, according to the class or species. The bones of quadrupeds and birds contain a much greater proportion of earthy matter than those of reptiles or fishes. In some fishes, the earthy matter is so small, that the cartilage continues, during the whole life of the animal, soft, flexible, and elastic, as the spine of the lamprey, or a little more indurated, as in the bones of the skate and shark. These fishes have been termed *cartilaginous*. Even in those fishes which are termed osseous, the cartilage bears a much greater proportion to the earthy matter than in quadrupeds.

Bone is readily reproduced, in small quantities, especially in youth. In the case of fracture, the periosteum inflames and swells, the crevice is filled up by a cartilaginous basis, abounding in vessels, and the earthy matter is at length deposited, giving to the fractured part, in many cases, a greater degree of strength than it originally possessed.

In animals of the deer kind, the horns, which are true bone, are annually cast off; a natural joint forming at their base, between them and the bones of the cranium, with which they are connected. They are afterwards reproduced under a skin or periosteum, which the animal rubs off when the new horns have attained their proper size.

In some cases of disease, the earthy matter is again absorbed into the system, the cartilaginous basis predominates, and the bones become soft and tender. This takes place in the disease of youth, termed *Rickets*, and in a similar complaint of advanced life, known under the name of *Mollities ossium*. In other instances, bone is formed as a monstrous production, in organs which do not produce it in a state of health, as the brain, the heart, and the placenta *.

The ordinary colour of bone is white. This, however, is usually mixed with other colours in the teeth of herbivorous quadrupeds. The bones of some varieties of the common fowl approach to blackness, and in some fishes they are tinged with green. The nature of the food exercises a considerable influence over the colour, as demonstrated by the red tint which the bones of fowls acquire, when madder is mixed with their food.

The texture of bones exhibits many remarkable varieties. It is compact in some places, as the teeth and bones of the ear,—fibrous as the bones of the head in a fœtus,—or cellular as those of the head of the cetacea. In bones, there are likewise tortuous holes, termed *sinuses*, differing from those containing the bloodvessels or marrow, and communicating, more or less, directly with the exterior of the body.

The surface of particular bones is irregular, presenting eminences which are termed *processes*, (apophyses.) When these processes are united to the bone by the intervention of cartilage, which, with age, becomes ossified, they are termed *Epiphyses*.

As intimately connected with bone, we may here take notice of *Cartilage*. This can scarcely be said to differ in its nature, from the cartilaginous basis of the bone. It is of a fine fibrous structure, smooth on the surface, and remarkably elastic. It covers those parts of bones which are exposed to friction, as the joints, and is thickest at the point of greatest pressure. By its smoothness, it facilitates the motion of the joints, and its elasticity prevents the bad effects of any violent concussion. It is intimately

* MONRO's Outlines of Anatomy, p. 63.

united with the bone, and can scarcely be regarded as different from an elongation of the cartilaginous basis. Where it occurs at a joint with considerable motion, it is termed *articular* or *obducent* cartilage. In other cases, it occurs as a connecting medium between bones which have no articular surfaces, but where a variable degree of motion is requisite. The ribs are united to the breast-bone in this manner. Between the different vertebræ, there are interposed layers of cartilage, by which the motions of the spine are greatly facilitated. As these connecting cartilages are compressible and elastic, the spine is shortened when the body remains long in a vertical position, owing to the superincumbent pressure. Hence it is that the height of man is always less in the evening than in the morning. All these cartilages are more or less prone to ossification, in consequence of the deposition of earthy matter in the interstices. To this circumstance may be referred, in a great measure, the stiffness of age, the elasticity of the cartilages decreasing with the progress of ossification.

Cartilage occurs even unconnected with bone, in parts of the system where strength and elasticity are required.

II. Articulations of the Bones.

The manner in which the bones of the skeleton are united to one another, exhibits such remarkable differences, in respect to surface, connection, and motion, that anatomists have found it difficult to give to each kind of articulation an appropriate name, and a distinguishing character. Without attempting to enter into the details of the subject, it will be sufficient for our purpose, to enumerate the more obvious kinds of articulation, and the motions each is destined to perform; not confining our attention to the bones of the vertebral animals, but including the modes of junction between the hard parts which supply the place of bones in the lower animals

The different kinds of articulations naturally admit of a division into two classes. In the first may be placed the *true* joints, which possess articular surfaces, surrounding ligaments, and a lubricating liquor. In the second, there are no articular surfaces, the lines of junction either uniting by contact, or by the intervention of some connecting substance. The joinings are secured by a continuation of the periosteum of the one bone, until it unites with that of the other. The first class of articulations is usually termed *Diarthrosis*, (δια per αρθρον artus) ; the second has been long known by the title of *Symphysis*. As the articulation by symphysis is the most simple in its structure, we shall consider its characters in the first place.

1. *Articulation by Symphysis*.—This kind of junction takes place either without, or by means of a connecting medium. The articulation which has no connecting medium, is usually termed *Synarthrosis* ; that in which there is a connecting medium *Amphiarthrosis*.

Synarthrosis is a mode of junction which admits of no motion, the connecting surfaces coming into close and permanent contact. When two flat bones join each other by their edges, the line of division is called the *suture*. It is termed *serrated* when the edges are jagged, and when the projections of the one edge are received into recesses in the other. The bones of the human skull afford fine examples of this kind of junction. It is usually obliterated in old age by the crevices being filled up with osseous matter. The *dove-tail* in architecture, is merely an imitation of the serrated suture. When the edges of the bones are even, and come in contact without indentation, the suture is said to be *harmonic*. Examples of this kind occur in the bones of the head in quadrupeds and birds. It is admirably exhibited in the junction of the plates of the crust of the *sea urchin*, (Echinus esculentus, L.) When the thin edge of a flat bone covers that of another, the suture is called *squamous*. It is very common in the bones of the heads of fishes. In these three kinds of sutures, where they occur in vertebral animals, the line of separation is most distinct in youth, and gradually disappears in advanced life, by the two bones becoming ossified into one. In the kind of junction we are now to notice, this never takes place.

When a pit in one bone receives the projecting extremity of another, the junction is called *Gomphosis*, (γομφος *cuneus*) because the one bone is inserted like a wedge into the cavity in the other. The teeth in man and other quadrupeds, and the hooks in the snout of the saw-fish, afford good illustrations of this kind of junction. It is unknown among the invertebral animals.

There is another kind of immoveable articulation, exhibited in the claws of cats, seals, and many other animals. It consists in the one bone having a cavity, with a protuberance at its centre, which receives another bone or hard part, sheathing it at the margin, and projecting into its interior at the centre. The claws are fixed in this manner on the last phalanges of the fingers and toes of the above mentioned animals; and thus have a bony cone and duplicature, to give greater stability to their base.

Amphiarthrosis—is a mode of junction between bones, by the intervention of a substance, softer and more pliable, by means of which, a degree of motion is permitted, limited by the extent of its flexibility. Where two bones are united by the intervention of cartilage, as is most frequently the case, the junction is termed *Synchondrosis*, (συν cum χονδρος cartilago.) In this manner the ribs are united with the breast-bone, and the two sides of the lower jaw with each other, in the vertebral animals. The motion in this case is limited by the dimensions of the intervening cartilage, sometimes by its nature; for, where it serves to connect the vertebræ of quadrupeds, it is much more flexible and elastic than it is in the other parts. Where two bones are united, by the intervention of ligament, or even skin, the junction may be termed *Syndesmosis*, (συνδεσμος *ligamentum*.) It is *harmonic* when the even edges of thin bones are prevented from coming in contact, or from receding, by the intervening soft part. In this case, the extent of motion is limited by the degree of flexion or extension of which the connecting medium is susceptible. The shells of the Bernacle (Lepas) afford good examples of this kind of connection. In the *squamous syndesmosis*, the thin edge of one bone covers the surface of another, in such a manner as to permit both bones to slide upon each other. The gill-flaps of many fishes, some of the bones of the head of the lamprey, and the dorsal plates of the *chitons* exhibit this kind of articulation. In the syndesmosis, by *inclusion*, the bones or hard parts are in the form of tubes, united to each other at the extremities by a common membrane passing from the one to the other. When the common membrane is long, flexible, and elastic, almost every kind of motion may be performed. When the tubes are nearly in contact, a lateral movement only can be executed, by a part of the circumference of the one tube penetrating the interior of the other. When the one tube is smaller than the other, it may be withdrawn into the cavity of the larger, or extended, at the will of the animal. There are frequently processes and cavities in the contiguous margins of the tubes, which regulate the direction and extent of the motion this kind of junction is designed to perform. Numerous examples of articulation by inclusion, occur in the limbs of insects and crabs.

When bones are united by the intervention of flesh, the junction is termed *Syssarcosis*, (σὺν cum σαρξ caro.) Of this sort is the union of the scapula with the ribs. In some cases, particularly in the junction of the upper mandible with the head, in many birds, there is an intervening thin bony plate, which being flexible, admits of a small degree of motion, and thus constitutes an example of *osseous* amphiarthrosis.

2. *Articulation by Diarthrosis*—presents several varieties differing in the forms of the articular surfaces, and the kind of motion of which they are susceptible. The articular surfaces are enveloped with cartilage, remarkable for the smoothness of its free surface, and its intimate union with the bone, of which it forms a protecting covering. The periosteum is not continued over the surface of the cartilage, but is prolonged like a sheath over the joint, until it joins that of the opposite bone. It thus forms a close bag at the joint, into which nothing from without can enter, and from which nothing can escape. Into this bag the lubricating liquor termed *synovia* is conveyed. It is secreted by a mucous membrane on the interior, on which, as it in some cases appears like little bags, the term *bursa mucosa* has been bestowed upon it.

The nature of synovia has never been investigated with much care or success; even the experiments which have been detailed, afford results so very different, that no accurate conclusions can be drawn from them, although the synovia of the joints of oxen has been exclusively employed. The two following analyses may here be exhibited.

Water, - -	80.46	98.3
Fibrous matter, - -	11.86	
Albumen, - -	4.52	.53
Gelatine and Mucilage, -		.93
Muriate of Soda, - -	1.75	.23
Soda, - -	.71	a trace.
Phosphate of lime, -	.70	——
	100.00	99.99
	Margueron *,	*J. Davy* †.

Besides the sheath formed by the continuation of the periosteum, and which is too slender to retain the bones in their proper place, the joints are furnished with *ligaments*. These are membranes of a dense fibrous texture, flexible, elastic, and possessed of great tenacity. They have their insertion in the periosteum and bone, with which they are intimately united. When they follow the course of the periosteum, and form a sheath round the joint, they are termed *capsular*. When they occur as broad belts on the sides of the joint, they are called *lateral*; and when they pass from one articular surface, like a cord, into the other, they are called *central*. Their composition has not been investigated with care. It is probable that they approach the nature of cartilage. They serve to secure the joints from dislocation—an accident that can scarcely happen, unless the ligaments be previously lacerated. They likewise, in some cases, furnish a basis for the origin or insertion of muscles.

The motions which joints of the kind are capable of performing, may be reduced to three kinds—flexion, twisting, and sliding. In *flexion*, the free extremity of the bone which is moved, approaches the bone

* Annales de Chimie, xiv. 127.

† Monro's Outlines of Anatomy, i. p. 82.

which is fixed, describing the segment of a circle, whose centre is in the joint. In *twisting*, the bone which is moved turns round its own axis, or round an imaginary axis, passing through the articulation. In *sliding*, the free extremity of the bone moved, approaches the bone which is fixed, in a straight line.

The form of the joints, in order to admit of these different motions, is exceedingly various. When a large spherical shaped head, called a *condyle*, is received into a deep cavity, there is formed the ball and socket joint, or the *Enarthrosis* of anatomists. In this articulation, both flexion and twisting are permitted, and it is the kind of joint employed in those members where great freedom of action is required. When the cavity is shallow, and the head flat, or when both surfaces are nearly plain, the articulation is termed by anatomists *Arthrodia*. Flexion cannot be performed by this kind of joint to any extent, and the motion by twisting is likewise limited. It is principally employed where a great freedom of action is not necessary, but where flexion, twisting, and sliding are all in a small degree requisite, as is the case with the bones of the hand and the foot. Where sliding only is required, the articular surfaces are either flat, as is the case with the bones which join the orbital septum in ducks, and unite the central branches of the palatine arch with the os quadratum, or, there is a groove formed by the one bone to slide along a ridge on another. This appears to be the case with the central branches of the palatine arch in the solan-goose, which are ossified, and cover the inferior edge of the orbital ridge like a saddle. These articulations are necessary to regulate the motions of the upper mandible in birds.

When the articulating surfaces are semicylindrical, the one convex and the other concave, or when both are partly convex and partly concave, the joint is called a hinge or *ginglymus* (γιγγλυμος cardo.) This kind of articulation may be either simple or compound. It is simple when the convex surface of the one bone is received into the concave surface of the other. The under jaw of quadrupeds is articulated in this manner with the head. In the badger, this hinge-joint is so perfect, that the jaw remains in its place, even after all the ligaments and cartilages have been removed, and cannot be separated from the head without breaking the margin of the joint. In the compound ginglymus, both articulating surfaces mutually receive and are received. In this kind of joint, motion can only be performed by flexion or extension.

In attending to these different articulations in the animal system, it will readily be perceived, that they pass into one another by insensible gradations, according to the kind or the extent of the motion which is required to be performed. Hence, the same bone is not similarly articulated at the same part in different species, nor at its different extremities in the same species.

The different bones which occur in the body of an animal, when united, are termed its *skeleton*. They serve as a support and protection to the soft parts, and to determine, to a certain degree, the forms, magnitudes, and motions of the individual. The bones of a skeleton are usually divided into those of the head, the trunk, and the extremities.

The bones of the *head* are present in every animal furnished with a skeleton. They consist of those various plates which form the *cranium*, for the reception of the brain; those of the *face*, which, including the orbits of the eyes, and the cavities of the nose; and those of the *mouth*, including the jaws. All these bones vary according to the species, in number, form, articulation, and size. That part of the body where the head is placed, is denominated the anterior extremity.

The bones of the *trunk* consist of the vertebræ, ribs, and sternum. The *vertebræ* form what, in common language, is called the back-bone. They vary greatly in number, and even structure. They serve to protect the spinal marrow, and to give stability to the ribs and the extremities. They obtain different names, according to the situation in which they occur, as cervical, dorsal, lumbar, sacral, and caudal. But, in many animals, these distinctions can scarcely be retained with propriety, in consequence of the absence of the parts from which they have derived their names.

The ribs are wanting in several vertebral animals, as the *frog*, *shark*, and *ray*. When present, they are either united by the one extremity to the vertebræ, and by the other to the sternum, forming *true ribs;* or they are only fixed by a joint at one of their extremities, either to the vertebræ or sternum, and are then called *false ribs.*

The *sternum*, or breast-bone, is wanting in many animals, as serpents; and, where it does exist, it varies greatly in its form and dimensions.

The extremities, where perfect, as in man, are four in number, two arms and two feet; termed by anatomists the anterior and posterior extremities. In quadrupeds they are termed fore and hind legs; in birds, wings and legs; and in fishes, pectoral and ventral fins. The bones of the anterior extremity consist of those of the *shoulder*, the *arm*, the *fore-arm*, and the *hand*. Those of the posterior extremities, are the *hip*, the *thigh*, the *leg*, and the *foot*. In some animals, as serpents, the extremities are wanting; in whales, the posterior extremities cannot be distinguished from the tail; and in many fishes they are wanting.

Besides these bones, which are considered as more immediately connected with the skeleton, there are many others, such as those of the tongue in quadrupeds, and of the fins in fishes, which are not intimately united with the skeleton, but which serve to give strength and support to the parts where they occur.

In taking a rapid sketch of the different parts of the skeleton, we may perceive, in the more perfect animals, that, while some bones, such as those of the head and the vertebræ, are always present, others, as the sternum, ribs, and extremities, are only found in the species of particular classes or orders. The animals in which a skeleton is found, even in its simplest forms, are few in number, in comparison with those whose hard parts are either connected with the integuments, as we have already noticed, or form solid pieces of support, differing from bone in composition, and destitute of articulation. These circumstances have given rise to a division of animals into two classes, the *Vertebral* and *Invertebral*; the former possessing a vertebral column, of which the latter are destitute. The honour of forming these divisions, is claimed by M. LAMARK, in his " Systême des Animaux sans Vertebres," 8vo, 1801. The same groups, however, have been known to naturalists since the days of ARISTOTLE, the vertebral having been termed *Sanguineous*, and the invertebral *Exanguineous*. Among the vertebral animals, are included Mammalia, Birds, Reptiles, and Fishes; and, among the invertebral, those which were termed by LINNEUS, *Insects* and *Worms*.

This primary division of animals into vertebral and invertebral, is natural; and the distinction, after a little acquaintance with the species, is obvious, and of easy application.

It may, however, be observed, that, among the invertebral animals, there are hard parts, which, though differing from bone in composition and structure, answer the same purposes in the animal economy. They all appear to be formed after the manner of shell, or horn, by the addition of layers of earthy matter on one of their surfaces, and not by the extension of a cartilaginous frame-work. In composition they likewise bear a close resemblance to shell, consisting principally of carbonate of lime, with a variable proportion of animal matter. In some cases, there are traces of phosphate of lime. They are either connected with the digestive system, as the teeth in the stomach of the lobster, or the jaws and teeth of the Sea-Urchin, (Echinus,) or they constitute a central support for the soft parts, as the bone of the Cuttle-fish, or the stalk of the Sea-pen.

CHAP. VIII.

ON THE MUSCULAR SYSTEM.

HAVING considered, in the two preceding chapters, the integuments which invest, and the bones which support, the animal frame, we are now prepared to consider the organs by which motion is executed, constituting the muscular system.

The investigation of this class of organs, unfolds the most singular mechanism of parts, and an infinite variety of movements, enabling the animal to perform the internal and external actions necessary to its existence.

The muscles appear in the form of large bundles, consisting of cords. These, again, are formed of smaller threads, which are capable of division into the primary filaments. Each muscle, and all its component cords and filaments are enveloped by a covering of cellular membrane, liberally supplied by bloodvessels and nerves.

The primary filaments appear to be parallel, and to be of very small demensions. PROCHASKA * found them nearly $\frac{1}{40000}$ of an inch; while Mr BAUER estimates them at only $\frac{1}{2000}$, or the size of the globules of the blood, when deprived of their colouring matter †. Differences of

* "In meis experimentis maxima globula rubri sanguinis diameter comparata cum diametro fili carnei se habere videbatur ut 8 vel 7 ad 1; unde concluditur filum carneum a globulo rubro sanguinis superari crassitie sua vicibus inter quadraginta et sexaginta."—Operum Minorum, pars 1. p. 198.

† Phil. Trans. 1818, p. 175.

opinion, equally remarkable, prevail among observers, with regard to the appearances of these primary fibres.

LEEUWENHŒCK found the filaments of the muscles of a frog transversely striated. When recent, they were cylindrical; but, upon drying, they contracted, and exhibited a groove on one side *. Mr BAUER describes them as moniliform †, and apparently constructed of a series of globules. These globules Sir E. HOME is inclined to consider as the remains of the globules of the blood, from which they have originated, adhering in a line; an opinion which the facts of the case by no means warrant ‡.

At the extremities of the muscular fibres, where they are attached to the more solid parts, there are usually threads of a substance, differing in its appearance from the muscle, and denominated *Tendon* or *Sinew*. The tendons are, in general, of a silvery-white colour, a close, firm, fibrous texture, and possess great tenacity. The threads of which they consist, are attached on the one extremity to the surface of a bone, or other hard part; and on the other, they are variously interspersed amongthe fibres or bundles of the muscle. They are considered as destitute of sensibility and irritability, and form a passive link between the muscle and the bone, or other point of support.

The muscles, after being freed from those substances which adhere to them, such as bloodvessels and cellular tis-

---

* Oper. Om. vol. ii. p. 58. f. 4. & p. 59. f. 5.

† By the term *Moniliform* naturalists in general refer to an organ, consisting of globules lineally and closely arranged, like the beads of a necklace. When the globules press so closely as to alter their shape, and render the line of separation less distinct, the term *Jointed* is usually applied. On the other hand, when the globules are a little distant from one another, with an apparent connecting thread, the term *Perfoliate* is substituted.

‡ Phil. Trans. 1818, tab. viii. f. 4, 5, 6.

sue, consist of *fibrin*, whose properties have been already detailed. The tendons, on the other hand, consist chiefly of gelatine. Hence, when meat is boiled, the tendons are dissolved by the warm water, and the muscles separate readily from the bones. Although the presence of tendons in the muscles of testaceous mollusca cannot be perceived by the eye, yet, as boiling water destroys their connection with the shells, the presence of gelatine, and hence of tendon, may be inferred. In quadrupeds which leap much, as the Jerboa, and in birds which walk much, the tendons become in part ossified; so that the points of support of the muscle are thereby increased.

Muscles are said to be either simple or compound in their structure. In the *simple* muscles, the fibres have a similar direction; and are either formed into a long round bundle, thickest in the middle, termed *ventriform*; or they proceed from an extended base, and converge to a small tendon, when they are termed *radiated*; or resemble a feather, and are called *penniform*. In these last, the muscular fibres are arranged like the barbs of a feather, along a middle line of tendon, resembling the shaft. In the *compound* muscles, the bundles of fibres and tendons are variously interwoven.

Muscles are the most active members of the animal frame. They alone possess the power of irritability, and execute all the motions of the body. The causes which excite them to action, may be reduced to two kinds. In the first, the will, through the medium of the nerves, excites the irritability of the fibres; and in the second, the action is produced by the application of external objects, either directly, or by the medium of the nerves.

The division of the muscles into *voluntary* and *involuntary*, is sufficiently accurate to convey distinct ideas of these two classes of exciting causes in ordinary circumstances. In some cases, it is true, we can acquire by habit a controul even over the involuntary muscles; and the muscles which are voluntary in one animal, may be involuntary in another. We may, however, add, as peculiar characters by which they may be distinguished, that the involuntary muscles are more durable in their action, more easily excited, and retain, even when separated from the body, their facility of irritability for a great length of time; while the voluntary muscles become fatigued by continued action, and require intervals of rest to recruit their exhausted energies.

When the fibres of a muscle are excited to action by any irritating cause, a simultaneous movement is performed. They become shorter, more rigid, and in many cases appear somewhat angular. In consequence of this contraction, the two extremities of the muscle approach, bringing along with them the parts to which they are attached. In what manner the particles of the muscular fibre arrange themselves during contraction, anatomists have been unable to discover. That they approach each other, is obvious; and that their cohesive power suddenly increases, appears evident, from the small force which prevents a muscle from contracting, compared with the prodigious resistance which it offers when contracted. Dr WOLLASTON is of opinion, that each muscular effort, apparently single, consists in reality, of a great number of contractions, repeated at extremely short intervals; so short, indeed, that the intermediate relaxation cannot be visible, unless prolonged beyond the usual limits, by a state of partial or general debility. He arrived at this conclusion, by attending to the sounds perceived in the ear upon the insertion of the tip of the finger. It is probable, however, that these sounds are occasioned partly by the air in the ear, and partly by the motion of the bloodvessels *.

The changes which take place in the tenacity of muscles after death, are very remarkable. The same force which they could resist with ease, in a living state, is sufficient to tear them to pieces after the vital principle has departed. The tendons, which are sometimes lacerated by the violent contractions of the muscle during life, are much stronger than the muscle itself after death. These circumstances serve to confirm us in the belief of a vital power. If it does not exist, it remains with the mechanical and chemical philosophers to determine, what strengthening principle has departed with life, and whence the weakness of death.

Muscles, from their mode of action, may be divided into circular and longitudinal. In the circular muscles, the fibres are so arranged, that they contract the part to which they are applied in all its dimensions, as the muscles of the heart and tongue; or they merely contract the extremity of a tube, so as to close it. In this last case, they are denominated *sphincters*. The longitudinal muscles may be compared to ropes, drawing towards them the objects to which they are attached in a direction influenced by the form of these objects; and the other muscles acting in concert. Where the one extremity of the muscle is attached to a hard part, and the other to a soft, the mechanism is usually of the simplest kind. But where both extremities are fixed to hard parts, which are destined to be used as levers, more complicated machinery is requisite. We have already adverted to the different forms of the joints of the bones suited to different movements. The muscles are fixed to these bones at various angles, and various distan-

* Phil. Trans. 1810, p. 2.

ces from the joints. The positions assigned them are not always the most favourable for the movement of the levers; but they could not be otherwise, without altering the external shape, and giving to it an appearance the very reverse of those round or tapering forms so commonly exhibited in organized bodies. But the disadvantages attending the obliquity of the muscles, with respect to the bones which they are destined to move, are abundantly compensated by other means. The heads of the bones are frequently larger than the body to which the tendons are attached; so that these, by passing over the convexity to the place of their insertion, form a more obtuse angle with the lever, than if the head did not exist. The origin and insertion of a muscle are frequently at some distance; and the muscle passes through holes in other bones, or is bound down by annular ligaments; so that the motion is performed without destroying the proportion of the part. Although muscles, which thus move the bones as so many levers, may be compared to ropes in the effects which are produced, there is a remarkable difference in the manner of their action. As the muscles contract in length in every part, and, by the contraction, change their shape and become more rigid and angular, every bundle may be said to assist another; so that in the very muscle itself, there may be formed both levers and fulcra; thus giving to it powers, which, were it acting like a cord, it could not have in consequence of its position.

Let us now take a view of the muscles, as concerned in the production of the different motions exhibited by animals. These are exceedingly various in their extent and duration, and in the organs employed to perform them.

It is obvious, that as animals are exposed to the vicissitudes of the elements, the fluctuation of the atmosphere and the waters, they could not perform the ordinary functions of existence, unless possessed of faculties fitting them for resisting such disturbing forces. Let us therefore contemplate the provisions made for enabling animals to remain at rest, previous to an examination of their displays of the faculty of locomotion.

1. *Proneness.*—Many animals protect themselves against the disturbing movements of the air and water, by placing their bodies in a *prone position*. They thus diminish the extent of their resisting surface. To give still greater efficacy to this protecting attitude, they retire to valleys, woods, or dens, on the earth, or to the deepest places in the waters; and are thus able, by the weight of their own bodies and the advantages of their position, to outlive the elemental war. The Zetlandic fishermen have repeatedly assured me, that cod-fish swallow stones before a storm, to enable them to rest more securely at the bottom of the sea, during the continuance of the agitated waves.

But there are other animals, which, while they are equally cautious to make choice of proper situations for their safety, employ in addition, peculiar organs with which they are provided, to connect themselves more securely with the bases on which they rest.

2. *Grasping.*—The most simple of these expedients, *Grasping*, is displayed by bats, birds and insects, in the employment of their toes, with their claws, in seizing the subjects of their support. In birds, the assumption and continuance of this attitude is accomplished by a mechanical process; so that there is no expenditure of muscular energy. In every case of this kind, the claws are so admirably adapted to the station of the animal, that the detention of the body in the same spot during this state of rest, is accompanied with little exertion.

3. *Suction.*—The third method of fixing themselves, employed by animals, is *suction.* The sucker, which acts in the same manner as the moistened circular piece of leather, with a cord fixed to its centre, and applied to the surface of a stone, known to every school-boy, varies greatly in its form, and even structure. In the limpet, and other gasteropodous mollusca, its surface is smooth and uniform; and the adhesion appears to depend on its close application to every part of the opposing surface. In other animals, as the leech and the sea-urchin, the sucker is formed at the extremity of a tube; the muscular motions of which may serve to pump out any air which may remain, after the organ has been applied to the surface of the body. In a third class, the sucker is more complicated in its structure, consisting of many smaller ones, so disposed as to act in concert, as on the head of the Remora, or the breast of the lump-fish. Among the vertebral animals, neither quadrupeds nor birds possess any sucker. It is found among a few reptiles and fishes. The extremities of the toes of many insects possess complicated suckers. Among the mollusca and zoophytes, there are few in which suckers in some form do not exist. By means of this organ, whose power of cohesion must depend, not only on the extent of its surface, but the strength of the muscles which produce the vacuum, these animals can remain in the same spot, although acted on by forces to which their own weight could offer no adequate resistance. PENNANT states, that he heard of a lamprey which was taken out of the Esk, weighing three pounds, adhering to a stone of twelve pounds weight suspended at its mouth *.

4. *Cementation.*—The fourth method, termed *cementation*, employed by animals to preserve themselves stationary,

* Brit. Zool. iii. p. 78.

consists in a part of their own bodies being cemented to the substance on which they rest. This takes place in the common mussel, by means of strong cartilaginous filaments, termed the *byssus*, united in the body to a secreting gland, furnished with powerful muscles, and, at the other extremity, glued to the rock or other body to which it connects itself. In other cases, as in the oyster, the shell itself is cemented to the rock. This method of resisting the action of the disturbing forces of the air and the water, is unknown among the vertebral animals. In the Mollusca, it occurs in those with shells termed *byssiferous*, and *fixed*. In the Annulosa, among spiders and some caterpillars, protecting threads are frequently employed. Among the Zoophytes, the adhesion between them and the substance on which they grow, is generally accomplished by means of a cement, connecting (as in nearly all animals fixed by cementation) their body for life to the spot where they first adhered.

Let us now attend to the muscular motions of animals, as displayed in the various positions assumed and actions performed, in the exercise of the locomotive powers. The first of these which demands our attention, is,

1. *Standing*.—In this position the body is raised above the ground, and supported on its legs. The facility with which this attitude is assumed, and the length of time in which animals can remain in it, depend on a variety of circumstances, connected with the figure and density of the different parts of the body, and the form, position, and strength of the limbs.

In order to enable an animal to support itself firmly on its limbs, it is necessary that these parts be so disposed, as that the centre of gravity of the whole body fall within the space which they occupy, and that the muscles have sufficient power to counteract those movements which might displace the body from that position. It is obvious, that the more numerous the limbs, and the more equally they are distributed on the inferior side of the body, the more securely will the centre of gravity be retained within the space which these feet include. In the annulose animals, termed *myriapoda*, the feet, which are numerous, and very strong in proportion to the size of the body, are so placed, that the centre of gravity of the body can never fall beyond the surface occupied by them; and consequently, these animals may be considered as possessing, in the greatest degree of perfection, the qualifications necessary for assuming and maintaining a standing position. In many of the true or hexapodal insects, the same provision for preserving the body in a standing posture may likewise be observed. The feet are well qualified, by their strength, for supporting the body, and when stretched out, the space which they include is, in general, many times larger than its bulk.

Among the vertebral animals, standing is an attitude practised by the Mammalia and Birds only. Among the mammalia, all, except Man, stand on four feet. The body, in this case, rests on an extended base; but as the head projects more or less in front of the trunk, to which the fore-legs are attached, the centre of gravity falls nearer these than the hind-legs. They are on that account stronger than these last, to enable them to support this additional burden, unless in those cases where the hind-legs are used for particular kinds of motion. The head is supported by the cervical ligaments, and by muscles, which vary in strength with the weight of the head, and the actions which it is required to perform.

In Man, standing is performed on two legs only. The mechanical structure of his body enables him to do this without any singular exertion. The weight of his head, the weakness of his arms, and the inconvenient length of his legs, all prevent him from standing on his four extremities. On the other hand, the great breadth of his feet, the form of his toes, and the position of the muscles which move them, render them peculiarly calculated, not only for pressing the ground to a considerable extent, but embracing some of its inequalities. From the breadth of the pelvis, the legs are placed farther distant in proportion than in the other mammalia, and thus include a more extended base, while the trunk is able to rest upon it more securely. It is owing to these circumstances that the vertical position is easily assumed and maintained by man, while, among quadrupeds, it is painful. When they do assume it, they in general bend their hind-legs so that their buttocks rest on their heels.

In Birds, the body, during the position we are now describing, is seldom vertical with respect to the extremities. In some water-fowl, as the grebes, the feet are situated so far behind, and the weight of the body is so great anteriorly, that they could not stand with the body leaning forwards, like the generality of birds, without a peculiar arrangement of the feet. We accordingly find the body assuming nearly a vertical position. In some cases where birds stand in this attitude, the tail likewise assists in forming a more extended support, by resting on the ground, as may be seen in the cormorant. In some birds, however, as the rail, in which the legs are placed far behind, the toes are uncommonly long, so that the tendency of the fore part of the body to fall to the ground is thus prevented, by the resistance which they offer. The length of the toes, and the manner in which they are disposed, joined to the disposition of the body with regard to its centre of gravity, enable many birds to stand for a great length of time on one foot only, without much exertion. In the stork, "the surface of the femur that articulates with the tibia, has, in its middle, a depression which receives a projection of the latter bones. In bending the leg, this process is lifted out of the depression, and removed to its posterior edge. By this motion, the ligaments are necessarily more stretched than during the extension of the leg, in which the process remains in its socket. These ligaments, therefore, preserve the leg extended in the manner of some springs, without receiving any assistance from the muscles *."

2. *Walking.*—This action is defined by CUVIER, to be a motion on a fixed surface, in which the centre of gravity is alternately moved by one part of the extremities, and sustained by the other, the body never being at any time completely suspended over the ground. It is produced by the alternate flexion and extension of the limbs, aided by the motions of the trunk, advancing the position of the centre of gravity in the intended direction.

In animals with many feet, as the *myriapoda*, walking is performed by so uniform a motion, that the body may be said to glide along the surface. The feet do not move by pairs, but by divisions, containing from five to twenty, and upwards. In the insects with six feet, the anterior and posterior legs on one side, and the middle leg on the other, are moved at the same time, so that the body is always supported by two legs on the one side, and one leg on the other. The hair on the rings of caterpillars, likewise serve as feet in assisting progressive motion. In animals with four feet, "each step is executed by two legs only; one belonging to the fore-pair, and the other to the hind-pair; but sometimes they are those of the same side, and sometimes those of the opposite side." The latter is that kind of motion in horses, which grooms term a *pace.* The right fore-leg is advanced so as to sustain the body, which is thrown upon it by the left hind-foot, and at the same

* CUVIER's Comparative Anatomy, Lect. vii. *a*, 1.

time, the latter bends in order to its being moved forward. While they are off the ground, the right hind-foot begins to extend itself, and the moment they touch the ground, the left fore-foot moves forward to support the impulse of the right foot, which likewise moves forward. The body is thus supported alternately by two legs placed in a diagonal manner. When the right fore-foot moves, in order to sustain the body, pushed forward by the right hind-foot, the motion is then called an *amble*. The body, being alternately supported by two legs on the same side, is obliged to balance itself to the right and left, in order to avoid falling; and it is this balancing movement which renders the gait so soft and agreeable to women and persons in a weak state of body *.

In walking with two feet, as takes place in man and birds, the whole weight of the body rests on each leg alternately; so that the movement is in a zig-zag direction. These undulations are rendered more conspicuous in some birds, by the form of their feet and toes, by which they are prevented from moving them in the direction of the mesial line; but are, at each step, compelled to move the leg outwards, in a semicircular manner. In some quadrupeds, as the seal, and in many reptiles, the limbs are placed perpendicular to the mesial line; and, in the progressive motion, the body is dragged along the ground, as the flexion and extension of the limbs are unable to elevate it above the surface. Such animals are said to *crawl*.

But, besides these modes of walking, which we have now stated, there are others, in which the progressive motion is similar, but the means employed to accomplish it are different. The principal of these may be termed the *Serpentine* Motion. It consists in bringing up the tail towards

* CUVIER's Comp. Anat. Lect. vii.

the head, by bending the body into one or more curves, then resting upon the tail, and extending the body, thus moving forward, at each step, nearly the whole length of the body, or one or more of the curves into which it was formed. In serpents this motion is well displayed; and, in some cases, it would appear that they are assisted in it by means of their ribs, which act as feet. Among the mollusca, and many of the annulose animals, the same kind of motion is performed by alternate contractions and expansions, laterally and longitudinally, of the whole of the body, or of those parts which are appropriated to progressive motion. In some cases, these actions are so minute that the body seems to glide along the surface with a uniform progress. Many of the annulosa are assisted in their progress, by the hairs or spines with which their bodies are furnished, entering the inequalities of the surface, and preventing a retrograde movement. In some, the body is so soft and pliable, as easily to accommodate itself to the inequalities of the surface over which it glides, and derives assistance from these in its progress. In others, there is a viscous substance secreted from their bodies, which, in the slug, may enable it to attach one part of its body more firmly to the surface on which it is moving, while it drags up the remainder to a new position.

But there is another mode of moving analogous to walking, performed by means of the *suckers* to which we have already referred. Where the organ of motion is a uniform extended surface, as the foot of the limpet, the manner of advancing must resemble the motion of serpents. Part of the foot will be detached from the surface, and form arches; while the remaining points will adhere by suction, until the others reach new points of support. Where the suckers are numerous, as on the belly of some caterpillars, they act by alternate adhesion and separation, with an intervening motion resembling the action of feet. Where the organs of adhesion are double, one at each extremity, as in the leech, the mouth adheres to one part of the surface, while the tail is brought up towards it, and is then fixed, the body being at this time like an arch. The head then quits its hold, the body extends itself, and, when at full length, the head is again attached, and the tail brought up. By these alternate movements, the leech, at every *step*, advances nearly the length of its own body. In some of the intestinal worms, the adhesion of the head takes place by means of reversed spines or hooks, as in the Echinorinchus.

3. *Leaping*.—In the action of *leaping*, the whole body rises from the ground; and, for a short period, is suspended in the air. It is produced by the sudden extension of the limbs, after they have undergone an unusual degree of flexion. The extent of the leap depends on the form and size of the body, the length and the strength of the limbs. The *myriapoda* are not observed to leap. Many of the spiders and insects leap with ease, both forwards, backwards, and laterally. In those which are remarkable for this faculty, the thighs of the hind-legs are in general of uncommon size and strength. Among reptiles, the leaping frog is well known, in opposition to the crawling toad. Among quadrupeds, those are observed to leap best, which have the hind legs longer and thicker than the fore legs, as the kangaroo and the hare. These walk with difficulty, but leap with ease.

The motion of leaping is not confined to animals furnished with legs; nor, even among these, is the action confined to the legs. Serpents are said to leap, by folding their bodies into several undulations, which they unbend all at once, according as they wish to give more or less velocity to their motion. The jumping maggot, found in cheese, erects itself upon its anus, then forms its body into a circle, by bringing its head to the tail; and, having contracted every part as much as possible, unbends with a sudden jerk, and darts forward to a surprising distance. Many crabs and poduræ bend their tail, or hairs which supply its place, under their belly, and then suddenly unbending, give to the body a considerable degree of progressive motion.

4. *Flying.*—Flying is the continued suspension and progress of the whole body in the air, by the action of the wings. In leaping, the body is equally suspended in the air, but the suspension is only momentary. In flying, on the contrary, the body remains in the air, and acquires a progressive motion by repeated strokes of the wings on the surrounding fluid.

The centre of gravity of the bodies of flying animals, is always below the insertion of the wings, to prevent them falling on their backs, but near that point on which the body is, during flight, as it were suspended. The positions assumed by the head and feet are frequently calculated to accomplish these ends, and give to the wings every assistance in continuing the progressive motion.

The action of flying is performed by animals belonging to different classes. Among the Mammalia, bats display this faculty, by means of wings, formed of a thin membrane extending between the toes, which are long and spreading, the fore and hind legs, and between the hind-legs and the tail. In Birds, the wings, which occupy the place of the anterior extremities in the mammalia, and are the organs of flight, consist of feathers, which are stronger than those on the body, and of greater length. Among Reptiles, the flying lizard may be mentioned, whose membranaceous wings, projecting from each side of the body, without being connected with the legs, enable it to fly from one tree to another in search of food. A few Fishes are likewise capable of sustaining themselves for a short time by means of their fins; these are termed *flying-fish*. The power of flight, among the invertebral animals, is confined to insects with six feet. Here the wings are distinct from the feet, and vary in number from two to four. They are membranaceous in their structure, and are, in general, covered with hairs or scales. Spiders are able to move in the air by means of their threads.

In the action of flying, the tail, especially in birds, is of great use in regulating their rise and fall, and even their lateral movements.

5. *Swimming*—is the same kind of action in water, as flying is in air. The organs which are employed for this purpose, resemble the oars of a boat in their mode of action, and in general possess a considerable extent of surface and freedom of motion. The former condition enables them to strike the surrounding fluid with an oar of sufficient breadth to give progressive motion to the body; and the latter permits the same organ to be brought back to its former position for giving a second stroke, but in a different direction, and without offering so great resistance. The centre of gravity is so placed, that the body, when in action, shall rest upon the oars or swimmers, or be brought by certain means to be of the same specific gravity with the water.

The animals which are furnished with oars or swimmers, and are capable of performing this action, are chiefly confined to the vertebral, and to those with articulated limbs among the invertebral animals.

Swimming, however, is not confined to those animals which are furnished with oars or swimmers. Many animals move with ease in the water, by means of repeated undulations of the body, as serpents, eels, and leeches; or by varying the form of the body by alternate contractile and expansive movements, as the medusæ.

In these different displays of voluntary motion, the muscles are only able to continue in exercise for a limited period, during which, their irritability diminishes, and the further exertion of their powers becomes painful. When thus fatigued, animals endeavour to place themselves in a condition for resting, and fall into that state of temporary lethargy, denominated *sleep*.

The positions assumed by animals during sleep, are extremely various. In the horse, they even differ according to circumstances. In the field he lies down, in the stable he stands. Dogs and cats form their bodies into a circle, while birds place their heads under their wings.

The ordinary time of sleep is likewise exceedingly various in different animals, and in the same animal is greatly influenced by habit. It in general depends on circumstances, connected with food.

It is probable, that all animals, however low in the scale, have their stated intervals of repose, although we are as yet unacquainted either with the position which many of them assume, or the periods during which they repose.

There are many animals in which the muscular filaments cease to be perceptible, as in many zoophytes; yet, when we see all the actions of the muscular fibres performed, such as contraction and expansion, we admit their existence from analogy, and repose the utmost confidence in our conclusions.

In the classification of animals, the organs of motion are very extensively employed. They, however, do not aid us in forming divisions of the highest kind, but they assist in the construction of subordinate groups. This may appear obvious by an examination of the facts stated in this chapter. Among Quadrupeds, for example, many walk, some fly, others swim,—among Reptiles, some possess feet, others are destitute of them,—among Insects, the same species, in some genera, can either walk, fly, or swim at pleasure.

But although the characters furnished by the muscular system, are thus various and indefinite with regard to classes, they serve to mark the characters of species or genera with wonderful precislon.

The three systems of organs which we have considered in the three preceding chapters, when viewed in connection, qualify the animal frame for the higher purposes of existence. The insensible parts of the skin, bones, and cellular membrane, are the rudiments of the singular fabric, and occupy, in reference to their functions, the lowest place in the scale. The muscles, though under controul, and occupying a subordinate situation, exercise functions of a higher kind. The quality of irritability which they possess, permits them to be excited to action by the application of certain stimulants, and likewise enables the nervous system to exercise its absolute controul over them. If all the motions of the body be performed by means of the muscles under the guidance of the nervous system, we must regard its functions as of a still higher kind; and a closer attention to its various characters, will induce us to assign to it the dominion of the body. To this system, as to a second class of functions, we now propose to direct the attention of the reader.

CHAP. IX.

ON THE NERVOUS SYSTEM.

As the nervous system contains the organs of sensation and volition, and distinguishes animal from vegetable structures, it has long occupied the attention of anatomists, and is now employed by naturalists as the basis of systematical arrangement. The discoveries with which these investigations have been rewarded, are indeed numerous; but much yet remains to be done, in order to ascertain the structure and actions of various parts of this system, and to reconcile the contradictory statements of different authors. In taking our view of the subject, it will be necessary to avoid all minuteness of detail, and to attend chiefly to those circumstances which characterise the different races of animals, instead of investigating the peculiarities of particular species.

1. *Structure of the Nervous System.*—The nervous system, as it appears in its most perfect form in the vertebral animals, consists of the Brain, the Spinal Marrow, and the Nerves.

The BRAIN occurs in the anterior part of the body, surrounded by the bony covering of the skull. Between the brain and the skull, there are three membranes which are considered as the *integuments* of this part of the nervous system. The exterior of these is termed the *Dura mater*, and may be considered as the inner periosteum of the skull. It forms various processes for dividing and supporting different portions of the brain, and contains, within its duplicatures, tortuous cavities for the reception of blood, which are called *sinuses*. Underneath the dura mater, is the *Arachnoid coat*; so named from its resemblance to a spider's web in thinness. Like the former, it is extended merely over the surface of the brain, without entering its various convolutions. The third membrane of the brain is termed *Pia mater*. It adheres every where to its surface, following the course of all its irregularities, and lining its different cavities. It is thin and vascular, containing numerous bloodvessels, which likewise penetrate into the brain beneath.

The brain itself, in the more perfect animals, appears in the form of a soft, compressible, slightly viscous mass. It exhibits such differences in its texture, as to induce anatomists to consider it as composed of two distinct substances, to which they have applied the terms Cineritious and Medullary.

The *cineritious*, or cortical substance, as it is likwise called, is of a greyish colour, usually tinged with red, semitransparent, and, to the eye, appears to be homogeneous. It is soft and vascular in its texture. Injections penetrate its substance, and exhibit the existence of bloodvessels. "Its quantity, (says Cuvier,) with respect to the rest of the brain, decreases in the cold blooded animals. It is proportionally greater in man than in any other animals."

The *Medullary* matter is of a white colour, opaque, and of a firmer consistence than the cineritious. When examined with a glass, it appears to consist of fibres disposed in different directions. It possesses few bloodvessels, and injections do not penetrate to all its parts.

These two substances constitute not only the brain, but the spinal marrow and the nerves. They differ from each other in relative situation, according to circumstances. In the brain the cineritious matter is chiefly peripheral, while in the spinal marrow it is central. They cannot be distinguished, with certainty, as existing separately in the nervous system of the white blooded animals.

The chemical analysis of brain has hitherto made us but imperfectly acquainted with the ingredients of which it consists, or their peculiar state of combination. According to Vauquelin it is composed of

Water, - - - -	80.00
White fatty matter, - -	4.53
Reddish fatty matter, - -	0.70
Albumen, - - - -	7.00
Osmasome, - - - -	1.12
Phosphorus, - - - -	1.50
Acids, salts, and sulphur, -	5.15
	100.00 *

In professor JOHN's examination of the cineritious matter of the brain of a calf, he detected the following salts:—muriat of soda, a sulphat,—the phosphats of lime, soda, ammonia, and magnesia, with a trace of the phosphat of iron, and silica †.

The brain in consequence of its structure, is usually divided into two portions, termed Cerebrum and Cerebellum.

The CEREBRUM occupies the whole frontal, coronal, and a great part of the occipital portion of the skull. Its peripheral portion consists of cineritious matter; its central, of medullary, here and there mixed with cineritious. Its surface is marked by numerous furrows, of various depths, and which are convoluted in different directions. It is divided longitudinally and vertically, by a deep fissure, which receives a duplicature of the dura mater, into two nearly equal portions, which are termed *Hemispheres*. These are convex externally, irregular beneath, and flat on the opposing surfaces. Each hemisphere is divided into three *lobes*; the anterior, which rests on the orbital plate of the frontal bone; the middle, occupying the cavity formed by the sphenoid and temporal bones; and the posterior, which occurs nearest the neck, and rests on the cerebellum. In descending between the two hemispheres, towards the centre,

* Annals of Phil. iii. p. 27. † Ib. vii. p. 54.

a transverse hard and white membrane termed *Corpus callosum*, appears, of considerable thickness, convex, with its anterior and posterior extremities incurved. At each of these extremities, the hemispheres are united by a medullary cord, termed the anterior and posterior *commissures*. Underneath the corpus callosum, is the less dense partition, termed *Septum lucidum*, the inferior arched side of which is termed the *Fornix*. On each side of this division, the substance of the brain admits readily of a separation, exhibiting two cavities, termed the *Lateral Ventricles* of the brain, which are said to communicate with each other, near the middle of the fornix. In the lower and interior parts of each ventricle, there is a curved medullary prolongation, termed *Cornu Ammonis*. In the anterior part of each ventricle, there is a pyriform eminence, of a cineritious colour, and striated when divided. These are termed *Corpora striata*. Behind these, lie two other eminences, of a whiter colour, united by a soft transverse plate of medullary matter, forming what are termed *Thalami nervorum opticorum*. Between these, there is another cavity, termed the *Third Ventricle* of the brain. The communications between the different ventricles, and the fluids which they contain, are still subjects of controversy. Immediately behind the thalami, lies the *Pineal Gland*, remarkable for containing small concretions, like sand, consisting of phosphate of lime. Beneath this gland, are situated four medullary eminences, termed *Corpora quadrigemina*, the two superior and anterior of which are called *nates*; the inferior and posterior, *testes*. In each lateral ventricle, there is a plexus of minute arteries and veins, which are expanded upon the optic thalami, the pineal gland, and corpora quadrigemina.

The CEREBELLUM occupies the inferior cavities of the occipital bone, and is covered by the posterior lobes of the cerebrum. Its surface is marked by transverse furrows, which are parallel and contiguous. It may be considered as of a firmer consistence than the cerebrum, and containing a greater proportion of cineritious matter. When divided vertically, the medullary matter appears disposed throughout the cineritious, like the branches of a tree. This has given rise to the appellation *Arbor vitæ.* The cerebellum is divided longitudinally into two lobes or hemispheres, which are united with each other by means of a medullary part, termed the *vermiform process*, or third lobe of the brain. On the inferior surface, where it rests upon the commencement of the spinal marrow, there is a cavity, termed the fourth ventricle of the brain. It communicates with the third ventricle, by a passage which has been termed the *Aqueduct of Silvius.* In the bottom of the fourth ventricle, there is an angular impression, bearing a resemblance to a writing pen, and denominated *Calamus scriptorius.*

2. The SPINAL MARROW may either be considered as taking its rise from the brain, or as terminating in its substance. If we view it as originating from the brain, it consists of four cords, which are termed *Crura*, two of which proceed from the cerebrum, and two from the cerebellum. Those from the cerebrum arise between the anterior and middle lobes. The body formed by the union of these four crura, when contained within the cavity of the skull, is called *Medulla oblongata.* It is separated from the mass of the cerebrum by a medullary sheath, striated transversely, termed *Pons Varolii* or *Tuber annulare*, and is marked with a longitudinal furrow in the middle, and one on each side. Within each lateral furrow, there is a slight eminence, denominated *Corpus olivarium;* and, between this and the middle line, there are some longitudinal fibres, termed *Corpora pyramidalia.* In its progress downwards, in the canal formed in the vertebræ for its reception, it is

enveloped by the same integuments which we have already noticed as belonging to the brain itself. It differs from the brain, however, in the arrangement of its component parts, the medullary matter here occupying the surface, while the cineritious is disposed towards the centre in a cross-like form. It is divided longitudinally into two equal halves, by a channel which is very obvious on the dorsal aspect, and on each side there is likewise a groove. By some anatomists, it is considered as consisting of two medullary cords only, divided in the direction of the mesial line; while others consider it as consisting of four cords, two anterior and two posterior. Considerable support seems to be afforded to the last opinion, by the circumstance of its originating from four medullary cords, two from the cerebrum, and two from the cerebellum. The two former may be considered as forming the anterior, and the two latter the posterior portions. These cords, however, are brought into intimate union by numerous filaments, which pass from the one to the other. Mr SEWELL, of the Veterinary College, London, has ascertained the existence of a canal in the centre of the spinal marrow in the horse, bullock, sheep, hog, and dog. It extends, uninterruptedly, from the calamus scriptorius to the cauda equina; is lined by a membrane, resembling the tunica arachnoidea, and contains a transparent colourless fluid, like that which is contained in the ventricles of the brain *.

3. The NERVES may be regarded either as originating from, or terminating in, the brain or spinal marrow. We shall consider them, for the present, in the former point of view. The integuments of the nerves resemble, in appearance, those of the brain, closely invest them on all sides, and

* "A Letter on a Canal in the Medulla Spinalis of some Quadrupeds."—Phil. Trans. 1809, p. 146.

preserve their form. Besides these, there is a delicate cellular membrane, containing bloodvessels, which penetrates the interior, and forms sheaths round the component filaments. This membrane is termed *Neurilema*. It may be exhibited by dissolving the substance of the nerve in caustic potash; or the existence of filaments may be shewn by removing the neurilema by acids. In this last case, they may be observed anastamosing in various ways.

Some of the nerves appear to have a simple origin; but in general, several filaments, from different parts of the brain or spinal marrow, unite to form the trunk of a nerve. This trunk again subdivides in various ways; but the ramifications do not always exhibit a proportional decrease of size. It frequently happens, that the branches of the same or of different nerves, unite and separate repeatedly within a small space, forming a close kind of net-work, to which the name *Plexus* has been applied. Sometimes filaments pass from one nerve to another; and, at the junction, there is usually an enlargement of medullary matter termed a *Ganglion*. Numerous filaments, from different nerves, often unite to form a ganglion, from which proceed trunks frequently of greater magnitude than the filaments which entered. Thus nerves, very different in their origin, form communications with one another; so that the whole nervous system, may be considered as a kind of net-work, between the different parts of which, an intimate connection subsists. In consequence of this arrangement, it is often matter of very great difficulty to ascertain the *origin* of those filaments which unite to constitute the trunk of a nerve. In some instances, they appear to arise from the surface of the brain or spinal marrow; in other cases, from the more central parts.

The manner in which the nerves *terminate* in the different organs to which they are distributed, is still considered by many authors as undetermined. Mr CARLYLE, however, seems disposed to consider the question as of easy solution, and states the result of his investigations in the following terms:—"The terminal extremities of nerves have been usually considered of unlimited extension. By accurate dissection, however, and the aid of magnifying glasses, the extreme fibrils of nerves are easily traced, as far as their sensible properties and their continuity extend. The fibrils cease to be subdivided, whilst perfectly visible to the naked eye, in the voluntary muscles of large animals; and the spaces they occupy upon superficies where they seem to end, leave a remarkable excess of parts unoccupied by those fibrils. The extreme fibrils of nerves lose their opacity; the medullary substance appears soft and transparent; the enveloping membrane becomes pellucid; and the whole fibril is destitute of the tenacity necessary to preserve its own distinctness. It seems to be diffused or mingled with the substances in which it ends. Thus the ultimate terminations of nerves for volition, and ordinary sensation, appear to be in the reticular membrane, the common covering of all the different substances in an animal body, and the connecting medium of all dissimilar parts *." When it is considered that the nerves of sensation and volition exercise functions so very different from each other, they may be expected to exhibit corresponding differences in their connections and terminations. To detect these, the scalpel and the microscope are necessary, under the guidance of a mind habituated to observation, and cautious in its inductions.

The differences which may be observed in the nervous system of the vertebral animals are numerous, and have long occupied the attention of physiologists. But the ob-

* Phil. Trans. 1805, p. 9.

servations which have been made on this subject, are too imperfect to enable us to determine, with precision, the peculiar functions of the different parts, or the effects which a modification of the form or situation of these parts is calculated to produce. Before taking any farther notice of this part of the subject, let us attend to the actual differences of form and structure exhibited by the different classes of animals.

II. *Varieties of Structure in different Animals.*

From an examination of the brain of quadrupeds, birds, reptiles, and fishes, M. Cuvier draws the following conclusions, which mark the peculiar features of each of these classes.

"1. The character which distinguishes the brain of Mammalia from that of the other red-blooded animals, consists,

"*a.* In the existence of the corpus callosum, the fornix, the cornua ammonis, and the pons Varolii.

"*b.* In the tubercula quadrigemina being placed upon the aquæductus Sylvii.

"*c.* In the absence of ventricles in the optic thalami, and in the position of these thalami within the hemispheres.

"*d.* In the alternate white and grey lines within the corpora striata.

"2. The character peculiar to the brain of Birds, consists

"*a.* In the thin and radiated septum, which shuts each anterior ventricle on the internal side.

"3. The character of the brain of Reptiles depends

"*a.* On the position of the thalami behind the hemispheres.

"4. The character belonging to the brain of Fishes, consists,

"*a.* In the tubercles of the olfactory nerves, and the tubercles situated behind the cerebellum.

"5. The three last classes have, in common, the following characters by which they are distinguished from the first.

"*a.* Neither corpus callosum, nor fornix, nor their dependencies.

"*b.* Some tubercles, more or less numerous, situated between the corpora striata, and the optic thalami.

"*c.* The thalami containing ventricles, and being distinct from the hemispheres.

"*d.* The absence of any tubercle between the thalami and the cerebellum, as well as the absence of the pons Varolii.

"6. Fishes have certain characters in common with birds, which are not to be found in the other classes. These are,

"*a.* The position of the optic thalami under the base of the brain.

"*b.* The number of the tubercles placed before these thalami, which are commonly four.

"7. Fishes and reptiles have, for a common character, distinguishing them from the two first classes, the absence of the arbor vitæ in the cerebellum.

"8. All red-blooded animals have the following characters in common.

"*a.* The principal division into hemispheres, optic thalami, and cerebellum.

"*b.* The anterior ventricles double; the third and fourth single; the aquæductus Sylvii; the infundibulum; and a communication between all their cavities.

"*c.* The corpora striata, and their appendices, in the form of a vault, called hemispheres.

"*d.* The anterior and posterior commissures, and the valve of the cerebrum.

"*e.* The bodies named Pineal and Pituitary Glands.

"*f.* The union of the great single tubercle or cerebellum, by two transverse crura, with the rest of the brain, which gives origin to the two longitudinal crura of the medulla oblongata.

" 9. It also appears, that there exist certain relations between the faculties of animals, and the proportions of their common parts.

" Thus the intelligence they possess, appears more perfect in proportion to the volume of the appendix of the corpus striatum, which forms the vault of the hemispheres.

" Man has that part greater, more extended, and more reflected than the other animals.

" In proportion as we descend from man, we observe that it becomes smaller and smoother on the surface, and that the parts of the brain are less complicated with each other, but seem to be unfolded and spread out longitudinally.

" It even appears that certain parts assume, in all classes, forms which have a relation to particular qualities of animals: for example, the anterior tubercula quadrigemina of *carps*, which are the most feeble and least carnivorous of fishes, are proportionally larger than in the other genera, in the same manner as they are in the herbivorous quadrupeds. By following these inquiries, we may hope to obtain some knowledge of the particular uses of each of the parts of the brain *."

In the animals without vertebræ, the brain is destitute of the protecting bony covering, which forms the head and backbone in the vertebral animals. The brain itself is much more simple in its structure. The cineritious and medullary parts, can scarcely be perceived as distinct; and all that appears to be present is a cerebellum, with one or more tubercles, resembling a cerebrum, and nerves furnished with ganglia.

Independent of very remarkable differences in the struc-

* Comparative Anatomy, vol. ix. p. 8.

ture of the nervous system, in the different genera of invertebral animals, there may still be perceived two models, according to which, the organs belonging to it are arranged.

In the first, the brain is situated upon the œsophagus. It presents different forms, according to the species. It appears more like a ganglion, than the brain of the vertebral animals. It sends off several nerves to the mouth, eyes, and feelers. One on each side passes round the œsophagus; these uniting below, form a ganglion; in some cases larger than what is considered the true brain. From this ganglion, nerves are likewise sent off to different parts of the body. The animals in which this nervous system prevails, belong to the great division termed MOLLUSCA.

In the second, the brain is situated as in the mollusca, sends out nerves to the surrounding parts, and likewise one nerve on each side, which, by their union, form a ganglion, from which other nerves issue. This ganglion produces likewise a nervous cord, which proceeds towards the extremity of the body, forming throughout its length ganglia, from which small nerves proceed; this cord, at its commencement, is, in some cases, double for a short distance. It has been compared to the medulla oblongata and spinal marrow of the vertebral animals. This kind of nervous system is peculiar to the Annulose Animals. There are usually ganglia on the nervous cord, corresponding with the number of rings of which the body consists.

III.—*On the Nervous System considered in Action.*

This is a subject unquestionably the most interesting in the whole range of zoological science. Yet it is still involved in much obscurity, and will probably continue to be so, unless new methods of observation shall be devised, and more rigorous induction practised.

1. *The Brain.*—That the brain is the organ to which the impressions produced by external objects are conveyed, and from which the excitements to motion in the different parts are propagated, has been demonstrated by observation and experiment. We have already stated the complicated structure of this organ, and the variety in the texture and situation of its different parts. What, then, are the uses of each? Physiologists have always been greatly divided in their opinions on this important subject.

Some, supposing that the seat of the intellectual operations must exist near the centre of the brain, have considered the pineal gland as the *common sensorium*. Others have bestowed the same honour on the corpus callosum, corpora striata, pons Varolii, and medulla oblongata. SŒMMERING considers the aqueous fluid, with which the ventricles of the brain are, in general, in part filled, as the common centre of sensation. According to GALL and SPURZHEIM, the various operations of sensation and volition, are performed in particular parts of the brain, every faculty or feeling having a distinct organ in which it is generated. The fore part of the brain, they consider as subservient to intellect; the middle to sentiments, and the back part to propensities. According to Mr WALKER *, the cerebrum is the organ of sensation, or the centre to which all the impressions are communicated, and in which deliberation is practised, while the cerebellum is the organ of volition. The nerves which terminate in the cerebrum, and the anterior columns of the spinal marrow, convey impressions to the mind; and the nerves which arise from the cerebellum and the posterior columns of the spinal marrow, execute the purposes of volition.

---

* Annals of Phil. vol. vi. p. 26.

For the confirmation or refutation of these opinions, various methods have been employed, by physiologists and anatomists.

The dissections of the brain have, to a certain extent, pointed out its structure, and some of the modifications which it exhibits; but there are few who will venture to assert, that the peculiar functions of the nervous system of any animal may be determined, by the appearances exhibited by the brain. How vastly superior are the intellectual powers of man to the monkey; yet the brains of both bear the closest resemblance to each other. By comparing the brain as it appears in the different classes of animals, we certainly witness very remarkable modifications of form. As we descend towards fishes, the cerebrum diminishes so much in size, that its total absence may be inferred in the lower classes. Observation confirms the supposition. It can scarcely be detected in the mollusca, and it is wanting in the annulosa. Now, if the opinions with regard to the uses of the cerebrum and its different parts were correct, we ought to find, in the animals which are destitute of the organ, a total want of the functions which it is destined to perform; (for we can scarcely suppose, that any of the other organs of the body can supply its place). But still we find, among insects, for example, not merely sensation and volition, but instincts, propensities, and deliberation, which, when they occur in the higher classes, are considered worthy of having peculiar organs set apart for their production. But the cerebellum still exists in these mollusca and annulosa; and may, therefore, be considered as exercising the functions of sensation and volition. Let us descend, therefore, to the inhabitants of the Corals or to the Hydræ; in these, neither brain nor nerves can be perceived. Yet they evidently possess both sensation and volition, and as evidently want a cerebrum and cerebellum.

In order to arrive at accurate conclusions on this subject, it would be of importance to examine the brain in a variety of species, with whose manners we are intimately acquainted. But as we know the characters or dispositions of few animals, with any degree of accuracy, except those which have been *domesticated*, our observations, for the present, must be imperfect, and our general conclusions premature.

Attempts have been made to determine the functions of the different parts of the brain, by attending to the effects of disease or injuries of its different parts *. There are, however, peculiar difficulties attending this method of investigating the subject. Between the different parts of the nervous system, there is an intimate connection by means of what are termed *sympathies*. Hence, when we witness any derangement of the functions of the brain, it is, in many cases, difficult to determine the seat of the disease, or the various causes which combine in order to produce the effects.

2. *The Nerves.*—In considering *the action of the nerves*, it is necessary to attend to their effects in the production of sensation, of voluntary, and of involuntary motion.

a. Sensation.—In general, sensation is produced by an impression made on the external organs, and conveyed by the nerves to the common sensorium. That this is the ordinary course of sensation, is rendered obvious by experiment. We can protect the external organs from receiving the impression, or when received, stop its progress along the nerves by a ligature or section. It may therefore be asked, has any substance entered the nerves from the ex-

* See an enumeration of a number of very remarkable cases of this kind, in Sir Everard Home's "Observations on the Functions of the Brain" Phil. Trans. 1814, p. 469.

ternal object? The answer is obtained from an attention to a well established fact, that people who have had their limbs amputated, often experience sensations indicating the existence of the lost organs, and the presence of external objects making an impression upon these. Here is a sensation conveyed to the sensorium without an impression on the external organ, and arising, obviously, from a change in the condition of the remaining part of the nerve. The same fact demonstrates, that sensation is not produced by a vibration of the nerves, arising from the concussion of external objects, for it is excited without the external impression or concussion.

A sensation may likewise be produced, in certain cases, by means widely different from those by which, in ordinary cases, it is excited. Thus a blow on the eye, or the contact of two pieces of metal, zinc and copper for instance, one piece being placed under the upper lip, and another under the tongue, make us perceive a flash, in the same manner as if light had really struck the eye. This can only take place in consequence of a change in the optic nerve, similar to that which light itself produces.

The susceptibility of the sensitive faculty for receiving impressions, is liable to considerable variations, depending on the influence exercised over it by different circumstances. Thus, different medicines increase or diminish its sensibility, and inflammation frequently heightens it to a painful degree. In these cases, it appears obvious, that the causes which are considered as in operation, produce a change in the substance of the nerve, or its connections.

But the most obvious changes which take place in the sensitive faculty, are indicated by diminished energy, in consequence of continued action. This state of fatigue or exhaustion, must be known to every one who has attended to the sensation produced by a feeble impression, immediately after the same organ has been acted upon by one of greater intensity. In this case, the feeble impression scarcely excites a perceptible sensation. Thus, weak sounds are not distinctly heard, when they follow those which are much louder; and feebly illuminated objects are scarcely visible, when the eye, immediately before, has been directed to those which are placed in a stronger light. If we look at a dark object on a white wall, and then direct the eye to another part of the wall, we shall still continue to observe the figure of the dark object, but now become apparently more bright than the wall itself. That part of the eye on which the image of the black spot fell at first, experienced a kind of repose, while the other parts on which the image of the white wall fell, were in action. The part, therefore, which was inactive, is now able to receive a stronger impression from the colour of the wall, than the other parts already fatigued by its influence *.

b. Volition.—The action of the nervous system in the functions of *volition*, bears a close resemblance to its operations as a sensitive faculty. In the case of sensation, however, the action excited in the nerve was communicated to the common sensorium; in this of volition, the action excited in the nerve, is communicated to the muscles.

The origin of volition may be traced to the impressions communicated to the brain by the sensitive faculty, or to changes taking place in the brain itself, independent of the action of external objects. In the former case, the volition follows the sensation, in some instances, after an observable interval, in others, it appears almost instantaneously. In the one there is an interruption of the circuit, if I may so speak, by the interference of the mind; in the other, the mind appears to exercise little or no controul. But whether

* CUVIER's Comp. Anat. sect. ix.

the volition be generated in the brain itself, or produced by the sensitive faculty, the action excited in the nerve appears to be the same. How this effect is produced, we are ignorant. There is no reason to consider it as the effect of the concussion of the brain on the nerves, producing vibration, but there is some reason for regarding it as the result of a change in the substance of the nerve, occasioned by the action of the brain.

Various substances applied to the brain, weaken its power of exciting this action in the nerves. But the same action may be excited in the nerves, independent of the brain. Thus, after the nerve has been separated from the brain, the galvanic fluid will so far excite its energies, as to throw the muscles, to which it is distributed, into as violent motions as if these had been produced by the ordinary process of volition. M. Humboldt has employed this method in distinguishing the nerves from the small vessels. He uses two needles, one gold, another silver. A point of one is applied to the muscles, and a point of the other to the filament, the nature of which he wishes to discover, while the other extremities of these instruments are brought in contact. If the filament be a nerve, contractions immediately take place in the muscular fibre.

In whatever manner the impression is produced, it is propagated along the nerve in a direction opposite to that of sensation, being from the sensorium to the external organs. Its progress, however, is liable to be interrupted by the same causes as those which obstruct the propagation of sensation, namely ligature or section.

The powers of volition appear, like those of sensation, to be liable to fatigue. The continued effort of the will becomes painful, and if prolonged beyond a limited time, the nervous energy becomes exhausted, and the death of the individual ensues. If fatigued only in a moderate degree, the nervous energy is recruited by repose. There are stated intervals, which differ according to the species, in which the powers of volition are allowed to rest, or the individual is said to *sleep*. This condition, we have seen, is likewise requisite, in order to recruit the peculiar irritability of the muscles employed for the purposes of voluntary motion.

The action which may thus be excited in the nerves subservient to voluntary motion, is communicated to the muscles. As it exists in the nerves, it is invisible in its operations; but when communicated to the muscle, its effects are obvious. The different particles of the muscle, enter into a more intimate union; it contracts, and its tenacity is increased to such a degree, that it is capable of suspending weights which would have torn it asunder when separated from the body. The act of the will by which the action has been generated in the muscle, can be suddenly altered, and the muscle permitted to return to its relaxed state, in the absence of the exciting cause. We are, however, ignorant of the nature of the change in the condition of the muscle on which the nerve has thus exerted its influence, and we are equally ignorant of the nature of that influence, or its mode of communication.

c. Involuntary Motion.—Besides those nerves which are employed in sensation and volition, there are others which are concerned in the production of *involuntary motion*. These are distributed to the various organs of digestion, circulation, and respiration, whose continued functions are necessary to existence. These nerves neither communicate to us distinct sensations, nor act immediately under the influence of the will. When the parts, however, with which they are connected, become inflamed, they then communicate to us painful sensations; and when the will is thrown into violent action, corresponding motions are excited in those nerves which, in ordinary cases, are not under its controul. Substances irritating the muscles to which these nerves belong, excite them to action, but the nerves themselves resist the stimulus of galvanism or electricity.

Though they do not communicate these sensations to the sensorium, they appear to be, in some degree, excited to action by the contents of the vessels to which they are distributed. Thus, the blood excites the nerves of the heart, and the food those of the stomach and intestines. These nerves, in their turn, act upon the muscles of these organs, and enable them to execute the requisite movements.

The nerves of involuntary motion, chiefly take their rise from the *ganglia*. Hence we perceive some of the properties of these medullary knots, in the changes which they have produced on the nervous filaments in their passage through them. These nerves become incapable, in ordinary cases, of conveying sensations to the brain, or of executing the purposes of volition, and they have become less capable of being excited to action by the electric fluid. But the most remarkable change consists in their becoming capable of continued action through life, without being exhausted, or exhibiting any symptoms of fatigue. These circumstances favour the supposition, that the ganglia are destined to remove the nerves, to which they give rise, beyond the direct controul of the will, to bestow upon them new energies, and to form individual systems, capable of exercising, to a limited extent, independent powers *.

But although the nerves appear thus to occupy different situations, and to perform different functions, they are connected, and form one system, in which the different parts

* See "Essays on the use of the Ganglions of the Nerves," by JAMES JOHNSTONE, M. D. Phil. Trans. 1764, p. 177, and 1767, p. 118, and 1770, p. 30.

are more or less intimately related. This connection is very conspicuously displayed in those actions which arise from what is termed *sympathy*. Thus, an injury on the toe will sometimes bring on lock-jaw; certain smells will occasion sickness; terror will cause the heart to palpitate, the cheek to turn pale, and augment the secretions of the intestines and bladder. The sight of a good meal to a hungry man *makes his mouth water*.

This intimate connection between the different parts of the nervous system, is likewise very strikingly displayed in the case of poisons. Many kinds of poisonous substances, when applied to the stomach, speedily bring on death, by annihilating the energies of the nerves of that organ, and indirectly the vitality of the whole nervous system. The removal of the brain is likewise followed by the loss of power in the nerves.

But this intimate connection, between the different parts of the nervous system, becomes scarcely perceptible in the lower orders of animals. Where the bulk of the brain is greatest in proportion to the nerves connected with it, as in man, we find this union most intimate. As the bulk of the brain decreases in proportion to the bulk of the nerves, the connection ceases to be so close. In reptiles and fishes, this is so conspicuously displayed, that it becomes difficult to induce death. The brain, or the spinal marrow, may be removed, and yet the other functions of life still proceed for a considerable time. Among the mollusca, an equal want of sympathy among the different parts, is well known to prevail. As we descend still lower, to those animals in which the nervous system, instead of appearing in the form of brain, nerves or ganglia, is uniformly diffused, we observe scarcely any dependent connection between the different parts. When portions of the body are removed, they are speedily reproduced, and the detached fragments even begin to enjoy an independent existence.

The power of the nervous system to repair the injuries it receives, or supply the loss of abstracted portions, has been sometimes called in question on very insufficient grounds. Its generation at the commencement of life, is surely more surprising than its production at an after period. But the subject has assumed a character different from mere conjecture. The numerous experiments of SPALLANZANI and others, to prove the power of reproduction in the cold-blooded animals, establish at the same time, the capability of the nervous system, in these animals, to reproduce abstracted parts. But this power is not confined to cold blooded animals. According to the experiments of Mr CRUICKSHANKS * and Dr HAIGHTON †, portions of the abstracted nerves of a dog were speedily regenerated, and the nerve restored to its ordinary functions. It is probable that portions of nerves are generated and destroyed periodically in those ruminating quadrupeds which have deciduous horns.

Are we to conclude, from the view which we have now taken, that the parts of the nervous system are homogeneous, and susceptible of a certain number of similar functions; that the apparently functions of each nerve depend on the organ with which it is connected,—to accessary circumstances, and not to the nature of the nerve itself? The opposite of all this appears to be the case. The nerves which are employed in sensation, obviously differ in their mode of action, from those employed for the purposes of volition, and hence we may reasonably conclude, that there is a corresponding difference in structure and composition. The same nerve may execute both functions; but in that case,

* Phil. Trans. 1795, p. 177. † Ib. p. 190.

it probably consists of different filaments, each acting their peculiar part. This appears to be demonstrated in those cases where a nerve, upon being cut, has afterwards united, and, while capable of executing the action of volition, has become unfit for that of sensation.

It would be to no purpose to enquire into the nature of that action which is excited in the nerve, either in sensation or volition, because the subject is yet in obscurity, and its elucidation, perhaps, impracticable. The rapidity with which the functions of the nervous system are executed, have induced some to consider its action as performed by means of some fluid similar to electricity, secreted by the medullary matter, and restrained by the tunics of the brain and nerves. All this may be true, but it is without proof. Others, from contemplating the effects of electricity, on the parts of dead animals, have concluded, that the nervous and electrical fluids were identical. There is, however, one experiment, easily performed, which proves the fallacy of this conclusion. The nervous energy is suspended or destroyed by the compression or section of the nerve, while the electrical matter is not arrested in its progress, provided, in the latter case, the cut ends of the nerve are brought in contact *.

The effects of electricity on many of the organs concerned in the vital functions, in exciting them to action, may, at first sight, favour the supposition of its identity. Thus, the action of the lungs, heart, and stomach, may be continued for a short time, after the natural nervous influence has been removed. In these cases, however, although electricity can act on the irritability of the muscles, it is probably through the intervention of the nervous filaments, and may be occasioned by exciting the languishing energies of the injured nerves, to expend the remainder of their strength.

* See MONRO's Anatomy. vol. iii p. 113.

Of the elementary nature of this exciting power, we indeed know nothing. Physiologists have termed it a secretion of the nervous system, without perceiving that, in the manner of its operations, it is essentially different from any other secretion in the system. That it results from organization, is disproved by the phenomena of death; that it is of an electrical or magnetical nature, is contradicted by the totality of its phenomena.

Having thus examined the structure of the nervous system, and attended to its functions of sensation and volition,—let us now take a view of the *mind*. The intimate connection which subsists between this mysterious part of the animal frame and the nervous system, points out this place as the most suitable for the investigation of the phenomena which it exhibits. But, in order to give to this subject the requisite illustration, it is necessary to examine more particularly, the nature of our different sensations, the organs employed in their production, and the kind of information which they convey to the mind, with regard to the properties of external objects.

CHAP. X.

ORGANS OF PERCEPTION.

In the numerous references which we have hitherto made to the faculty of Sensation, as a display of the operation of the nervous system, we have considered it as indicating merely the presence of bodies, and as giving no information respecting their character. If we attend more minutely to this faculty, we shall find, that all the sensitive parts of the body, are not equally capable of warning us of the presence of the same kind of objects. The rays of light make no impression upon the tongue or the fingers, indicating their presence, while they act with energy on the eye. The vibrations of the air make no impression on the eye, the mouth, or the nose, while they instantly act upon the ear. Sensation, therefore, is a generic term, intimating the capability of the nervous system to receive impressions of external objects; and it includes as many species as there are impressions calculated to act on one organ, and not upon another, distinguished by this common property, that they intimate the presence of objects.

The number of impressions which may be regarded as distinct species, is more extensive than is generally imagined, and would justify us in considering the term Sensation as the index of an order or class, rather than of a subordinate division. Philosophers, however, have agreed to reduce our sensations to five kinds, namely, those of Touch, Sight, Hearing, Taste, and Smell, to which I have ventured to add Heat.

When an impression is produced on any one of these senses, there is an intimation given to the mind, of the presence of an object with which that sense is more immediately connected. Thus, the sense of touch warns us of the contact of a body with our skin. Taste, smell, hearing, sight, and heat, of the presence of sapid, odorous, sonorous, luminous, and calorific bodies. The first impression of an object on the senses, conveys, therefore, little information, except that of presence or existence; and if the body producing the impression, be speedily withdrawn from the excited organ, our notions of its character will be very imperfect. But, if the object continues to act upon our sense, we begin to analyze the impression. By comparing it with some former impression, we discover the extent of resemblance or diversity; and, by varying the condition of the object, or the sense directed to it, we attempt to discover the circumstances, which give to the impression its common or peculiar character. Our knowledge of objects, therefore, in reference to any particular sense, will be in proportion to the number of impressions, continued a sufficient length of time, which have been made upon that sense. In early life, we cannot analyse an impression, with any degree of accuracy, because we have few former ones with which to compare it. As we advance in life, our impressions increase in number, and consequently, our capability of analysing any one of them. In youth, we know but few of the relations of an impression; and, after a rapid and superficial examination, we turn away to a new object. In more advanced life, we can trace, in any impression, a much greater variety of relations to other impressions, and in doing this, we occupy much longer time in its examination. To this circumstance, we are disposed to refer the fickleness of children. When a new object is presented to them, they are much gratified; but as they can speedily trace its more obvious relations to all their former superficial impressions, it speedily loses its interest. If these observations be correct, we never could acquire an accurate knowledge of a solitary impression produced on any one of the senses; all our accuracy, on this subject, being the result of comparison.

When the senses are thus employed, in investigating the nature of impressions,—the objects which produce them, and the circumstances by which they are modified, they then become, what the leaders in the science of Mind have been pleased to denominate—ORGANS OF PERCEPTIONS. The knowledge we acquire of the relations of an impression, is usually termed *idea*, sometimes *perception.* The last phrase, however, is frequently used indiscriminately, to denote the presence of the impression,—the attempt to analyse its character,—or the knowledge which results from the examination. We are disposed to use Sensation to express the first of these meanings, and Idea the last, and could we succeed in restricting Perception, to express the inclination, to examine, and the means which we employ, no inconsiderable portion of ambiguity would be removed from the statements of moral science.

In giving a general view of these organs of Perception, we shall offer a few remarks on their structure, mode of acting, and the kind of information each is calculated to convey to the mind.

1. Sense of Touch.

The organ of Touch is widely distributed over the surface of the body. In the higher classes of animals, as quadrupeds, birds, and reptiles, it is supposed to be seated in the villous surface of the skin, where the cutaneous nerves terminate. In those places of this part of the corium, where the greatest inequalities prevail, the sense of touch is observed to be most acute.

In many animals, there are parts of the skin to which no nerves are distributed, as the hair, feathers, and horns, which, in consequence, are termed insensible. In such animals, the organs of touch are limited in their extent But even in those whose skin possesses few of these insensible appendices, there are particular parts of the body where the skin is more sensible to the impressions of external objects, than in other parts; and where the sense of touch is considered more particularly as residing. Thus, in Man, the hand and foot convey to us the most accurate information connected with the sense of touch. The lips and tongue are likewise employed for a similar purpose. Among Quadrupeds, the monkeys have their sense of touch, in many respects similar to that of man; in others, it is seated in their lips, snout, or proboscis. In Birds, whose bodies are covered, to a great extent, with feathers, the sense of touch must be confined to the feet and head. We accordingly find, that the villous surface of the skin is very conspicuous in the soles of the feet, particularly of rapacious birds, which use their toes for seizing their prey. In those birds with long bills, as snipes and woodcocks, which search for food among mud, the extremities of the mandibles are usually of a softer texture than the base, and evidently appear to possess an exquisite degree of sensibility.

But, even in those animals whose skin is considered as destitute of a villous surface, the presence of the sense of touch may be distinctly ascertained. In Fishes, the surface is covered with insensible scales; yet the contact of an object with the body is readily felt. About the mouth and head, however, of many species, there are places destitute of scales, usually more or less raised above the surface, which are usually considered as subservient to the purposes of touch, although the evidence of their utility in this respect is far from satisfactory. These are soft in their texture, pliable, and capable of various motions. They obtain different names, according to their situation. When placed upon the lips, they are termed *cirrhi*, on the head *tentacula*, and on other parts of the body *fingers*. In the Mollusca, the sense of touch more particularly resides in the tentacula. We may observe the application of these organs to the examination of the surface of bodies in the common slug. Among Insects, the antennæ, or *feelers*, as they are termed, are organs of touch, possessing, in some species, very great sensibility. Even among the Zoophytes, the sense of touch is present. By means of the cirrhi which surround their mouth, they are warned of the presence of their prey.

In the examination of the appearance of the sense of touch in the imperfect animals, it is difficult to ascertain, whether the information obtained by these cirrhi is confined to sensation, (or a knowledge of the presence of objects), or if it includes, likewise, intimations of their qualities.

In order to excite to action the organs of touch, it is necessary that the body which communicates the impression, be brought into *contact* with them ; and, in general, the more close and intimate the contact, the more correct is our information with regard to those qualities which it possesses, and which it is the business of touch to examine. Pressure appears to excite the nervous action of the organ, and when removed, that action ceases. It is obvious, therefore, that the perfection of any part, as an organ of touch, will depend, not only on its sensibility, but its pliability, or power of applying itself to the unequal surfaces of bodies, and to its extent, to enable it to touch a large surface at once.

The information communicated to the mind by the medium of touch, is more varied and extensive than that which is obtained by the aid of the other senses. It makes us acquainted with the *dimensions* of bodies,—their *form*, whether round or angular,—the condition of their *surface*, whether rough or smooth,—their *structure*, whether hard or soft, fluid or solid,—their *connection*, whether moveable or immoveable,—their *gravity*, whether heavy or light,—and their *situation*, whether near or distant.

These qualities, some of which are primary, and others secondary, are ascertained, by the organ of touch, in so correct a manner, that we are scarcely ever deceived with the knowledge thus acquired.

The accuracy of this sense is much improved by habit, and, in certain cases, where the other senses have been injured, as sight, this has acquired so great a degree of sensibility, as, in a great measure, to supply their loss. Thus, blind men are able to perceive their approach to a wall, by effects produced on their skin by the air. These may arise from the resistance of that fluid, or other causes with which we are but imperfectly acquainted. Bats appear to possess this exquisite sensibility of touch, independent of habit or experience. SPALLANZANI has observed these animals, even after their eyes had been destroyed, and ears and nostrils shut up, fly through intricate passages without striking against the walls, and dexterously avoid cords and lines placed in their way. The expanded membrane of their wings is probably the organ which, in such cases, receives the impressions produced by a change in the resistance, motion, or perhaps temperature of the air. The susceptibility of being acted upon by these agents, which exercise a feeble influence over us, is a condition natural to these animals, and necessary to enable them to find their way in the dark caverns in which they dwell. It cannot therefore be regarded as one of those resources employed in time of need, in the case of bats, although it appears to be such in blind people. Man naturally moves but little in the dark, so that he does not require such susceptibility. It is, however, fortunate that his body is susceptible of acquiring it in the time of need.

The sense of touch appears, in man, to be able to obtain nearly all the information, with regard to external objects, which it is capable of receiving. In a few instances, the lower animals surpass us in the delicacy of this sense, as the bat, which is warned, indirectly, by its aid, of the presence of bodies, previous to coming in contact with them. The feelers of insects are likewise better adapted for exploring the condition of the surface of bodies, than any organ which we possess. But, in all these cases, the sensibility of touch is limited to particular qualities, or confined within narrow bounds. The human hand, on the contrary, by its motions, the pliability and strength of the fingers, and the softness of the surface, is the most extensive and perfect organ of touch possessed by any animal.

II. Sense of Heat.

The sense of touch is exclusively occupied with the examination of the conditions of resistance. Contact, therefore, is indispensably requisite for enabling the organ to act upon the object, and muscular exertion, to examine its condition. Neither of these are necessary to enable the sense of heat to act. Calorific rays emanate from a heated body, though at a distance, and, in order to ascertain their direction and intensity, no muscular effort is required. When the heated body happens to be in contact with us, we in like manner examine its conditions, in reference to temperature, without any muscular exertions, or rather, we try to avoid them. Thus, when I lay my hand upon the table, to examine its hardness or smoothness, I make an obvious muscular effort with my fingers; but when I lay my hand on the table to examine its temperature, I endeavour to check all motion, so as to keep my hand in the same position.

These qualities of the sense of heat, sufficiently distinguish it from that of touch, with which it has been confounded, and justify its establishment as a distinct power of perception.

The organ of heat, like that of touch, is seated in the skin, and appears to be co-extensive with those portions of it where the cuticle is thin and destitute of appendices.

By means of the sense of heat, we are enabled to judge of the relative quantity of caloric in our own bodies and in surrounding objects. When no sensation is produced, we conclude, that our body, and the object with which it is compared, have the same temperature. When a sensation is felt, it arises from the difference of temperature, and the changes which the caloric is experiencing to bring about an equilibrium. When the body feels cold, the caloric is passing out of it into the neighbouring object; when the body feels warm, it is receiving heat. As the body is thus constantly either giving or receiving heat, in consequence of being surrounded with a variety of objects, changeable in their temperature, the information which it can communicate to the mind, in reference to its own temperature, or that of other bodies, is merely relative. Thus, an object which would feel warm to the body in one state, would feel cold in another. But though the information communicated to the mind by this sense, be not equally accurate with the indications of the thermometer, it exercises a powerful influence over our volition, and is intimately connected with our comfort. It gives warning of the approach or retreat of certain objects, when no other sense could have indicated their presence or change. It is a faculty common to all animals.

III. Sense of Sight.

The organ of Sight, or the Eye, exhibits, in the different classes of animals, so many modifications of form and structure, that it is difficult to assign to each part its proper function, and to form a just estimate of the sensations which, under different circumstances, may be excited, or the ideas which may be communicated to the mind.

1. Structure of the Eye.—Among all the Vertebral Animals, including Man, there is a considerable resemblance between their organs of sight. Hence a very general description of their structure will suffice, to enable us to comprehend their peculiar functions, and to estimate the

relative importance of the same organs, as they appear to be modified in the inferior tribes.

(1.) *Coats of the Eye.*—There are two coats which may be considered as constituting the integuments of the Eye, and within which, the other essential parts are contained. These are, in the form of two cups, applied to each other at the margins, and hence the eye-ball approaches a globular form. The cup, which is situated with its convexity inwards, is termed the *Sclerotic coat.* Its texture is cellular, and is composed of filaments, which are interwoven in every direction. It possesses considerable opacity, tenacity, and flexibility. It determines the shape of the eye. Hence, in those animals whose eye-balls are more or less depressed, the sclerotic coat acquires a proportional degree of strength, to enable it to retain its form. In birds, for example, its anterior part is divided into two plates, between which, a circle of thin hard bones is interposed.

The sclerotic coat is perforated posteriorly, for the passage of the optic nerve. By some, it is considered as a continuation of the dura mater, forming a sheath for the protection of the nerve, and afterwards expanding to support its terminal enlargement, denominated the Retina. On its internal surface, a thin blackish membrane adheres, which has been considered as a prolongation of the pia mater.

The sclerotic coat not only preserves the form of the eye, but furnishes a firm support for the insertion of the muscles destined to regulate its motions. It is supplied with few bloodvessels, and possesses little sensibility.

The sclerotic coat protects the posterior part of the eye, but it is wanting on the anterior surface. Its placc, however, is supplied by another membrane, termed the *Cornea,* which may be compared to the glass of a watch, the case resembling the sclerotic coat.

The *Cornea* is composed of thin concentric plates, united together by a compact cellular substance. It is thickest in the middle, becoming thinner towards the edges, wher it unites with the sclerotic coat. The line of junction exhibits considerable differences. In man, the edges are bevelled, and the cornea slides under the sclerotic coat. In the hare, the sclerotic coat divides at the edges, and embraces, like a forceps, the margin of the cornea. In the whale, the fibres of the sclerotic coat pass into the substance of the cornea, in the form of very delicate white lines.

The cornea is transparent. It possesses few, if any, bloodvessels or nerves. It is, however, porous, and readily admits the escape of the aqueous humour, as appears by the eye speedily becoming flat after death.

In insects, this membrane is hard and scaly, and supplies the place of the crystalline humour; while in the cuttle-fish it is regarded as absent.

These two coats may be considered as forming the case of the eye. There are other membranes on their central aspect, which are usually classed along with them, which we shall now consider.

Immediately within the sclerotica, is the *Choroides*, a thin delicate membrane of a vascular structure. It lines the concave surface of the sclerotic coat, with which it is connected by cellular tissue, and by the nerves and bloodvessels which terminate in its substance. On its concave surface, its substance is compact and uniform in its texture, and has been regarded as a particular membrane, and termed *Membrana Ruyschiana*.

Towards the anterior part of the sclerotic coat, near its termination, the choroides is more intimately united with it, by means of a circular band of dense cellular substance, moistened by a whitish mucus, termed the Ciliary Ligament. From this band, numerous processes arise, which project into the cavity of the eye-ball, and form a broad radiated ring. These projecting laminæ are termed *Ciliary Processes.*

After forming the ciliary ligament, the choroides still proceeds along the concave surface of the sclerotica, and parallel with it, until its junction with the cornea, when it forms another projection into the cavity of the eye-ball, termed the *Uvea.* It is here continuous, and constitutes an annular veil, perforated in the centre. Its anterior surface is covered, and intimately connected, with a membrane of a spongy, fibrous texture. This membrane is termed the *Iris*, and is well known as the seat of the colour of the eye. It is sensible to the impressions of light, and readily expands or contracts, according to circumstances ; thus, enlarging or diminishing the size of the central perforation. The central aperture is termed the Pupil, and varies greatly in its dimensions and forms in different animals.

On the central surface of the choroides, including the ciliary processes and uvea, there is a slight villosity, to which there adheres a mucous pigment. This is usually dark coloured, and termed the *Pigmentum nigrum*, but, in some cases, it is light coloured approaching to white. This pigment is insoluble in water, but soluble in alkalies and the stronger acids. LEOPOLD GMELIN considers that it approaches the nature of indigo. On the temporal side of the bottom of the cavity of the eye, there is a small space destitute of this pigment, through which the colours of the membrana Ruyschiana appear. This spot is termed the *Tapetum*, and is peculiar to quadrupeds.

(2.) *Humours of the Eye.*—The cavity of the eye, formed by the coats which have been described, contains Fluids differing in consistency, form, and situation. They are divided into three kinds, termed the *Vitreous*, *Crystalline*, and *Aqueous Humours*.

The *Vitreous Humour* occupies the posterior and lateral parts of the cavity. It is convex behind, and concave before. It is invested by a delicate thin transparent membrane, termed the ***Hyaloid.*** At its anterior surface, it may be separated into two laminæ, between which, air may be introduced. This is the *bullular canal* of PETIT. The interior of this humour is divided by the same kind of membrane, into numerous cells. These are filled with a fluid, of the consistence of the white of an egg, which does not readily escape from them, even when the external membrane is punctured. This fluid consists chiefly of water, with a small quantity of albumen and gelatine.

The *Crystalline Humour*, or Lens, as it is frequently called, occupies the centre of the cavity of the eye. It is doubly convex, its posterior side resting in a concavity of the vitreous humour. It is enclosed in a membranaceous capsule, which is soft and transparent, and with which it is but loosely connected. The lens itself is denser at the centre than towards the circumference. It becomes indurated by boiling and by alcohol, and then exhibits its peculiar structure. It consists of an infinite number of concentric laminæ, formed from delicate fibres which proceed from two centres, situated at the two extremities of the axis. At each axis, there is the appearance of a membrane disposed in rays, from which the fibres originate. These rays vary in number, in different animals, from two to five. The rays of the one axis, are placed opposite the interstices of the other. The matter of the lens coagulates by boiling, and consists of water and albuminous matter, with a small quantity of cellular substance.

The *Aqueous Humour* occupies the remaining part of the cavity of the eye, in front of the crystalline. It is divided by the iris into two chambers. The anterior, which is the largest, occupies the space between that membrane and the cornea; and the posterior, between the uvea and ciliary ligament, which is very small, and whose existence is even denied by some. The aqueous humour agrees with the vitreous in chemical composition, but it is less viscid. It is not lodged in cells, and hence it readily flows out when the cornea is punctured *.

3. *Nerves of the Eye.*—The position of the optic thalami, and their relation to the cerebral portion of the brain, have been already explained. From these thalami, the optic nerves take their rise. They are two in number, and in some animals, the nerve which proceeds to the right eye, originates in the left side, and the nerve of the left eye originates in the right. Each nerve is divided, internally, into a great number of canals, formed by the neurilema, which contain the medullary matter. This structure is displayed when the medullary matter is removed by maceration, and the nerve inflated and dried.

The optic nerve penetrates the sclerotica and the choroid coat, and becomes expanded on the concave surface of the latter, in the form of a delicate transparent membrane, without, however, adhering to any part of it. It extends to the ciliary ligament, terminating at the base of the processes. The internal surface, next the vitreous humour, is of a firmer consistence than the external, contiguous to the choroides, as it contains numerous minute bloodvessels. In some animals these surfaces can be separated, and the dense central lamina is termed the Arachnoid.

The retina is perhaps the most delicately sensible membrane of the animal frame. It is readily affected by the rays of light, which, when too intense, excite very painful sensations.

* The relative position of the different parts of the eye, is displayed most distinctly when the eye-ball is in a frozen state.

4. *Muscles of the Eye.*—In the vertebral animals, each eye is placed in a socket, situated in the bones of the face, and termed the Orbit. The eye-ball does not fill this cavity at its posterior part, but the remaining space is occupied with fat or gelatinous matter. The ball rests upon these, and is thus able to move, without being injured by friction. As the fat, in lean and old animals, is in a great measure absorbed, the eye appears, in such cases, sunk in the orbit. In some cases, the ball of the eye is supported on a foot-stalk, which enables the muscles to act on a longer lever, and admits of a greater extent of motion.

The muscles by which the eye is moved, have their origin in the walls of the orbit, and their insertion in the sclerotic coat. Tendinous fibres, however, pass on to the cornea, with which they become incorporated. In man, the four straight muscles of the eye can be separated from the sclerotica, and the external layer of the cornea demonstrated to be a continuation of their tendons. In birds, the same connection between the tendons and cornea prevails *. The muscles vary in number in different animals.

5. *Glands of the Eye.*—These bodies vary in form and number in the different classes of animals. In man, they consist of three kinds, the Lachrymal, and Meibomian, and the Caruncula Lachrymalis.

The Lachrymal gland is situated between the ball of the eye and the upper arch of the orbit, a little towards the temples. It consists of a number of small whitish granular bodies collected together into lobes. From these proceed several canals which descend through the substance of the upper eye-lid, and open on its internal surface. The fluid which is secreted is the *Tears*. This fluid, according to the observations of FOURCROY and VAUQUELIN, con-

* Phil. Trans. 1795, p. 11. and 263, and 1796, p. 1.

sists of water, mucus, muriat of soda, soda, phosphat of lime, and phosphat of soda. Its taste is perceptibly saltish, although the saline ingredients do not exceed a hundredth part of the whole. The use of the tears is to lubricate the surface of the eye-ball, and protect it from dust and the drying influence of the air. They are not secreted in those animals which live in water. When the eye-lids close, the tears are pressed towards the internal angle of the eye, from whence they are conveyed to the nose through the lachrymal sac. The opening into this sac is either by a single fissure, or by two small pores, the mouths of two canals which are situated at the nasal angle of the eye, and termed *puncta lachrymalia*.

The *Meibomian Glands* are situated in the substance of both eye-lids. They consist of small follicles, arranged in vertical lines, terminating in small round holes on the edge of each eye-lid. The albuminous substance which they secrete, covers the margins of the eye-lids, and while it prevents the tears from flowing out, preserves the tarsi from adhering to each other.

The *Caruncula Lachrymalis* is situated at the internal angle of the eye-lids. It is a roundish reddish mass, secreting a thick whitish humour, which is supposed to protect the lachrymal pores. The Harderian gland, which occurs nearly in a similar situation, and secretes a fluid somewhat similar in its properties, although not found in man, is observable in many quadrupeds, and in birds.

6. *Coverings of the Eye.*—The external coverings of the eye are all derived from the common integuments, more or less altered in their texture. They may be regarded as three in number, and described under the names, Eye-lids, Nictitating Membrane, and Conjunctiva.

The Eye-lids, when present, are generally two in number, the one protecting the upper half of the eye, and termed the Upper eye-lid; the other, which covers the inferior portion, is termed the Under eye-lid. These meet in a line, which usually observes a horizontal direction. The margins of both eye-lids are thickened by the cells of the meibomian glands, and rounded, so that when they meet on the eye-ball, there is a conduit formed for the tears within. The *Tarsus*, as the margin is termed, is still farther strengthened for supporting a row of hairs, called Cilia, or eye-lashes. The eye-lids exhibit considerable difference in their motions. In some animals, both eye-lids are capable of approaching each other, while in others, only one eye-lid is able to move, either the under rising upwards to join its antagonist, or the upper descending for the same purpose. In one fish, (Tetraodon mola or Sun-fish,) the eye-lid is single, circular, with a perforation in the centre, the aperture contracting or enlarging according to circumstances.

The Nictitating Membrane, or third eye-lid, as it is sometimes called, is of a more delicate texture, and more liberally supplied with bloodvessels than the eye-lids, and is even transparent in some animals. In many animals it is single, situated at the nasal angle of the eye, within the eye-lids, or on the interior of the under eye-lid. In others it is double, situated at each corner of the eye. In some cases it is destitute of motion, while in others it is capable of covering the eye-ball, by extending in a horizontal or vertical direction, according to its position.

The Conjunctiva forms a permanent and continuous covering on the eye-ball itself. In man, it adheres so closely to the cornea, that it cannot be separated unless by maceration. In other cases, where there are no eye-lids, and where the skin passes directly over the eye, the adhesion is very slight. In almost all animals the conjunctiva is transparent where it passes over the cornea, but is usually thickened and coloured when it covers the sclerotic coat, forming what is called the White of the eye. In some animals, however, (as the Myxine glutinosa of Linn.) this membrane is so thick and opake, as it passes over the eye, scarcely differing from the skin on the rest of the body, as to render the eye beneath useless as an organ of sight.

As connected with the eye, we may here take notice of the *eye-brows* or supercilia. These form a screen to protect the eye from too much light,—intercept the sweat flowing down the forehead,—and greatly contribute to the expression of the countenance.

The observations which we have hitherto made, illustrative of the structure of the organs of vision, relate to the more perfect animals. Among the Gasteropodous Mollusca, the eye is too minute to admit of accurate dissection. It appears as a black spot, convex, however, on the surface, and furnished with a nerve from the cerebral portion of the brain. Among the Annulose animals, black spots are observed in many species, but to which no nerves have been traced; while in others the nervous filaments have been detected. Even in the apparent absence of the nerves, these spots have retained the name of Eyes, and the analogy on which their claim to be so denominated rests, is far from remote. Such black spots in the annulose animals are termed *simple*, in opposition to other organs which are termed *compound* eyes. The surface of the compound eyes is convex, and, when viewed through a microscope, appears to consist of a number of hexagonal facets, slightly convex, forming a hard elastic membrane. Each facet is concave internally, but always appearing thicker in the middle than at the edges. Behind this external compound plate, which may be regarded as a cornea, or aggregation of lenses, there is a close covering of an opake substance. On the posterior part of this last substance, there is a fine delicate membrane of a black colour, having behind it an expansion of the optic nerve, which may be considered as the common retina. Nervous filaments proceed from this retina, penetrate the black membrane, which has been regarded as the choroides, and proceed to the concave surface of each facet, between which the opake matter only is interposed. We may consider the retina in these animals, in the light of a ganglion, and the individual filaments as the separate branches of the retina.

Some of the annulose animals have only these compound eyes, others only the simple ones, while many species are in possession of both kinds.

2. Functions of the Eye.—We come now to consider this important organ in action.

Rays of light emanate from luminous bodies in all directions, and the eye may be regarded as an optical instrument destined to act on these rays, and produce an impression on the retina, indicative of their colour, intensity, and direction.

These cones, or pencils of rays, falling upon the convex and transparent cornea, have their direction changed, and are made to converge. The effect is increased by the three humours of the eye through which the rays pass, so that they meet at a point beyond the vitreous humour, which, by opticians, is termed the *Focus*. Here the retina is spread out to receive the impression, and to communicate the same to the mind.

No object is visible to the eye, unless the angle formed by its extreme points exceeds thirty-four seconds of a degree. In order to render the impression distinct, it is necessary that all the rays which proceed from any one point of a body, should be collected in one point of the retina, and that all the points of union thus formed, should be disposed in that organ, in the same relative position as in the body from which they emanated. For the accomplishment of this purpose, the humours of the eye are so adjusted, in their form, density and refractive power, as to prevent any dispersion or decomposition of the rays. They thus act in a similar manner to the compound object-glasses of an achromatic telescope.

As animals reside in different media, it is obvious that the eyes of each must possess different refractive powers. In the land animals the cornea is usually convex, and the aqueous humour abundant; while in aquatic animals the former is flat, and the latter in small quantity. In land animals, the aqueous humour possesses great power of refracting rays, passing to it through air, aided likewise by the convexity of its surface. But its refractive power in water would be comparatively weak. This defect, however, is supplied by the spherical form and great refractive power of the lens, as may be seen in whales, diving-birds, and fishes.

When we look at the image of an object in the focus of a convex lens, or in that of the natural eye of a recently slaughtered bullock, prepared by removing the coats on its posterior side, and thrown upon white paper, we observe a picture formed, but in a reversed position,—the rays of light emanating from the upper part of the object forming the lower part of the image, and those from the right proceeding to the left. If the retina, in the living eye, be considered as occupying the place of the white paper in these experiments, it must follow, that the image of any object thus painted on the retina will be inverted. How comes it to pass, therefore, that we see every object in its natural upright position? All this difficulty originates in a misconception of the nature of the retina, and the impressions which it receives. There is no white screen in the eye, on which the image of an object can be painted. The retina is translucent, and the choroid behind it is black. The retina is not, therefore, acted upon by the reflected rays of the inverted image, as our eye is, when looking on the picture formed on the white paper, but by the direct rays from the object passing through its substance. We do not, therefore, see the *picture* of the object, but the object itself. And as we see the object, or any part of it, in the direction of those rays which proceed from them, and which produce the sensation, it follows, that the eye really sees objects in their natural and relative situation.

It is well known that the eye discerns objects placed at different distances. As the rays of light, which reach the eye from a distant object, are nearly parallel, they will converge into a focus nearer the humours than those rays which proceed from near objects and which are more divergent, and, consequently, will unite in a focus still more remote. In order to obtain distinct vision in these different circumstances, either the retina must recede or approach, according to the focal distance; or, if we suppose the retina stationary, the lens must move, or experience a change in its refractive powers, by an alteration of its form or density; or, in viewing near objects, those rays only may be admitted which are nearest to the axis, and which are consequently the least diverging. But physiologists are by no means agreed in their opinion of the means employed by nature for this purpose.

The sclerotic coat is considered by some as subservient to this end. They suppose that the muscles compress it, and that the humours are thus pressed forward, to encrease the convexity of the cornea, and enable it more readily to converge the rays. But while the sclerotic coat is flexible in man, whose eye is globular, and easily retains that form by the humours pressing equally in all directions, it is nearly inflexible in many animals whose eyes possess this adjusting power, but which are more or less removed from a spherical form. It is indeed pretty obvious, that the use of the sclerotic coat is to preserve the form of the eye, to furnish points of insertion or attachment, for the muscles which move the eye-ball, and to support the delicate membranes which line its central surface. At the same time, the action of the straight muscles on this organ, when flexible, must have a tendency to alter the form of the contained humours, and in this manner part of the desired effect may be produced.

The cornea has likewise been considered as the instrument, by which the eye is able to exercise the power of seeing objects at different distances. This effect has been supposed to take place by the elasticity of the laminæ of the cornea, acted upon by the straight muscles, with which it is so intimately united. But the effect produced in this manner, is considered by many so small as to be incapable of accounting for the display of the power we are now considering. More recently, Mr Crampton has demonstrated the existence of a muscle in birds, capable of changing the form of the cornea, and which he considers as the organ employed to alter the convexity of the eye *.

When we look first at a near and then at a distant object, or the reverse, we feel that a muscular effort is required for the adjustment of the eye, to the change in the distance of the object to which it is directed. This exertion becomes very evident, when we look at a spot on one of the panes of glass in a window, and then look through the glass in the direction of the spot to some distant object, as a tree or house. What are those muscles which are called into exercise? The manner in which the effort is made, leads irresistibly to the belief, that the straight muscles are excited to action, and may either act upon the sclerotica or cornea, according to the conditions of these objects, or upon

* Annals of Philosophy, i. p. 173.

both, and thus alter the form of the humours, or change their relative position with regard to the retina.

The crystalline lens has also been regarded as possessing the power of changing its form, and varying the focal distance of the eye, either in consequence of the action of the ciliary processes, or of a change in the internal arrangement of its parts. Its structure, as displayed by LEUWENHOEK *, YOUNG †, and others, after coagulation by heat or alcohol, is considered as muscular. But as it contains no fibrin, and is even soluble in water, with the exception of a small portion of extremely pellucid membrane, its muscular power is denied by some. Its increasing density towards its centre, rather indicates a cellular structure, the cells being filled with pellucid matter of different degrees of concentration. It may be added, as a still more decisive proof that this power of varying the focal distance is not seated in the crystalline, that when the lens is extracted in the disease termed the Crystalline Cataract, the limits of distinct vision suffer no diminution.

When we look at objects within the limits of distinct vision, the iris expands, so that the aperture of the pupil becomes contracted. In this manner the least diverging rays only are admitted, and distinct vision obtained.

By this arrangement, it is probable that the eye accommodates itself to objects at different distances, within the limits of distinct vision, as it is known to do by the same means in reference to the quantity of light.

The evidence in favour of this function of the iris appears to me to be conclusive. The enlargement and contraction of the pupil, are confined, indeed, within narrow limits, but so is the extent of distinct vision within the ordinary limits. There is a particular range of *minute in-*

* Oper. Om. p. 73.

† Phil. Trans. 1793, p. 169.

spection, within which the pupil enlarges and contracts, according to circumstances. Within this range vision is distinct; beyond it, at either extremity, it is obscure. When, however, we are able to assist the iris by any contrivance, the range of distinct vision is enlarged. Thus, in reading a book, the page is kept at least six inches from the eye; and when brought much nearer, the rays entering the eye to form the image of the letters possess two great a degree of divergence, and vision is obscure. When, however, we look at the letters through a pin-hole in paper, held close to the eye, we prevent the rays of greatest divergence from entering the pupil, admitting only those which are nearly parallel, and the resulting image is so distinct, that we can read print brought within an inch and a half of the eye. The pin-hole in paper is therefore the simplest kind of microscope, but the eye is fatigued by using it. When an object, beyond the limit of minute vision, is viewed through a smaller aperture than the pupil, a more obscure image is formed than when the eye is uncovered, and the same object ceases to be visible at a shorter distance *.

---

* The following experiment by Mr DUNGLISON, countenances the view taken above: "A portion of the newly prepared extract of Belladona, for the sake of experiment, was inserted between, and applied to the eye-lids; in consequence, in the space of about twenty minutes, the pupil was so much dilated, that the iris was almost totally invisible. From the time the pupil attained to three times its natural dimensions, objects presented to this eye, with the other closed, were seen as through a cloud; and as it proceeded to the point of extreme dilation, this effect gradually increased, so that minute and near objects, as letter-press, &c. could not be at all distinguished. By means of a double convex lens, the focus of this eye was found to be at twice the distance of that of the sound eye; the iris, however, dilated upon the sudden admission of light; and although the pupil approached by almost imperceptible degrees for six days to its natural size, yet, at the end of that time, it was dilated to twice the size of the other; and, in proportion as the contraction took place, the sight became more distinct, and the

When it is considered that the supporters of these various opinions have discovered in each of the parts of the eye which have been referred to, a provision for enabling it to see distant and near objects, it seems not unreasonable to conclude, that the same part may not exercise the same function in all animals, but, in the different classes, be assisted or superseded by those with which it is connected. It is likewise probable, that the necessity of this power of adjustment may not exist to the extent which has been supposed, and that the limits of distinct vision are included within a narrow range. The human eye sees objects most distinctly at the distance of from six to ten inches. When these are removed to a greater distance, we do not perceive so clearly the shades of colour, or the inequalities of the surface, and this indistinctness of vision increases with the distance. The action of the straight muscles, however, serves in some degree to correct the defect. But, in looking at distant objects, we are assisted greatly in our perceptions by our former experience; so that it may often happen that the praise which we bestow on the sightfulness of the eye, is due to the readiness of the recollection. Microscopical inquiries are seldom prosecuted so habitually as to furnish the same aid to the unassisted eye, when viewing objects within the range of minute vision.

Some physiologists have been disposed to conclude, that the formation of a perfect image on the retina, is not essential to distinct vision; and the following experiment of M. DE LA HIRE, has been brought forward in support of the opinion: "If a small object placed at that distance from the eye, at which vision is most distinct, be viewed through

---

focus nearer the natural. In the open air, all objects, except those near, were distinctly seen, but immediately on entering a room, all was again enveloped in mist." Annals of Philosophy, vol. x. p. 432.

three pin-holes, so disposed, that the interval between the most distant of them shall not exceed the diameter of the pupil, the object will be seen single; but if the object be brought either within or beyond the limits of distinct vision, it will be seen multiplied as many times as there are holes in the card, and each of the three images will be as perfect as the single one *." If the statements here made be correct, it must follow, that objects may be seen distinctly by rays which do not accurately converge on the retina. These three images are formed from three pencils of rays, which, as they possess different degrees of divergence when they strike the eye, must converge at unequal distances behind it, and be intersected by the retina under different circumstances. But upon repeating the experiment, I obtained very different results. Upon looking through three pinholes, placed on a line, the distance of the lateral ones from each other, not exceeding the diameter of the pupil, at a small dot made upon white paper, lying six inches distant from the eye, the dot appeared single. But when I brought the dot within three inches of the eye, and viewed it through the perforated card, keeping the central hole opposite the optical axis, the images of three dots appeared; differing, however, in their distinctness, the central image being clear, the lateral ones obscure. It is obvious, in this case, that the cone of rays which entered the middle pin-hole, possessed least divergence, and, consequently, converged nearly at the focus of parallel rays, the ordinary station of the retina. But the cones of rays which entered the lateral holes possessed greater divergence, and consequently met in points beyond the retina, or were truncated previous to their convergence into a focus. The images which they formed, were therefore ill defined. When one of the lateral holes

---

* Annals of Philosophy, i. p. 171.

was brought into the optical axis, the image formed by the rays which passed through it, was distinct, the image of the rays of the central hole was rendered obscure, and the rays of the remote lateral hole not entering the eye, only two dots appeared.

When I placed the dot at the distance of a foot from the eye, and then interposed the perforated card, only *one* image was formed, nor did more appear when the eye was removed to a greater distance, or brought a short way within the limits of distinct vision. In this case, the rays which entered the different holes were nearly parallel, and consequently converged at one point.

When I placed the dot and the perforated card, as in the first experiment, within the limit of distinct vision, where three images appeared, and gradually removed the dot beyond that limit, the three images still continued to appear even at that distance, where only one would have been visible, had the perforated card been there interposed for the first time. In like manner, when the dot, viewed through the perforated card, beyond the limits of distinct vision, and appearing single, is brought nearer the eye, it will still appear single within that distance; at which, if viewed for the first time through the card, three images would have been formed. These appearances are similar to those which take place in the eye, in ordinary circumstances. We see a dot upon paper at a greater distance, if first viewed at the limit of distinct vision, and then gradually withdrawn, than when the eye is directed towards it for the first time in its remote station; and the same thing takes place with objects held close to the eye. In these cases, the action in the retina is continued by a weaker impression than is requisite for its first excitement. These results warrant the conclusion, *that the rays which do not accurately converge upon the retina, do not produce a distinct image.*

The power of seeing objects beyond or within the ordinary limit of distinct vision, is greatly strengthened by habit. Thus, a sailor will discern the masts of a vessel appearing at a distance in the horizon, where nothing is visible to the eye of a landsman. In like manner, a botanist will detect a Lecidea on a rock, or the entomologist a fly, where an ordinary observer would perceive no trace of organized existence.

Some animals are destined to perform the functions of vision in the full light of day, while others are confined to the obscure light of the evening or night. In the animals of the former class, termed *diurnal*, the mucous pigment of the eye is of a dark colour. The purpose which it is supposed to serve, is that of absorbing the rays of light, after passing through the retina, and of preventing any reflection of the rays taking place in the lateral parts of the eye, and disturbing the image of the objects so contemplated. Where the tapetum exists, however, a portion of the rays must be reflected; but the reflection in this case may be so regulated, as to assist rather than disturb the action of the retina. In animals which seek their food in the dark, the eye is usually of a large size; the pupil is wide, to admit a greater number of rays; and the pigment and tapetum pale coloured, approaching to white. In these animals, as the cat, for example, whose eyes are so constructed, that the choroides reflects, instead of absorbs the rays of light, it is difficult to determine whether the reflected rays act upon the retina, and excite vision in their passage outwards, as is generally supposed; or pass through the retina outwards, without exciting any action, to be thrown on the object, in order to increase the distinctness of its image, by an increase of its light. It is not, indeed, probable, that both surfaces of the retina are equally adapted for receiving impressions of external objects ; and, judging from analogy, it is probable, that the rays in their passage inwards alone produce the image *.

In some individuals of certain species of quadrupeds and birds, the mucous pigment is entirely deficient ; so that the choroid coat is visible through the iris. This deficiency is always congenital, and is connected with a defect of the secreting organs of the colouring matter of the hair and feathers. Such animals are called *Albinoes*. Their eyes are tender, and impatient of light †.

Many animals can only see an object with one eye at a time. But in other animals, as man, both eyes may be directed at once to the same object, so as to produce an image in the retina of each eye. Still, however, we see objects simple ; and this single vision has, by some, been ascribed to habit. It is, however, probable, that vision is always single, when the images fall on precisely the corresponding points of both retinæ, and only double where this condition does not exist. Were this not the case, the compound eyes of insects would exhibit objects multiplied to an extent

* Phil. Trans. 1799, p. 1.

† Blumenbach, in reference to this subject, offers the following interesting observations. " It is well known that this pigment is entirely, or for the greatest part, deficient in the eye of the *albinoes* or *chacrelas;* which strange variety occurs, not unfrequently, in the human race, and in several other mammalia and birds. I know, however, no instance of an albino among cold-blooded animals. This anomalous deficiency is always congenital ; and is connected with a want of the colouring principle of the skin, and of the hair and feathers. It is hereditary in some mammalia, so as to form a constant breed of white aniamls, viz. in the rabbit, mouse and horse, (which latter are those called *glass-eyed.*) I cannot believe that any whole species of warm-blooded animals should originally want this pigment ; and, therefore, I consider the ferret, (Mustela furo,) to have descended from the polecat (M. putorius.") *Comparative Anatomy, Trans. Note to p.* 363.

which no habit or experience during their limited existence could reduce. I have observed, that children, from the time that they are capable of fixing their eyes steadily on any object, direct both of them towards it; and this effort they do not seem capable of making, until the iris has acquired the power of dilating and contracting.

3. Knowledge obtained by the sense of sight.—The information communicated to the mind by means of the sense of sight, is, perhaps, more varied than that of touch, but it is less accurate. The qualities and conditions of objects primarily ascertained by this sense, may be restricted to colour and direction merely. It is true, that we rely on the information which it communicates with regard to the distance, form, size and condition of the surface of bodies; but in these cases we are apt to be deceived, unless aided by the recollections of the sense of touch.

The eye is the only organ of the body, which is fitted to examine the quantity, quality, and motions of the rays of light. Hence we owe to it all our ideas with respect to the *colour* of bodies. We readily perceive the limits of different coloured spaces, and thereby ascertain their shapes and degree of illumination. Aided by the experience and the sense of touch, we speedily judge of the boundaries of objects themselves, by the distribution of colours, and their distance by their brilliancy. But in order to preserve the eye in a condition capable of perceiving correctly the differences among bodies with respect to colour, it is necessary that it be prevented from looking long on any one colour at a time, least the retina become fatigued, and less easily excited to receive impression. When the retina is thus fatigued, the eye ceases to judge accurately with regard to colour, seeing those only which have been termed *accidental*.

Thus, as has been already stated, if we look steadily at a white spot, and afterwards turn the eye towards white bodies, a dark spot will be perceived by the eye. If we look at a red spot on a white ground, and then direct the eye to another part of the white ground, a green spot approaching to blue will be perceived. In the first case, the retina was fatigued by the white colour, and could not be excited by any other colour, having the rays which constituted it in its composition. The accidental colour was, therefore, black. In like manner, after looking at the red spot, the retina was insensible to the impression of a compound colour, having red rays in its composition; hence, the accidental colour consisted of the other rays of the prismatic spectrum, forming a colour destitute of red.

The following Table exhibits the Natural Colours, with their corresponding Accidental ones.

Natural Colours.	Accidental Colours.
Red,	Blue, with a small mixture of green.
Orange,	Blue, with nearly an equal mixture of indigo.
Yellow,	Indigo, with a considerable mixture of violet.
Green,	Violet with a mixture of red.
Blue,	Red, with a mixture of orange.
Indigo,	Yellow, with a considerable mixture of orange.
Violet,	Green, with a considerable mixture of blue*.

The eye, in general, communicates accurate information with regard to the *direction* of objects. The sources of error in this case are few; and seldom interfere with the wants or the enjoyments of the species. They arise either from the reflection or refraction of the rays which proceed from the

* Accidental Colours, Edin. Encyclop. vol. i. p. 90.

objects exciting the sensation. Rays reflected from a mirror, exhibit to us objects in a direction in which they do not exist; and a similar deception is successful, when we mistake the *shadow* for the substance. Rays which are refracted in passing through media of different densities, likewise exhibit objects in the direction in which they do not exist. The rays of light from an object in water, coming obliquely to our eye, in air, are deflected from the perpendicular, so that it appears nearer the horizon or higher in the water, than it actually is, as is seen by immersing the lower end of a pole in water. On this principle it is well known, that the bottom of a river appears nearer the surface, or the water seems shallower than it really is, a deception which has proved fatal to many who have ventured into streams with which they were unacquainted. In attempting to kill fish in the water with a spear, this source of error will prove very inconvenient, until corrected by experience. Herons, gulls, soland geese and other animals which dart from the air upon their prey in the water, must often miss their aim in youth, from this optical illusion. When rays of light pass from an object in the air, to an eye in the water, they are bent towards the perpendicular, and the object appears nearer the zenith than it actually is. Trouts and other fish, which occasionally catch insects flying over the surface of the water, must learn to avoid this source of error, or meet with frequent disappointments. There is one fresh water fish, a native of India, the *Chætodon rostratus*, or Shooting Fish, which is able to correct this error with wonderful dexterity. "When the Jaculator fish intends to catch a fly or any other insect, which is seen at a distance, it approaches very slowly and cautiously, and comes as much as possible perpendicularly under the object; then the body being put in an oblique situation, and the mouth and eyes being near the surface of the water, the Jaculator stays a moment quite immoveable, having its eyes directly fixed on the insect, and then begins to shoot, without ever shewing its mouth above the surface of the water, out of which the single drop shot at the object, seems to rise.*" No part of the mouth is seen out of the water; and it shoots a great many drops, one after another, without leaving its place. Another fish, termed *Zeus insidiator*, is known to exhibit the same habits. These errors of refraction thus corrected by the fish, are, perhaps, in a great measure, avoided by birds, and other fishes, by confining their attempts to seize their prey to a perpendicular direction, in which no refraction takes place.

In examining the colour and direction of objects, the eye is seldom assisted by the other senses; and the information which it communicates, is in general correct. But it is otherwise in judging of the magnitude, the distance, or the condition of the surface of bodies.

We judge of the *Magnitude* of an object, by the different colours which it exhibits, in comparison of the colours of those objects by which it is surrounded. When the same colour pervades the whole, we must remain in doubt, so far as the eye alone is concerned, and even in those cases where there is a great dissimilarity of colour, we are at a loss to determine, whether the difference is produced by the margin of the body itself, or only a portion of its surface. Even after we have determined that the limit of the particular colour marks the boundary of the object, we have still to ascertain the angle which its extreme points form with the eye. But as the size of the angle depends not only upon the real magnitude of the body, but its distance, it

* Phil. Trans. 1764, p. 89. Tab. ix.; and 1766, p. 186.

is obvious, that the sources of error are numerous, and that the information thus gained is of little value, unless corrected by the experience of the sense of touch.

As we judge of the magnitude of objects, by the angle which they form with the eye, so, in like manner, do we employ the same means to determine their *Distance*, when restricted to the organs of vision alone. But as bodies of the same dimensions, when placed at different distances, will form with the eye different angles, and as bodies of unequal dimensions may be so arranged, as to appear under the same angle, our notion of distance acquired by unassisted vision, must likewise be very imperfect. It is true, that in judging of the distance of objects, their degree of illumination is likewise attended to. But as objects are not always luminous in proportion to their proximity, it may often happen that this character, instead of contributing to an accurate result, shall rather generate error.

In judging of the condition of the *Surface* of any object, in order to ascertain its inequalities by means of the eye, we attend to the degree of illumination of the different parts, and form our opinions from the manner in which the light and shade are distributed. But as the eye has not the power of perceiving the difference between a dark ground and a shaded place, nor between an illumined and a white spot, it is obvious, that, without the assistance of the sense of touch, it could not be determined whether the surface was rough or smooth, pitted or even.

The eye assists us in determining the *Motions* of bodies; but, like its intimations with regard to size and distance, it frequently misleads us. Thus, when an object is moving in a straight line from us, it appears as if at rest. When sailing on a river in a boat, the objects which are stationary on the banks seem to be in motion, while the boat which is in motion appears to be at rest. In like manner, the sun seems to move, and the earth to be stationary. When bodies move with great velocity, as a cannon ball, they are invisible; or when the motion is very slow, as the index of a clock, it is likewise imperceptible.

Although the intimations of external objects obtained by the eye, are, in many cases, apt to mislead, yet, when aided by the sense of touch, and, in some cases, by that of smell, the information communicated to the mind by the organs of vision, is more varied and extensive than that which is derived from any of the other senses. While the errors of the eye may mislead the inferior animals, and man himself, when seeking to supply the wants of existence, yet they have been converted into blessings by the ingenuity of our species, and made subservient to the increase of the pleasure and improvement of civilized society. It must be understood, that we here allude to the deceptive art of *Painting*.

The value of the characters furnished by the organs of vision, in the systematical arrangement of animals, is seldom estimated at a high rate. The differences exhibited in the eyes of animals, otherwise nearly related in form and structure, are so great, that little reliance is placed on the distinctions which they exhibit in the construction of the primary divisions of animals. But in the formation of the inferior groups, the characters of the eye are frequently employed with advantage, as they are remarkable on account of their constancy and obviousness.

There are many animals in which no trace of organs of vision has been discovered. Thus, eyes are wanting in the pteropodous and acephalous Mollusca, many of the Annulosa, and all the Radiata.

In the vertebral and molluscous animals having eyes, these organs are always two in number. But in the annulose animals, they vary greatly in number, some having only one, while others have two, four, eight, or more.

The eyes of some animals are fixed; while in others, they are capable of a great extent of motion, resting on a cushion of fat or gelatinous matter, or seated on the summit of a moveable peduncle.

The organs of vision are uniformly placed on the head; or, in the absence of that part, on the anterior portion of the body. They are usually situated on both sides of the mesial line, and have always more or less of a dorsal aspect, unless in Man, whose eyes are directed forwards.

There is no proportion observed between the eye and the size of the body. Insects present the greatest ocular surface to the light, in proportion to the size of the body,—whales the least.

IV. Sense of Hearing.

The essential parts of the organ of hearing, consist of a gelatinous pulp, inclosed in an elastic membrane, into which the last branches of the auditory nerve penetrate. The vibrations of sonorous bodies are supposed to excite movements in this pulp, by which the nervous filaments are affected, and the perception of sound produced. As it is difficult to assign to each part of the ear its peculiar function, we shall confine our descriptions to the parts themselves, briefly stating as we proceed, the uses which they are supposed to serve.

1. *Structure of the Organs of Hearing.*—The most complicated part of the organ of hearing, is that in which the auditory nerve terminates, and which is termed the *Labyrinth.* In quadrupeds, it consists of several cavities, hollowed out in the petrous substance of the temporal bone,

containing a fluid, through which the nervous filaments are dispersed. These cavities are of two kinds. In the first, there are tubular semicircular holes, which are three in number, and denominated *semicircular canals*, which open into a cavity termed the *vestibule;* the second is termed the *cochlea*, which likewise communicates with the vestibule, and consists of a tube revolving round a conical axis, hollow like a turbinated shell, describing, according to the species, from one turn and a half to three turns and a half. These cavities are filled with a fluid which has been termed the Water of *Cotunnius*.

The *auditory nerve* takes its rise from the inferior surface of the fourth ventricle of the brain, and penetrates the cavities of the labyrinth by traversing an infinite multitude of perforations, which it fills by its minute and anastomosing filaments.

In Birds, the labyrinth consists of nearly the same parts as in quadrupeds. In these, however, the cochlea is less perfectly developed, and merely consists of a short hollow bony process, divided into two scalæ. The semicircular canals are not hollowed out of an os petrosum, but consist of tubes, united by cellular bone. In Reptiles the cochlea is still more imperfect; and in many species it can scarcely be said to exist. The semicircular canals are apparent; and there is a bag containing gelatinous pulp, in which the filaments of the nerves are distributed. In this sac there is a substance of the consistence of starch, which may be regarded as occupying the place of the cochlea in the higher classes. In Fishes, the semicircular canals still exist; but there is no trace of a cochlea. The sac of gelatinous pulp, contains three cretaceous bodies, varying in hardness in the different races. The nervous filaments embrace these bodies. Among the cephalopodous Mollusca, the labyrinth consists of a simple sac, inclosing the pulp and a single cretaceous body. There are no semicircular canals. In the Crustacea, the sac consists of a scaly cylinder; the one extremity closed by an elastic membrane, exposed to the action of the element in which the animal resides; and the other extremity open for the entrance of the nerve. Neither semicircular canals nor cretaceous bodies are obvious.

In Quadrupeds, the labyrinth occupies the inner part of the ear; and on its external side there is a cavity, termed the *Tympanum*, or barrel of the ear. This cavity is irregular in its form, according to the species. It is closed on all sides; behind by the labyrinth, laterally by the osseous parietes, and externally by an elastic membrane, termed *membra tympani*, or drum of the ear. The inner wall of this cavity, which is formed by the labyrinth, contains two openings; the one termed the *fenestra ovalis*, which communicates with the vestibule and semicircular canals; and the other termed the *fenestra rotunda*, which opens into the cochlea, but whose orifice is closed by a membrane. From the floor of the barrel, a tube, osseous at its commencement, and cartilaginous in the remaining part, takes its rise, which terminates in the back of the mouth in the palate, called the *Eustachian tube*. Other holes may be observed in different species, leading into adjacent cells. The external membrane, by which the tympanum is closed, is attached to the osseous circle which forms the walls. It may be regarded as a continuation of the skin through the external ear, and of the integuments of the mouth through the Eustachian tube; it is dry, transparent and elastic. Within the barrel there are three bones, which form an immediate connection between the drum and the fenestra ovalis. The first, which is termed *malleus* or hammer, adheres to the drum; and is articulated with the central bone, *incus* or anvil, which is again united to a branch of the *stapes* or stirrup, whose base rests upon the fenestra ovalis.

These bones are susceptible of motion, and have muscles attached to them, by which they are supposed able to stretch the drum, to compress the fluids in the labyrinth, and communicate the vibrations of the drum to the pulp of the canals and cochlea.

In Birds, the tympanum bears a close resemblance to that of quadrupeds. Its cellular openings, however, are more numerous, and the cells themselves of greater dimensions. The fenestra ovalis and rotunda, are placed, the former above the latter, and separated by an osseous bar. The Eustachian tube is osseous throughout its whole length. In the barrel there is but one bone connecting the drum with the labyrinth, branched where united with the drum, proceeding by slender stalks through the cavity of the barrel, and forming a plate which closes the fenestra ovalis.

In Reptiles, the tympanum exhibits very remarkable differences. In some, it can scarcely be said to exist, while in others, it is membranaceous, exhibiting the Eustachian tube, and one or two bones in the barrel. The membrane, or drum, is wanting in some cases, its place being supplied by the common skin, while in others, it exhibits its usual characters of dryness, transparency, and elasticity.

In Fishes, particularly those with free branchiæ, there is no tympanum. In those with fixed branchiæ, the rudiment of a tympanum may be observed in the form of a minute tube leading from the labyrinth to the skin.

Between the drum and the surrounding element in Quadrupeds, the *External* Ear is situated. It consists of a tube continued from the drum to the surface of the body. It varies greatly in the length, direction, and texture of its walls. When arrived at the surface, this *meatus auditorius externus*, as it is called, either terminates in a simple hole, or it is expanded into a cartilaginous arch, termed the *concha.* In Birds, the tube is short, and there is no expansion of the skin to constitute an external ear. In Reptiles, there is no tube, nor any external opening, (unless in the crocodile where the skin forms a kind of lip). The skin passes directly over the tympanum, exhibiting no change in its direction, and becoming only a little more transparent.

Although the organs of hearing, in Insects, have not been satisfactorily demonstrated, their existence may reasonably be inferred from the circumstance, that many species are capable of producing sounds, and others of acting under their influence.

From the preceding review of the organs of hearing, it appears, that in some animals the action excited in the air or water by vibrations of sonorous bodies, is communicated directly to the auditory nerve, by the medium of the common integuments. In other cases, there appears a complicated apparatus, to collect the vibrations, and to transmit them by means of cavities and tubes, variously arranged, to the auditory pulp. These differences in the form and structure of the parts, must occasion corresponding modifications in the impressions produced; but with regard to the nature of these, we are still in a great measure ignorant.

2. *Knowledge obtained by the sense of Hearing.*—The information communicated to the mind by the organs of Hearing, is far from being so important and diversified as that which is derived from the sense of touch or sight.

Sound is produced by the motion of the parts of a particular body, or by the friction of one body against another. In both cases, a tremulous motion is communicated to the surrounding medium, which extends in all directions like the waves produced in the water by a stone falling into it. These undulations affect the organs of hearing, and excite the sensations of sound. These undulations vary with the condition of the body which excites them, and when carefully attended to, give us intimations regarding the distance, direction, velocity, connection, and composition of sonorous bodies.

In judging of the *distance* of objects by the sense of hearing, we attend chiefly to the strength of the sounds which they emit, considering it as bearing some proportion to their proximity. But as this depends on the quantity of motion excited, and the resistance which is opposed, we may be led to conclude that a body is distant, because the sound emitted is weak, while it is actually near, but the extent of its vibrations limited.

The *position* of sonorous bodies is intimated to us by the direction in which the vibrations are communicated to the external organs. In many cases, our conjectures are verified by experience, but in others, we are deceived by the change produced in the direction of the sonorous waves by various obstacles, so that we mistake the reflected for the radiated vibrations,—echo for the direct sound.

In estimating the *velocity* of moving bodies by means of sound, we attend to the changes which take place in its strength. When the transition from loud to weak, or from weak to loud, is rapid, we infer, that the vibrations have a corresponding velocity, in proportion to the change of place in the body which produces them. We are more rarely deceived in this case than in the two former, although it sometimes happens that a change in the direction of the sonorous vibrations, may produce corresponding variations in their strength, without any alteration having taken place in their velocity.

When two bodies are rubbed or struck against each other, we are in many cases able, from the differences of sound emitted, to recognise their composition or structure.

Thus, we can in general distinguish between the sound of wood or metal, water or air, and hollow or solid bodies.

In judging, however, of any one of these qualities or conditions of bodies, we call to our assistance all the information which is communicated by the different characters of the sonorous vibrations, so that the conclusion at which we arrive, is frequently the result of a very complex, though rapidly executed mental operation.

To animals in general, the sense of hearing gives warning of the approach or retreat of their foes. In some cases, it is likewise the medium of communication between the individuals of the same species, in the expressions of their feelings of joy or grief, love or hatred. The human ear, judges of all the qualities of sounds. Some of the quadrupeds and birds can, however, perceive sounds which are inaudible to us, and perhaps can recognise more readily than we are able to do, some of the modifications of a particular quality. But our knowledge of the peculiar functions of the different parts of the ear, is still too limited to enable us to speculate, with any prospect of success, even were our information concerning the habits and feelings of the lower animals more extensive and precise.

In the systematical arrangement of animals, the characters furnished by the organs of hearing, are scarcely ever employed, unless in the exhibition of generic or specific distinctions. Even for such purposes, the attention is exclusively directed to the external ear, and to the form and position of the entrance; the internal ear being concealed from view, difficult to dissect, and furnishing characters which can scarcely be expressed by words.

V. Sense of Smell.

In those animals which possess organs of Smell, sufficiently developed to be obvious to the eye, the Nose, or entrance, is placed on the anterior part of the head. It is divided by a partition which varies in its breadth and position, and gives to the nostrils a great variety of character.

1. *Structure of the organs of Smell.*—In the vertebral animals which respire by lungs, the partition which is observed externally, is continued to the posterior opening into the throat, forming what is termed the *septum narium*. This division is formed by the vertical plate of the *ethmoidal* bone and the *vomer*. The cavities on each side are, in part, filled by the turbinated processes of the ethmoidal bone which occupy the vault, and the inferior turbinated bones, which adhere to the maxillary bones, and project into the middle of the cavity.

The olfactory membrane lines these cavities of the nose, covering its septum, walls, and projecting laminæ. It is merely a continuation of the external skin, which is attached by cellular substance to the periosteum of the bones, and which unites posteriorly with the integuments of the throat and gullet. Near the external openings it resembles the skin, but towards the interior it is of a red colour, which is derived from the numerous bloodvessels with which it is supplied. Its surface presents numerous little eminences, which have been regarded, by some, as nervous papillæ, and, by others, as the termination of excretory canals. It is kept constantly moist by a watery fluid, termed *nasal mucus*, in general, a secretion from the whole membrane, in other cases produced in particular cells.

This membrane is supplied with nerves from the first pair or olfactory nerves. These take their rise in the anterior lobes of the brain. Each nerve proceeds to its corresponding nostril, and after dividing into numerous filaments, is lost in the substance of the olfactory membrane, where it covers the septum and windings of the ethmoidal bone. Besides the olfactory nerve, which is considered as essential to the sensation of smell, the nose is likewise supplied with a division of the ophthalmic branch of the fifth pair, which is principally distributed on the inferior part of the membrane, and termed the *nasal nerve.*

Connected with the cavities of the nostrils which we have already mentioned, there are numerous cells termed *Sinuses*, in some of the adjoining bones, which communicate by contracted apertures. They are termed *frontal*, *sphenoidal*, and *maxillary*, from the bones in which they are imbedded. They are covered with a continuation of the olfactory membrane. It is thinner in these cells than in the cavity of the nose, is not supplied with any branch of the olfactory nerve, although filaments of the nasal nerve may be traced into it.

In Fishes, the nose does not communicate with the mouth. The nasal cavity on each side is simple, the olfactory membrane is expanded on the walls, and kept moist by a secreted mucus. One or two openings externally lead to this cavity.

2. *Functions of the organ of Smell.*—In order to comprehend the functions of the nostrils, it is necessary to make a few remarks on *Odours*.

Many bodies allow excessively small particles to escape from their surface, and diffuse themselves through the atmosphere. These particles, which are termed Odours, are always emanating from certain bodies, producing a rapid decrease of weight in some, as ether, while in others, as musk, no sensible diminution takes place. Such particles are not given off by some bodies when in a particular state of combination, as ammonia, when united with the stronger acids, while other bodies require to be combined, previous to their escape, as arsenic or phosphorus with oxygen. Heat assists the escape of odorous particles from a variety of bodies which are termed volatile. Light likewise influences the emanation of odorous particles. Hence we find some plants give out their smell during the day, while others perfume the air only at night. Dampness, in many cases, assists the escape of such particles, as appears by the fragrance of a garden after a summer shower, or clay when breathed upon. These odours emanate from bodies in all directions, with greater or less rapidity, penetrate only where air can enter, and obey the motions which it experiences.

The air, in passing through the nostrils to the lungs, comes in contact with the olfactory membrane, which lines the narrow passages, and enables the odorous particles which it conveys to act upon the olfactory nerves. We are ignorant of the manner in which this action is produced. Has the olfactory membrane an affinity for odorous particles? Does the nasal mucus retain these by its adhesive power, or is it employed in attracting them? Are odorous particles deposited on the membrane, or do they merely act mechanically, as they move along its surface? These are questions of difficult solution, and hitherto treated in a superficial manner.

In the case of fishes, the water impregnated with odorous particles, traverses the cavity of the nose, urged by the muscular action of the apertures and walls, and brings them into contact with the olfactory membrane.

3. *Knowledge obtained by the sense of Smell.*—The sensations produced by the different kinds of smells, are classified with difficulty. Some are *agreeable*, such, in general, as those which serve for food; others are *disagreeable*, as the most of those substances which are deleterious. They

are likewise classified, according to the effects which they produce immediately upon the organ or the feelings in general; thus, we have pungent, nauseous, and fragrant smells. In many cases, however, we are compelled, in describing an odour, to compare it with that which is emitted by some well known substance. Thus, we say, sulphureous, vinous, alliaceous, musky, in reference to the smell of burning sulphur, wine, garlic, or musk.

This sense gives us information of the presence of odorous bodies, and, in many cases, of their position. It is, however, more extensively employed by animals, to distinguish one body from another when contiguous, especially the different kinds of food. It informs us of many of the changes which take place in bodies by heat, light, or moisture, and thus serves the purpose of a chemical test.

The sense of smell contributes greatly to our enjoyments, in our anticipations of food, and in the pleasure derived from perfumes. In man, the organs of smell are more developed in the savage than in the civilised state. In the latter, multitudes destroy the utility of this sense, as the mean of procuring information or pleasure. Other mammiferous animals derive great pleasure from this sense, as is exhibited by the fondness of dogs to putrid substances, against which they delight to rub themselves, and of cats to particular plants.

From the difficulty of characterising the different kinds of smells, and of recollecting the particular sensations which they excite, the information communicated by this sense, though varied, is seldom to be relied on with much confidence. The smell of one body may be disguised by the presence of another, even when small in quantity, as may be seen to a great extent in the deceptions of modern cookery.

In the Molluscous, Annulose, and Radiated animals, no organs of smell have been detected, analogous to those which exist in the more perfect animals. They, however, appear to possess this sense, and to be guided by its intimations in seeking out the proper kinds of food, their mates, and a situation to deposit their eggs. In some cases, they are deceived by the resemblance between the smells of substances very different in other qualities. Thus, some plants emit a cadaverous smell, similar to putrid flesh, by which the flesh-fly is allured, and deposits its eggs on parts of these which can furnish no food to the future progeny.

VI. Sense of Taste.

The sense of Taste resides in the mouth, and the organs destined for exercising it are analogous to those of touch. The skin, upon entering the mouth, becomes of a finer texture, and is supplied by a greater number of bloodvessels and nerves than on the outside of the body. It is soft, covered by numerous papillæ, and kept continually moist by the saliva. The sense of taste is not confined to any one part of the mouth; it exists in the cheeks, tongue, palate, throat, and perhaps gullet, being most sensible in those parts which are softest, and have the greatest number of papillæ.

The nerves which are distributed to the mouth, take their rise from different branches of the fifth, eighth, and ninth pair. The peculiar office of each of these nerves is not distinctly understood, nor has it been demonstrated that they are all conducive to the perfection of the sense of taste.

The bodies which excite the sensations connected with taste, require to be dissolved or mixed with the saliva of the mouth, and in this state applied to its integuments. It is not known how the particles of sapid bodies moistened produce their effect on the organs of taste. Some suppose that they act chemically, others mechanically.

The nomenclature of savours, though very imperfect, is more precise than that of smells, arising from the circumstance that the impressions of sapid bodies are more permanent than of odours.

We distinguish savours into agreeable and disagreeable, sour, sweat, bitter, hot, and cold; and in our descriptions of them, we frequently refer to well known bodies, as salt, sugar, vinegar, as standards of comparison. Sapid bodies appear to act with greater energy on some parts of the organ of taste than on others. Thus, some affect the throat, others the palate or tongue.

As different bodies act on the organs of taste in producing a variety of different feelings, this sense is extensively used in the discrimination of bodies, more particularly those which serve as nourishment. Indeed, this sense appears almost exclusively subservient to the digestive system, so that the pleasure we derive from the savour of bodies in the mouth, is intimately connected with their salutary effects in the stomach.

The information communicated by this sense is limited in its nature, in the inferior animals, to food. Man, however, employs it to ascertain the composition and relation of bodies; and, by experience, communicates to this sense a wonderful degree of sensibility, as we see in chemists, wine-dealers, and even epicures.

It is probable that all animals possess the sense of taste, to enable them to make choice of the proper kind of food.

Before concluding our remarks on the organs of perception, some general observations may be made on their nature and mutual dependence.

a. There is no determinate relation observed in the degrees of perfection of the senses in different animals. Even in the same class, and subordinate divisions of a class, one species or genus may have the organs of smell very fully developed, another those of hearing, a third those of sight, while the other senses may be in a less perfect condition. Even among individuals of the same species, such differences prevail.

b. In judging of the properties of bodies, we seldom rest satisfied with the information obtained by one sense, but employ the results furnished by the others, to correct or strengthen our conclusions. Thus, the organ of touch assists that of seeing, and sight aids the efforts of touch. Hence, as the senses mutually assist each other, it is difficult to assign to each of them the knowledge which it has exclusively communicated. In the case of disease or accident, where one sense has been destroyed, the other senses, by an increased sensibility, in a great measure supply the defect. Thus, hearing and touch, in many cases, supply the loss of sight, and exhibit striking displays of that compensating or repairing power, to which we have had frequent occasion to refer.

c. In many animals, where some of the senses exist, although we are unable to detect the organs in which they are seated, as is the case with hearing in the annulose animals, it is probable that the deficiency of one sense may be supplied by the sensibility of the remaining ones.

d. The same qualities in bodies do not produce the same effects on the appropriate organs of all animals. There is a striking difference in the degrees of sensibility; so that an impression which would be overpowering to one animal is scarcely felt by another, as appears in the case of certain sounds, smells and lights. Even sensations excited by the same bodies in different animals, are dissimilar in kind. This appears from the circumstance, that what is agreeable to and sought after by one animal, is often disagreeable to another, and carefully avoided.

e. From these considerations it appears to be difficult to determine the knowledge of external objects, possessed by any one species, from the developement of the organs of the senses. Nor does it appear that there is any regular connected gradation in the relative perfection of the senses, keeping pace with the increasing simplicity of bodily structure.

f. Although many animals have *some* of the senses more fully developed than Man, comparative anatomy furnishes a demonstration, that there is no animal in which they *all* exist in so great a degree of perfection. Hence we are led to conclude, that man is more intimately acquainted with the properties of the material world in general, than any of the inferior animals.

g. Some of the senses, such as taste, smell and touch, require the existing object to be brought into contact with the organs; while in others, as sight, heat, and hearing, the sensation is produced by means of media. Touch, which is most universally distributed, appears to be the sense into which the others may be resolved. Thus, light comes in contact with the eye, heat with the skin, and the vibrating air with the ear.

CHAP. XI.

FACULTIES OF THE MIND.

In treating of the organs of perception, we endeavoured to ascertain the functions of the six senses, and the kind of information concerning the objects around us, which they are qualified to communicate to the mind. It now remains that we attend to the sensations themselves, in reference to the processes performed upon them by the mind, with the view of becoming acquainted with the faculties of that mysterious part of our nature.

In conducting this inquiry, it is difficult to avoid the use of ambiguous phrases; for almost every term which can be applied to mind, has been loaded with a variety of significations. It will be our aim to render obvious the meaning of the terms here employed, by the descriptions which accompany them.

When we attend to the phenomena displayed by the mind, we perceive that it exhibits certain relations to the sensations of the body, changes resulting from their production, and consequent efforts of volition.

These different states or conditions of the mind have been observed to be regulated by peculiar laws, and to be subservient to particular purposes in the animal economy. They have had bestowed on them specific appellations, to express their characters; and they have generally been denominated the Attributes, Faculties, or (in consequence of volition producing change) Powers of the Mind. Trivial objections have been urged against the use of these terms, as implying that the mind is composed of different parts; and, consequently, not entitled to its prerogative of *unity*. Of the essence of mind we absolutely know nothing; and hence the various phrases, Unity, Indivisibility, Immateriality, and others, which have been employed to express the nature of this essence, are, in fact, expressions of our own ignorance and presumption. When we witness the mind capable of exciting action in matter, and of being excited to action by matter,—exhibiting its identity by its local residence,—variable in its relations to matter,—variable relatively to its own conditions,—capable of exercising different functions at the same time,—and, last of all, multiplying with an increase of population,—we feel overwhelmed with the incomprehensible phenomena which it presents, and admit the suitableness of an expression of our Divine Master, when applied to the present case, "Ye know not what manner of spirit ye are of."

When we attentively examine the peculiarities of the mental powers, they appear to admit of distribution into two classes, and several subordinate divisions. These classes we term the *intellectual* and *instinctive*, and now proceed to give a detail of their characters.

In the illustrations which are here offered on the mental phenomena, I have given the result of the observations which I have made upon my own mind, and the sentient objects around me, without being influenced by the received systems of philosophy. It would be of great advantage to moral science, were individuals to study the conditions of their own minds, and observe the phenomena which are so accessible, instead of hastily subscribing to the tenets of a particular school, or yielding to the influence of questionable authority. The various results from different minds could thus be compared, discordant statements submitted to more rigorous scrutiny, and the number and nature of the attributes of mind more satisfactorily established.

I. Intellectual Powers.

1. Faculties of the mind.—In order to enable the mind to become acquainted with the objects which produce an impression on the senses, certain conditions are necessary. The organs of perception must be kept steadily towards them a sufficient length of time, that they may be examin-

ed in all the various circumstances in which they present themselves. This power or effort of the mind is termed

1. *Attention.*—Unless this power is exercised on the organs of perception, the impressions which are produced are confused and obscure, and speedily vanish. But, by its means, the organs are brought into a condition suitable to receive an impression and determine its nature and duration, and preserved therein a sufficient length of time.

In the exercise of Attention, we seldom are capable of employing more than one sense at a time; but can, without much difficulty, fix the mind on several different kinds of information which one sense is capable of communicating. Thus, we can distinguish a variety of conditions in sound, and the impressions produced on the organs of sight. But this faculty chiefly displays its peculiar nature and its utility, in its *selective* operations. Thus, for example, in looking at a rose, I can either attend to its size, its subdivisions, colour or fragrance. The character on which I am said to bestow the greatest attention, is the one to which I direct the suitable organ of sensation with the greatest intensity, for the greatest length of time, and of which I am said to have the most distinct conception. Man employs this essential faculty of his mind, in every inquiry he makes, and in almost every action which he performs. Without its assistance and controul, no knowledge of external objects can be gained, no train of reflection can be pursued, no successful bodily effort can be made.

In the lower animals, this faculty not only exists, but displays itself to our observation in various ways. What is it but the exercise of attention, when we see a cat watching for a mouse, or a kestril hovering in the air? In both cases, the faculty now under consideration is exercising its controul over the organs of sight. When we witness the fox-hound engaged in the chace, we see attention regulating the organs of smell; and, regardless of the other perfumes arising from the ground, permitting only the scent of the cunning fugitive to make a deep impression. The dog who has lost his master in a crowd, practises the same restraint upon his organs of smell, sometimes, also, on his sense of hearing, as he is able to detect his master by his voice, even when others are speaking at the same time.

Reasoning from analogy, we may conclude, that the faculty of attention reaches as low in the scale of animal life as the organs of sensation. It is necessary in the more perfect animals, for the regulation of every impression; hence it is probable that it exists, wherever there are organs to receive an impression.

The ideas which we thus acquire of external things, by the help of the faculty of attention, are so disposed by the mind, as to be retained in such a manner, that they can be recalled at pleasure, even in the absence of the objects which excited them. There are two faculties employed in this process. In the one the idea is recalled in the condition in which it was first formed; in the second, a part only of the idea is recalled, in union with the part of some other idea. The former is termed *memory*, the latter *imagination*.

2. *Memory*.—A variety of circumstances enable us to recall the idea which a particular impression has formerly produced, and exhibit it in nearly its original condition. This process of recollection is performed by retracing the steps which led to its production, the time and place in which it was excited, and the consequences which followed its reception.

The tendency of the mind to have impressions recalled by particular circumstances, in preference to others, has been exalted into a separate faculty, denominated the *power of association*, exhibiting its operations in the production of of rhyme, puns, colloquial phrases, and alliteration. In many cases, however, ideas return in the mind without any exertion to recall them. In some instances we can trace back the process by which they were introduced, while, in others, we have no notion of their origin or connection. If the process is known, we perceive that it has been excited by some impression of *resemblance*.

Some confusion has arisen with regard to our notions of Memory, from the introduction of the term *conception*. This phrase, in numerous instances, refers to the original image or impression produced on the mind, through the medium of the senses, thus using it as synonimous with sensation and perception; while in others, it is applied to the combining efforts of the imagination, or the simple recollective exertions of the memory.

When an idea returns to the mind either spontaneously or by an effort of recollection, we employ the faculty of attention in examining its condition. We either view it as a whole, in relation to the whole or the parts of other objects, or we attend to the qualities of its particular parts in reference to one another, or to the whole, or the parts of some other object. This selective act of attention some have considered as a separate power, under the denomination of *abstraction*. But as we have seen attention practising abstraction, in regard to the operations of the senses, and as it acts in a similar manner with the ideas of recollection, we see no necessity for regarding this operation of the mind as a distinct faculty. Indeed it must appear obvious, upon a little consideration, that the terms *abstraction, examination* and *comparison*, merely express the single or combined acts of the attention and memory.

Ideas are excited by the memory in general, with a facility proportional to the degree of attention originally bestowed on the impressions, their intensity, and the frequency with which they have been recalled. But there are some ideas which can be recalled with difficulty, whatever exertions we may have made originally to render them permanent. Thus, the perceptions of sight and touch are more readily recollected than those of sound, taste or smell. In the latter, we seldom can recall a distinct idea; but, in the attempt, rest frequently satisfied with the recollection of the symbols by which it is designated.

In comparing objects with one another, we employ the powers of attention and memory in conjunction. Having, by means of attention, fixed the mind on one object, we can then turn it to another: the memory preserves the impressions which the former produced, so as to compare these with the qualities of the latter.

This faculty of memory is essentially requisite to our attainment of knowledge. Without it our intellectual improvement would be confined within very narrow limits. It exhibits itself in various degrees of excellence in different individuals, partly the result of constitutional arrangements, and partly of habit. In all cases its various properties are improved by exercise.

The existence of the faculty of memory in the lower animals, can scarcely be doubted, as instances daily occur of its display in our domestic quadrupeds, as the elephant, horse, and dog *. It is likewise exhibited by birds. We

* In illustration of the extent of the memory of the elephant, Mr CORSE, in his valuable observations on the natural history of that animal, states the following circumstances, to which he was an eye-witness:—" In June 1787, Fâttra Munqul, a male elephant, taken the year before, was travelling, in company with other elephants, towards Chittigong, laden with a tent and some baggage for our accommodation on the journey. Having come upon a tiger's track, which elephants discover readily by the smell, he took fright, and ran off to the woods, in spite of the efforts of his driver. On entering the wood, the driver saved himself, by springing from the elephant, and

have evidence of a peregrine falcon, which was lost in the month of March, recognizing its master, when retaken in the end of September. Indeed in all those animals which are capable of being tamed, there must exist this faculty to enable them to recognize former sensations.

The memory, in the lower animals, likewise performs its operations in the same manner as with us, by the help of what is termed the *Association of Ideas.* Thus we have

---

clinging to the branch of a tree under which he was passing. When the elephant had got rid of his driver, he soon contrived to shake off his load. As soon as he ran away, a trained female was dispatched after him, but could not get up in time to prevent his escape. She, however, brought back his driver, and the load he had thrown off, and we proceeded, without any hope of ever seeing him again. Eighteen months after this, when a herd of elephants had been taken, and had remained several days in the inclosure, till they were enticed into the outlet, there tied, and led out in the usual manner, one of the drivers, viewing a male elephant very attentively, declared he resembled the one which had run away. This excited the curiosity of every one to go and look at him; but, when any person came near, the animal struck at him with his trunk; and, in every respect, appeared as wild and outrageous as any of the other elephants. At length, an old hunter, coming up and examining him narrowly, declared he was the very elephant that had made his escape about eighteen months before. Confident of this, he boldly rode up to him, on a tame elephant, and ordered him to lie down, pulling him by the ear at the same time. The animal seemed quite taken by surprise, and instantly obeyed the word of command, with as much quickness as the ropes with which he was tied permitted; uttering, at the same time, a peculiar shrill squeak, through his trunk, as he had formerly been known to do; by which he was immediately recognized by every person who had ever been acquainted with this peculiarity."—Phil. Trans. 1799, p. 40. The same observer furnishes satisfactory evidence that another elephant, a female, taken in 1765, and which was turned loose in 1767, when retaken in 1782, recollected the customs of her former bondage, and lay down at the command of her driver. "He fed her from his seat; gave her his stick to hold, which she took with her trunk, and put into her mouth, kept, and returned it, as she was directed, and as she formerly had been accustomed to do."—Ibid.

seen a spaniel exhibit all the ecstacy of joy when he observed his master put on any article of dress which he was accustomed to wear during the hours of sport. These things recalled to him the enjoyments of the field, as distinctly as the sight of the gun. We see the same animal practising an attempt at recollection, by smelling at a stranger, and at last, after many efforts, recognizing him as an old friend.

We have not the means of determining whether former impressions arise spontaneously in the mind of the inferior animals, or whether all the acts of memory be brought about by the excitements of present sensations. We know very imperfectly the nature of those spontaneous efforts of the memory, as they take place in our own minds; so that our reasonings on this subject, in reference to the brutes, must be of a doubtful kind. Indeed it is probable, that these spontaneous recollections are not purely mental processes, but have their rise in some of those organical circumstances which excite and frequently regulate our trains of thought.

3. *Imagination.*—The proper business of imagination is to decompose our impressions or ideas; to exhibit a part of any one of these, divested of its original associations, or in union with a portion of some other impression or idea. In performing these analytical and synthetical operations, the powers of attention and memory are indispensably requisite. The former fixes the mind on a particular quality, the latter recalls our ideas of other qualities with which we wish it to be united. The combinations which we thus form, influence the mind in a similar manner to an original impression. We can fix our attention upon their characters, and store them up, so as to be able to recall them at a future period.

The ideas on which the imagination exercises its powers with the greatest success, are those which the memory can recall with the greatest distinctness, such as those obtained by the aid of the sense of sight and touch. But the ideas produced by the impressions of the other senses, such as hearing, taste, or smell, cannot be so readily recollected, and at best but obscurely; hence the imagination cannot operate upon them with any degree of success. The combination of the ideas resulting from these last impressions, with new objects, is seldom intimate. When, in works of fiction, they appear to be so, it is because we deceive ourselves with the signs of the sensations, which we are able to combine in any manner we please, while we are incapable of combining the ideas which they represent.

By means of imagination, thus exercised on former impressions, we are able to produce new ones, in a greater degree, perfect or imperfect, agreeable or disagreeable, than any which we have ever experienced.

But if we restrict the operations of imagination to the decomposition or combination of ideas which we have actually obtained by sensation, we render doubtful its claims to rank as a distinct faculty of the mind, and greatly limit its usefulness as an instrument in the acquisition of knowledge. It is not merely a retrospective, but a prospective power. By the help of memory, we recall past impressions; by the help of imagination, in the exercise of this second function, we, as it were, render future impressions present. It is true that these pictures of futurity are formed of the materials of past impressions, but they are always in a new state of combination with regard to time. Sometimes they are more disagreeable than those of the past, when the mind is torn with despair. In other instances, more agreeable, when they are generated under the cheering influence of hope. Some of the pictures of the imagination, as those of fiction, we never expect to see realized; while some others, which we know never existed, are yet confidently anticipated.

It is by means of imagination thus exercised, that we arrange our plans of conduct, or invent new schemes, for the employment of our time. It is the greatest source of our activity, and is the only intellectual faculty which accelerates our exertions to improve.

The justly celebrated Mr STEWART considers it as sufficiently evident, "that imagination is not a simple power of the mind, like attention, conception, or abstraction, but that it is formed by a combination of various faculties *." Of these simple powers which he has enumerated, we have already considered Conception as identical with Memory, and Abstraction with Attention; but it remains to be pointed out what those faculties are, which, when combined, produce imagination. Attention is exclusively occupied with present impressions or ideas, and memory with those which have been. Now, although the imagination makes use of the materials furnished by the memory, and employs attention while acting upon them, yet it forms from these ideas which never have existed, and never will exist; or it forms such pictures as by exertion may be realized, and bends all the faculties of the mind to their production. Were the efforts of the imagination thus confined to what had taken place, there might be room for considering it as a combination of attention and memory; but when it looks into the future it exhibits its peculiar and exclusive character. Indeed it is this power of anticipating the future, much more than of acting upon the recollections of the past, that gives to the imagination its rank as a distinct power of the mind.

* Elements of the Philosophy of the Human Mind, vol. i. p. 488.

Without this faculty we would never even attempt to know what a day might bring forth.

The mind appears to retain, and the memory to recall, both the retrospective and the prospective ideas which the imagination has formed ; and, what may appear surprising, the latter with the greatest readiness. This faculty, like those of attention and memory, is greatly improved by exercise, and appears in the greatest degree of perfection in those who have received a liberal education, and habituated themselves to attend to their intellectual operations. In the savage state, this faculty is probably chiefly occupied in prospective efforts to secure food or shelter.

In the lower animals, the faculty of the imagination certainly exists, although, from the imperfect communication which subsists betwixt us and them, its operations, as distinct from memory, cannot be traced with any degree of certainty. The pointer, who exhibits impatience to travel when his master takes his gun in his hand, recollects the pleasure of his former sports, and wishes to have them renewed. What is it but imagination that persuades him that they may return, and even points out the channel of their course ! A dog howls when his master is absent, and will anxiously look for his return in particular directions. Here is anticipation of a future event ; and action founded on the certainty of its occurrence. We have seen a dog, evidently entertaining suspicions that his master would prevent him being a companion in his journey, steal away unobserved, and wait on the road, at a considerable distance from the house. Here we have the anticipation of the master's going from home ; apprehension of being detained ; the prospect of gratification from the journey ; an expectation of his master's road ; and the success which would crown the plan ;—all of these efforts of the imagination. As we descend in the scale, these displays of the imagination can scarcely be perceived, unless in actions which suppose a succession of events similar to those which have occurred.

There is one very striking difference between this faculty, as it exists in man and in the lower animals. With us, it is frequently exerted on speculative truths: In them, on present or future sensations. With us, sometimes on things which we know will never happen: With them, on things which the probabilities of experience warrant.

II. Ideas of Reflection.

The knowledge which we obtain through the medium of our senses, by means of the three faculties whose nature we have been attempting to illustrate, is denominated *Experience.* In acquiring it, the mind employs the senses to collect, the attention to fix, the memory to recall, and the imagination to combine, the various impressions which external objects make upon us. In the course of this very complicated process, the mind, by degrees, perceives relations among external objects which were not discovered immediately after the first impressions, and obtains results, from an attention to the intellectual process, by which the limits of our enjoyment and power are extended, and the senses directed to proper objects of examination. In consequence of this habitual or occasional attention, we acquire a number of ideas, which do not resemble those which have been formed directly from the impressions of the senses. These are, therefore, properly termed Secondary Ideas, or *Ideas of Reflection.* In the formation of these ideas, the imagination exercises the greatest influence. They, indeed, arise from the unceasing combinations and decompositions of this faculty, and might, with propriety, be termed Ideas of the Imagination. We may likewise observe, that while they depend upon the condition of the primary impressions, and are subordinate to these, they are still inseparably connected with the constitution of the mind, and are necessarily formed in the course of its ordinary operations. We shall now offer a few observations on some of the more remarkable of these ideas of reflection.

1. *Personality.*—In the course of attending to the sensations which external objects excite, we speedily discover that the perceptions obtained by one sense, differ from those procured by the assistance of another, and that the perceptions of any one sense differ with the objects which excite them. As these different impressions frequently recur, the mind perceives their resemblance or dissimilarity, and begins to classify them by the aid of memory. In this manner, the primary and secondary qualities of matter,—the peculiarities of individuals,—the general character of groups,—the relation of one object to another,—and the notion of number, are apprehended.

Uncultivated minds have not very correct notions of these ideas of reflection: but, still they do arise even in such; and, in various ways, influence their future operations.

The inferior animals likewise possess similar ideas of reflection. It is true, that they appear to be equally acted upon by the impressions produced by the secondary, as well as the primary qualities of matter; nor have we any evidence that they act upon the distinction; but they know that there is a difference among the qualities. They are, to a variable extent, acquainted with them, and regulate their conduct by this knowledge which they possess. They readily perceive changes in the objects with which they are familiar. They are acquainted with individuals, or their identity, as the dog is acquainted with his master;—with groups, as in the case of the shepherds' dog, who is capable of marking the individual of a flock pointed out to him by his master, and steadily pursuing it. Even the notion of number is not unknown, as appears from the wolf uniting in bands in the chace,—and, while afraid of attacking men in company, appearing fearless of the resistance of a straggler.

In our intercourse with the world, we are much guided by our ideas of personality. We recognise objects which we have formerly examined, and, convinced of their identity, confidently expect them to be still possessed of the same qualities we formerly discovered them to be invested with. If a change has taken place in their appearance, we expect a corresponding change in their qualities. When we witness two individuals exhibiting similar characters, we expect to meet in both the same qualities. It is by means of this acquaintance with individuals, and this power of detecting resemblance or dissimilarity, that we obtain the greater part of our knowledge of ourselves and the world around us, and place confidence in the value of our systematical arrangements. But if much knowledge is thus gained, a fruitful source of error is established by the employment of what is termed *Analogy*.

Analogy may be safely employed in the prosecution of knowledge or the business of life, were we to regard *the probability of the truth of its deductions, to be directly as the resemblance of the condition of the objects compared.* By not attending to this test of the value of a deduction from analogy, much error has been introduced into science, and a legitimate instrument of philosophical research perverted and misapplied. In justification of the severity of our remark, we may here quote a few examples.

It is well known, that the age of trees can be determined with considerable certainty, by counting the layers of wood; their number being, in general, equal to the years of growth. This particular character of ligneous stems has given rise to attempts to discover the age of other substances which are composed of concentric layers. Thus, the number of concentric circles of the vertebræ and scales of fishes, have been regarded, in the absence of experience, as indicating the age of the animal to which they belong. Even the age of the Earth, apparently, in the opinion of some geologists, may be determined by regarding its strata, as analogous to the layers of growth in a tree.

An anatomist of eminence observed, that the cells of a piece of bee-comb which he examined were double *. The acute author of the article, "Vegetable Anatomy," in the Supplement to the Encyclopædia Britannica yields to the opinion, that the cells of plants are double, influenced by this observation of the condition of the cells of bees.

Since the windpipe of birds is employed to convey air to and from the lungs,—vegetable physiologists have ever been inclined to consider the spiral vessels of plants, which have a remote resemblance thereto, as likewise air-vessels,—and much vain reasoning and idle conjecture, have in consequence been displayed on the subject.

As the materials of the crust of the earth are lighter than the mean density of the earth, many geologists conclude, that the weighty materials of the centre must be metallic, without attending to the condition which must be produced by the force of gravitation, or the analogous arrangement of the atmosphere. Who ever risked the conjecture, that the lower and denser parts of the atmosphere, were of a more metallic nature than the higher and lighter parts?

On what foundation does the belief rest, that the other planets are inhabited, but on that of analogy; yet the circumstances of the other planets are all different with regard to the force of gravity, motion, temperature, and light. Their inhabitants, therefore, if such exist, must be unlike us in physical constitution at least,—neither bone of our bone, nor flesh of our flesh.

* Mem. Wern. Soc. ii. p. 259

I have thus exhibited a few examples of perverted analogy, amidst many others equally obvious which might have been produced. The science of zoology abounds with them. I have already referred to a few, and shall have occasion afterwards frequently to return to the subject.

The inferior animals are, in many cases, deceived by analogy, as in the case of *baits*,—considering the identity as perfect, when there are only a few points of resemblance.

2. *Time.*—It is probable that our notion of Time, is derived from a comparison of the succession of our ideas, as they spontaneously or voluntarily pass through the mind, with the succession of impressions produced by external objects. In looking at the commencement and conclusion of a change in one body, as the hour-hand of a clock, we find, that before it can exhibit a sensible change of place, several ideas can pass through the mind; but in the case of a cannon ball, an obvious change of place has been produced before a single idea can be contemplated by the mind. The one moves slower, the other faster than thought. Our ideas of time and motion, therefore, are cotemporary. The rapidity with which the mind performs its intellectual processes, differs among men in different individuals, and in the same individual at different times,—an interesting idea being contemplated much longer than one to which we are indifferent. The succession of ideas therefore is irregular.

The standard of time is sought for in the succession of changes of the material world. In an uncivilised state, man employs as his chronometer, the motions of the heavenly bodies,—the changes of day and night,—the seasons,—the moon,—and the tides.

The inferior animals evidently have a knowledge of time. Those which leave a particular dwelling at stated intervals, measure the distance they ought to travel, and return with regularity to their home. The sun appears to be their great regulator, as they are influenced by the changes which take place with his light and heat. Fishes, and other animals which live in the sea, or search for food on its shores, appear to regulate themselves by the motions of the tide. The regularity of the crowing of the cock has been long admired,—but it appears difficult to point out the measure of time by which it is governed.

Man alone has devised artificial plans for measuring time,—in the employment of dials, clepsydræ, sand-glasses, clocks and watches.

By attending to the succession of past events, and finding our anticipations frequently realised, we begin to form some notions with regard to futurity, and by imagining a continued succession, we acquire our ideas of eternity, if we can be said to have any distinct ideas on the subject.

The notions of future time, in all probability, exist wherever there are notions of past time, and bear a co-ordinate degree of distinctness. In this respect, man is unquestionably superior to the brutes, as he exercises himself more frequently in measuring time, and devises more complicated plans for his future comfort. But the brutes are not ignorant of future time, as many of their actions clearly testify, in the prospective efforts of their imagination, already noticed. A fox or a dog possessed of more food than is necessary for the supply of present wants, conceals the remainder until again urged by the calls of hunger. The ermine will conceal a number of eggs in a particular place, and return at intervals to its magazine.

3. *Power.*—Were we merely inactive spectators of the changes which take place in the world, it is probable that the ideas of reflection which would result from the con-

templation of these, would be limited to resemblance and succession. But as we begin to act upon the objects around us, and produce in them various changes, we acquire a knowledge of our own Power. When we see changes produced independent of us, we consider it as the display of some other power. These changes, and the efforts which have preceded them, excite our ideas of cause and effect, means and ends.

That the lower animals possess some notion of power and of cause and effect, may be inferred from various actions which they perform. Thus, for example, we have seen the hooded crow *(Corvus cornix,)* in Zetland, when feeding on the testaceous mollusca, able to break some of the tenderer kinds by means of its bill, aided, in some cases, by beating them against a stone ; but as some of the larger shells, such as the buckie, *(Buccinum undatum,)* and the wilk, cannot be broken by such means, it employs another method, by which, in consequence of applying foreign power, it accomplishes its object. Seizing the shell with its claws, it mounts up into the air, and then loosing its hold, causes the shell to fall among stones, (in preference to the sand, the water, or the soil on the ground,) that it may be broken and give easier access to the contained animal. Should the first attempt fail, a second or a third are tried, with this difference, that the crow rises higher in the air in order to increase the power of the fall, and more effectually remove the barrier to the contained morsel. On such occasions, we have seen a stronger bird remain an apparently inattentive spectator of the process of breaking the shell, but coming to the spot with astonishing keenness, when the efforts of its neighbour had been successful, in order to share in the spoil *. Ani-

* Pennant Brit. Zool. iv. p. 114, mentions similar operations performed by crows on mussels.

mals, in general, seem to have a tolerably correct notion of their own powers, as we do not often see them attempting to accomplish objects for which their strength is inadequate. Thus, we have seen a pointer which, if a hare was wounded, would pursue with the utmost keenness, but if otherwise, would witness her escape without exertion. It is the knowledge of the variety of power which sometimes makes a horse run away with a bad rider, when he would not even make the attempt with a good one.

4. *Truth.*—Though many have endeavoured to give a satisfactory answer to the question, "What is Truth?" few have succeeded in the attempt. The failure, we apprehend, has in a great measure arisen from the variety of meanings attached to the term, and the impossibillity of giving a definition which shall include the whole, independent altogether of our limited acquirements. Thus, truth is by some considered as opposed to falsehood, by others to ignorance, and by many to duty. At present, we shall consider *Truth* as expressing the actual existence of things.

Our knowledge derived from the impressions made on the senses, and from reflection, or experience, is to every one the standard by which he judges of truth. In consequence of the origin of this idea of reflection, a thing may be true, which, from experience, we cannot affirm; or false, which we cannot deny.

In the acquisition of truth, we are aided by Experience and Testimony.

(*a.*) *Experience.*—The value of the information furnished by experience, must necessarily depend on the successful employment of the various instruments of our perception and powers of reflection. Man appears to attend to the exercise of those powers with greater intensity, and to vary and repeat the operations which they perform more frequently, than the inferior animals. His knowledge of truth from experience must, consequently, be more extensive.

The belief which arises in the mind from the testimony of experience, is the most secure and permanent, and upon which all act with unreserved confidence. So rapidly does the mind judge in such cases, that the evidence is said to be intuitive. Here the memory renders powerful assistance.

The indications of experience, however, may deceive, since, as I have already pointed out, the senses frequently communicate impressions which do not correspond with the actual state of things, and the mind, by employing these materials, may arrive at very erroneous conclusions. In these cases, when there is no chance of deception, the confidence in experience is unlimited, and to these the phrase intuitive evidence ought to be restricted. In other cases, we do not entertain such firm belief, until by repeated trials we have satisfied ourselves that all sources of error have been avoided.

The inferior animals possess a knowledge of truth, but in a more limited degree than man. They do not direct their organs of perception to so many objects, or examine the same object under such a variety of aspects. But that they possess much knowledge derived from experience, all will be ready to admit, who have traced the operations of memory which they exhibit. Besides, we witness its display in the caution of an old horse, in comparison of a young one,—and of the older animals which we wish to ensnare, compared with the more ignorant ones which more easily fall into our hands. The absence of experience enables us to deceive the latter; its presence in the former, teaches them to avoid the error.

From the knowledge which the mind possesses of time and power, aided by experience, we confidently expect that the same event shall take place in the same circumstances; but when there is any change in the circumstances, we expect the event with less certainty, and our confidence is weakened with the extent of the change.

(*b.*) *Testimony.*—Experience is the result of our own efforts to attain truth: Testimony is the medium by which we become acquainted with the knowledge or truth of others. Our confidence in the truth of the testimony of others, depends on its conformity to our own experience, and the usual agreement of facts with the report of witnesses. Besides, the signs which we generally employ, represent the ideas which we have formed, and we suppose that others employ signs in the same manner.

In early life, we give ready assent to any statement, because, not being aware of the difficulties of ascertaining the truth, and ignorant of the process of dissembling, we draw conclusions from the testimony of others, in the same hasty and incautious manner as we do from the evidence of our own senses.

When experience has convinced us that our senses frequently give us false information, we become more cautious in admitting the evidence of others, until we are satisfied that they *were not deceived.*

In the course of our experience, we find that false appearances deceive ourselves, and, from various causes, we perceive that they are employed to deceive others. The natural result of this information, is a diffidence in testimony, until we are satisfied that there is *no attempt to deceive.* Still more extensive experience and reflection strengthen the diffidence, by making us acquainted with the ambiguity of the signs which are employed in communicating the experience of others.

When once we are satisfied that, having guarded against all sources of error, we have ascertained the truth, no opposing testimony will be admitted, as we presume that others have not exercised the same degree of caution. But if our belief is founded on evidence which appears imperfect, and the belief of others on evidence which would have been regarded by us in similar circumstances as complete, we are disposed, after exercising the necessary precautions, to admit their testimony, even in preference to our own experience. Our confidence in all cases in the testimony of others, increases with their number.

Much of the truth, with which we consider ourselves acquainted, is derived from the experience of others. Without the assistance thus obtained, our opinions would often be ill founded for want of a standard of comparison, and by neglecting to exercise sufficient caution in attending to the evidence of the senses.

Among the inferior animals, there are some species which, during life, are solitary and fixed to the same spot, as the common oyster. These can derive no information from the testimony of others. Their knowledge of external objects must be limited to the results of their own sensations. But in the case of all monogamous or polygamous animals, whether gregarious or otherwise, a considerable dependence is placed on the testimony of others in a variety of circumstances. Thus, in the case of wild geese or crows feeding in a field, the knowledge of approaching danger observed by one, is speedily communicated to the whole, who immediately act upon the information.

Errors in testimony, among the lower animals, are frequently committed, as the result of erroneous information or experience. Thus, a cock will often give warning of danger to the hens under his charge, if a pigeon flies rapidly over his head, mistaking it for a rapacious bird. In other cases, the sentinel may be deceived by false appearances, and, considering that there is no danger approaching, fail to do his duty *.

There are few instances of attempts to give false testimony among the inferior animals, which do not appear to arise from the impulse of the instinctive rather than the intellectual powers. The Fox, however, in a tamed state, will often scatter food within the reach of his chain, and then remain motionless until an unwary chicken approaches the cunning observer. This is an attempt to deceive, not so much by scattering about food, as by lulling asleep all suspicion by his quietness.

5. *Duty.*—In the examination of our idea of reflection with respect to Duty, moralists have not unfrequently mixed with their speculations the notions which they entertain regarding truth. Our ideas of reflection, however, in reference to truth, differ from those of duty, in the manner in which they are acquired, the standards by which we judge of their accuracy, and the kind of belief to which they give rise.

* Dr Edmondston in his "View of the Zetland Islands," gives a very striking illustration of this neglect of the sentinel, in his remarks on the Shag. "Great numbers of this species of the corvorant are sometimes taken during the night while asleep on the rocks, and the mode of accomplishing it is very ingenious. Large flocks sit, during the night, on projecting rocks of easy access, but before they commit themselves to sleep, one or two of the number are appointed to watch. Until these sentinels are secured, it is impossible to make a successful impression on the main body; and to surprise them, is therefore the first object. With this view, the leader of the expedition creeps cautiously and imperceptibly along the rock, until he get within a short distance of the watch. He then dips a worsted glove in the sea, and gently throws water in the face of the guard. The unsuspecting bird, either disliking the impression, or fancying from what he conceives to be a disagreeable state of the weather, that all is quiet and safe, puts his head also under his wing and soon falls asleep. His neck is then immediately broken, and the party dispatch as many as they choose." Vol. ii. p. 253.

Our notions of duty are, in but few instances, the result of experience, being in general derived from testimony. Our standard depends less upon the information of the senses than the dictates of others, and the belief, which is the consequence, may be easily overruled. The dictates of truth are universal and irresistible; those of duty are partial and variable.

In the prosecution of the objects of existence, and in attempting to supply the wants of our nature, we soon discover, that when we enter certain situations, or perform particular actions, we secure happiness or success, while by neglecting to attend to these conditions, we entail upon ourselves pain and disappointment. In many cases, our information respecting what will be hurtful or advantageous, is derived from the testimony given by others of their experience. In consequence of these suggestions of experience and testimony, we conclude, that we *ought* to observe the necessary degree of precaution. We likewise witness certain actions which we perform, hurtful or useful to others, and we conclude, employing our own feelings as the standard, that we ought to avoid those which are injurious, and execute, willingly, those which are beneficial. By degrees, we thus acquire our notions of duty, which greatly influence our conduct.

In the course of farther advances in the road of life, we find the scene changing so frequently, and the conditions in which we are placed so variable, that, for the accomplishment of the same end, we must employ means different from those we formerly made use of; or, in other words, the action which it was our duty, on a former occasion, to perform, has now become unnecessary, or, from the change of circumstances, unsuitable. When these changes occur, in reference merely to our own success or happiness, we consult analogy, our own experience, or that of others. When the happiness or success of others are involved, it is difficult to point out the principles by which we would be guided, unless the golden rule be considered as one of the natural standards of duty,—"All things whatsoever ye would that men should do to you, do ye even so to them."

To ascertain our duty would be an easy task, were the mind always to pursue so straight a course. But we often act in direct opposition to those rules of conduct which we have previously established. Experience tells us, that the conduct of others is equally consistent as our own. By persevering in this perverted course, we destroy that standard which our own experience had erected, and act upon another which we would formerly have considered imperfect and foolish. By habit or custom, our notions of duty are thus changed, converting, with respect to our conduct, evil into good, and good into evil.

Duty, according to the preceding remarks, consists in employing the most efficacious means to secure the greatest quantity of enjoyment, and to avoid every degree of pain. That the discovery of duty is an intellectual process, is obvious from the variety of opinions which have been entertained on the subject; the variable influence which these opinions exercise over the will; and the effort required to comply with them.

When a standard or rule of duty is proposed to us, we judge from experience, testimony, and analogy, of its suitableness,—and if embraced, we ever after condemn any conduct which is in opposition to it, (although we commit it,) and continue to do so, until a more perfect standard has been proposed and received.

By the help of memory, we acquire an astonishing quickness of perceiving whether we act conformably or oppositely to this received standard. We feel pleasure in the one case, and pain in the other; both of which are encreased according to our notions of the value of the standard by which we try ourselves. To the operations of the memory in such cases, theologians apply the term Conscience, and others, the Moral Sense.

The question, then, What is Duty? in reference to its rules or standard, is one of very difficult solution,—or rather, is one which unassisted reason cannot resolve. If we take the question in detail, and ask, What will be acceptable to the Supreme? the Hindoo devotee will say, to fall under the wheels of the moving temple of the god Juggernauth, and be crushed to death, or to drown himself in the Jumnah at its junction with the Ganges. What is the duty of children to their aged parents? Some will say, to nourish and comfort them, others to expose them to hunger and death. What treatment ought the mother to bestow on her new born babe? The wives of Madagascar will say, those that are born in the months of March and April, in the last week of every month, and on all the Wednesdays and Fridays of every week, ought to be exposed to perish with hunger, or cold, or be devoured by the wild beasts. We could easily swell the proofs of the variableness of the human standards of duty,—and although all are convinced that there is, or ought to be a standard, they differ with respect to its character. This display of a moral deficiency or want in our nature, is the strongest proof that can be urged for the necessity of a revelation. The Christian religion supplies this moral want,—and furnishes a standard which, if observed, would make all men in every condition happy, exalted and wise.

The divine original of Christianity may be almost demonstrated, from the circumstance of its containing an account of our own imperfections, whose existence human partiality would never have discovered, nor human pride acknowledged. What is it but this partiality whi chinduces many to believe in the existence of an active power, termed a Moral Sense, the origin of moral obligation?—

"the least violation of whose authority fills us with remorse;" "and the greater the sacrifices we make, in obedience to its suggestions, the greater are our satisfaction and triumph."

Were this representation true, it would exhibit a very gratifying picture of the moral dignity of our race. But, alas! if this moral faculty exists, it is in general too feeble in its operations to influence the conduct, neither prompting us to avoid this pictured remorse, nor to grasp at this imaginary triumph.

Among the lower animals, we do not observe any instances of their acting contrary to their experience. In a domesticated state, where laws have been imposed upon them, they obey from various motives; the prospect of reward, the dread of punishment, and ultimately habit. They are aware of the conformity or disagreement of their actions, to the standard by which they are tried. Examples of this kind of knowledge are daily exhibited in the Ox, Horse, and Dog.

6. *Deity.*—From the displays of our own power and that of others in the production of motion, we are led to attend to the changes which take place around us, as the marks of some other power; and by witnessing the variety of means which are employed in the accomplishment of these various alterations, and the regularity which they exhibit in their succession, we arrive at the conclusion, that a Being, superior to man in power and wisdom, exists, and continually exerts an influence on the surrounding world.

So simple is this effort of the mind, and so easily excited by the smallest degree of reflection, that the belief in a Superior Being may be considered as universal among mankind. Nations may be found who have scarcely devised signs to express their ideas on this subject, and over whom, their notions of Deity may exercise little controul; but we can scarcely believe it possible for man to exist in any stage of society, without being furnished, by the natural operations of his mind, with the first principles of religion.

Along with the progress of civilization, and consequent habits of reflection, we find this religious feeling extending itself, until an effort is made to hold intercourse with the Supreme by prayers and sacrifices.

There is no evidence to prove that the brutes have any idea of a Supreme Being.

II. Instinctive Powers.

The impressions which are made upon us by external objects, or the ideas of reflection suggested by memory, when they are the subjects of our intellectual powers, do not necessarily lead to any controul over the body in consequence of an act of volition. Between impression and action, there is always a process of thinking, varying greatly in its nature and duration, according to the subject, but absolutely necessary to connect the one with the other.

In the powers which we are now to consider, the case is very different. Here action follows impression *immediately*. There is no thinking,—no deliberation. There is likewise a difference in the nature of the action in the two cases. There is an *effort* required to perform that which is the result of the intellectual process,—whereas, the action which follows in reference to our instinctive powers, is *spontaneous*, or rather, it requires an effort to resist obedience to the impulse.

As the impressions, in the case of the intellectual powers, are variously modified by the thinking process, the corresponding actions exhibit, in their character, a great degree of variety. The impressions in the case of the instinctive powers, suffering no intermediate modifications, produce actions characterised by great uniformity.

In many cases, when an impression is produced upon us, which, as belonging to the instinctive powers, would have been followed by action, the intellectual powers interpose their controul, and the impression is surveyed before action is permitted. But, as might have been expected, such action is less varied, than when originally the result of an intellectual process, but more irregular than in those cases in which impression and action follow instantaneously. In the one case, the action is modified by the impression,—in the other, by the changes the impression has undergone by thought.

The powers to which we are now directing our attention, are usually denominated, by the writers on the science of mind, ACTIVE POWERS. To this appellation, however, there are strong objections. There are other powers which excite to action, inseparably connected with our constitution, which do not belong to this class. Some actions are produced by irritability, of which we are not conscious. Action is likewise produced by the information obtained by the senses, through the medium of thought,—or in consequence of the ideas of reflection which spontaneously arise in the mind. Hence the difference between the intellectual and instinctive powers, is not so distinctly marked in the acts of volition as in the manner in which these acts are excited. It is with a view to avoid all ambiguity on the subject, that we have ventured to substitute the term *instinctive*.

Much confusion has arisen by the vague use of the terms Instinct and Reason, and much vain speculation has been indulged, in consequence of no distinct and definite ideas being attached to them. No confusion, however, could arise, were we to consider *reason* as expressing the movements of our Intellectual powers,—and *instinct*, those which have hitherto been termed Active. The propriety of restricting the term Instinct, in the manner now done, will be rendered evident, in our enumeration of Instincts or Active Powers.

In attending to our instinctive powers, we shall follow the ordinary division into Appetites, Desires, and Affections. These include all the active principles, whose origin cannot be traced to those intellectual powers, or their results, which we have already enumerated.

Appetites.

The instinctive powers termed Appetites, are essentially requisite for the subsistence of the individual and the continuation of the race. They excite to action at particular periods,—are preceded by an uneasy feeling, yield much comfort during their gratification, and cease to act when the object has been attained. They may be regarded as three in number. Food,—Rest,—and Procreation.

1. *Appetite for Food.*—As the different parts of the body, from the moment of birth to the close of existence, stand in need of frequent repair; and as nourishment for this purpose is indispensably necessary, animals are not left to obtain a knowledge of these circumstances from observation, nor to supply their wants by the dictates of experience. They are impelled by feelings which arise at the suitable periods, to exert themselves to employ their peculiar organs in the most advantageous manner. This appetite, therefore, begins to act at birth, and, during the continuance of life, executes its movements nearly on a uniform plan.

Among the greater number of animals, the appetite for food is directed in every period to nearly the same substance, with this modification, that soft substances are preferred in youth. In the mammalia, the appetite is directed to liquids alone, the milk of the mother, in the first periods of existence, and, afterwards, to a mixture of solids and fluids. Yet this instinctive power regulates with equal success, in these different periods, the disimilar movements of sucking and chewing.

In many animals, such as the mammalia, the nourishment of the first period of life is provided by the parents, so that all the exertion necessary for this instinctive power to make, is confined to the proper employment of the food thus brought within the reach of the suitable organs. But, with many other animals, the case is widely different. They are brought forth in situations where there is no parent to assist this appetite, nor food provided for its supply. In such cases, this instinct must execute more complicated movements, and lead the individual to those places where food is to be obtained, and afterwards direct the choice. It is this influence that guides the caterpillar to the leaf,—the duckling to the pool, and the samlet to the ocean.

Under the controul of this appetite, and, prior to all experience, each species is directed to seek the kind of food which affords it the most suitable nourishment, and to shun that which would be deleterious. Thus, in looking at a pastured field, we observe that there are some plants which are left untouched, while others are cropped to the ground. But as the tastes of animals, in this respect, are exceedingly various, we observe that what is left untouched by one species, is greedily devoured by another. Nay, what is eaten by the goat, for example, with avidity, and with impunity by the horse or sheep, as the water-hemlock (Cicuta virosa,) is certain poison to the cow. Hence it has been called *water cowbane*, and we have heard a Fifeshire farmer, with a sigh, which intimated his experience of its effects, call it *deathen*.

In general, when animals have obtained a supply of food, they satisfy the cravings of appetite and retire. In other cases, the superfluous food of to-day is concealed to provide for the cravings of the morrow; or still more extensive arrangements are exhibited, in which the bounties of harvest are laid up in store against the scarcity of provision in winter.

In the other features of this appetite, there is a much greater variety of character. The *frequency* of the calls of hunger, varies according to the species or the habits of the individual. Thus, the caterpillar eats almost constantly, while the butterfly produced from it, scarcely ever seeks after nourishment. In almost all animals, however, there are stated intervals, during which the calls of hunger are felt, distinguished by a restlessness, or by the emission of particular sounds.

These *particular periods* when the calls of hunger are felt, appear in like manner to follow no general rule with regard to the species or even classes of animals. Some feed only during the day, and are termed *diurnal*, as nearly all the ruminating quadrupeds and land birds; others feed only in the twilight, as the bats and owls, and are called *crepuscular*; while many beasts of prey, aquatic birds, and others, prefer the darkness of the night, and are termed *nocturnal*.

The predilections of this appetite are equally anomalous: Some quadrupeds and birds preferring vegetable, others animal matter;—either in a fresh, a dried, or a putrid state. Some species feed on animals and vegetables only while these are living, others only while dead.

In the manner of obtaining food, this instinct, in different animals, acts in various ways. The spider weaves his web to entangle the fly: The ant-lion digs a pit-fall: The heron remains motionless at the margin of the pool, until the unwary fishes come within the reach of his long neck and bill: The cat kind take their prey by surprise; while the wolf trusts entirely to swiftness and strength. The manner of beginning the feast, even after the supply has been secured, is very different according to the species.

In a domesticated state, animals seem to lose the useful properties of this appetite, which are so essential to their existence in a wild state. Thus, cows which have been kept within doors during the winter, and supported chiefly on dry food, when turned out to pasture in the spring, devour, indiscriminately, every green herb, and frequently suffer for their indiscretion. LINNÆUS tells, that when he visited Tornea, the inhabitants complained of a distemper which killed multitudes of their cattle, especially during spring, when first turned out into a meadow in the neighbourhood. He soon traced the disorder to the water hemlock which grew plentifully in the place, and which the cattle, in the spring, did not know how to avoid *. We have been informed, that in Orkney many goslins die when first turned out into the hills, to pasture, in consequence of eating the leaves of fox-glove. In a wild state, however, this appetite directs animals with great certainty to the suitable objects of nourishment, and does not permit them to taste of those things which would injure or destroy.

By the *force of habit*, this instinct may be so modified, as that what was disagreeable at first, shall even become an object of desire. By this capability of changing taste, various animals can accommodate themselves to new situations,

* Lachesis Lapponica, ii. p. 136.

and preserve existence and its comforts under circumstances which, without it, would have been fatal to both.

But while the useful properties of this appetite disappear, in many cases, in consequence of domestication, in several instances they remain and exhibit themselves even when not necessary to the comfortable existence of the individual. Thus the dog, even when well fed, will often conceal the remainder of the food he is unable at the time to consume, although the calls of hunger seldom urge him to return to the hidden stores. The fox, so similar to the dog in many particulars, exhibits, (as we have already stated with another view,) in a wild state, the same disposition. When he has obtained more food than is necessary for the supply of his present wants, he buries the remainder, and returns to it when again pressed by hunger, often after an interval of several days. It is probable, that in a wild state, the dog obeys a similar instinct, of which, its appearance in a domesticated state is a strong proof. Even the love of hunting, which prevails among our race, almost universally, seems to be the remains of one of the original instincts of the species.

As intimately connected with this appetite, we may shortly notice the peculiar manner in which each animal disposes of its excrementitious or secreted matters. How carefully do the otherwise filthy swine, even in a domesticated state, void in a corner, and preserve the cleanliness of their bed. Cats are careful to bury their dung, while dogs exonerate upon stones. The uniformity of these peculiar modes of action in each species, indicates instinctive arrangements for their production, which the fastidious observer may despise, but which the enlightened naturalist will not overlook.

2. *Appetite for Sleep.*—We have already, while treating of the muscular and nervous systems, stated the neces-

sity of repose to recruit exhausted strength, and the different positions assumed by animals in accomplishing this object. But we are not left to discover this necessity by the slow process of experience; the appetite which we are now to consider determines our choice, and in each species regulates the different conditions which are necessary according to peculiar plans.

The *time*, in which animals sleep, is regulated by the instinct for food. Those animals which are diurnal feeders are nocturnal sleepers, while those which are crepuscular, sleep, partly in the night and partly in the day. In general there is but one sleep in the course of the natural day, and its length varies with the season of the year. In some animals, which enjoy sleep during the ordinary periods, the condition is assumed and continues permanent during the winter season. Such are termed Torpid animals. We shall afterwards have occasion to consider the circumstances attending their torpidity.

Sleep, in general, seizes the frame after the ordinary exertion of the body, or of the intellectual or active powers. Its approach is accelerated by any extraordinary exertion, although during the continuance thereof, at whatever period of the day, this instinct is overruled. All animals seem to be debilitated by exercise, and to require periodical intervals of repose for their renovation. There is likewise a tendency to obey the impulses of this instinct, whenever the calls of hunger have been satisfied.

The tendency to sleep is accompanied, in man, with a listlessness, and feeling of muscular weakness; the senses cease to convey impressions of external objects, so that the mind no longer holds its ordinary intercourse with them, —attention is relaxed,—memory ceases to recall,—and the imagination to arrange, and consciousness itself can scarcely be said to be in exercise. At the same time, every instinctive power is quiescent, and irritability alone regulates all the movements which prevail.

When in this condition of sleep, we find that all the powers of body have lost a portion of their sensibility, so that no impressions are produced, unless those which are very strong; in which case the state of sleep is interrupted and the individual awakes.

After continuing in this state of sleep for a definite time, our various powers again assume their wonted functions, —consciousness returns,—the senses again establish our relation with external objects,—the intellectual powers operate upon the materials which they collect,—and our instincts again exert themselves as active powers.

There is now a pleasing feeling of recruited strength, a sense of new vigour, and a disposition to activity.

But in some cases this instinct is so far influenced by circumstances that the sleep produced is imperfect, and we are said *to dream.* This happens when there is any organic derangement of the body, or when the irritability of the body, during the state of repose, is too much excited, (as when the stomach is overloaded or filled with matter difficult to digest), and calls into action the instincts by which it is accustomed to be controlled.

In one state of dreaming, the intellectual powers are in action, but the process of thought is conducted differently from the way in which it is carried on in our waking moments. We fancy that the senses are communicating to us information of external objects, while they are in a state of inactivity, so that we appear incapable of distinguishing between the impressions of perception and memory. We are equally incapable of perceiving the difference between the ideas of memory and imagination. Hence we form the most incongruous images, mixing the conditions of different ideas of memory, with regard to time, place and connection, without perceiving that it is the imagination chiefly that is in exercise. However confused the limits between the ideas of recollection and imagination, the attention seems sufficiently active to enable the memory to treasure up the distorted images.

The recollection of the efforts of our instinctive powers is likewise a common employment of the mind during this state of dreaming, and in some cases, in consequence of these instincts being in part awake, partial action is produced. This is exemplified in *Somnambulists*, or those who rise and walk during sleep, or those who speak. The former will sometimes rise out of bed and move from one room to another, or even go into the open air, without becoming awake; and the latter, while talking, if cautiously spoken to, may be made to keep up an irregular conversation. Somnambulism exhibits, however, this very singular character, that the memory retains no traces of the actions which were performed, or the feelings in which they originated.

That the same causes operate in producing imperfect sleep, in other animals, may, with considerable property, be inferred, even in the absence of direct proof. Indeed we can have no proof of such a state, unless the instinctive powers have been so far awake as to produce action. This proof, however, has been obtained in the case of the dog. This animal, during sleep, may sometimes be observed moving his limbs and tail with considerable velocity, and even uttering low and imperfect sounds, as if barking.

In preparing for sleep, some animals assume particular attitudes, such as we have already noticed, but remain in the open air. Others retire to shaded places or caves. Some sleep without any covering: others prepare for themselves a bed in order to keep up their standard of heat, which would become diminished by inaction.

3. *Appetite for Procreation.*—The two preceding appetites are necessary to the existence and well-being of the individual:—The one under consideration is subservient to the continuation of the species. As the exercise of the former cannot be dispensed with, in any period, without the decay or death of the individual, they continne to operate from the commencement to the close of life. It is otherwise with the present appetite: for it seldom begins to exert its influence until the body has nearly reached maturity, and usually ceases when it is beginning to decay. In many cases, the influence is only felt once during the continuance of life, as in many insects. In other cases, it operates periodically through a considerable portion of life. In all cases, however, it excites to action at such a period, that the young animals to be produced, shall enter life at a season of the year, when temperature, food, and all the other conditions of existence are in the best possible state for their comfort and increase.

But the most singular property of this instinct, is its selective quality, always guiding individuals to those of the same species. Even in the activity of this power, whatever be the age or the season, every animal is indifferent or averse to those of any other species. Hence there is in nature no mixture or contamination of species, every living thing seeking after its kind.

Domestication produces on this appetite effects equally striking as on the appetite for food. It usually accelerates the period when this power exerts itself,—increases the frequency of its demands,—and, in a few instances, destroys the selective quality.

II. Desires.

The instinctive powers which are usually denominated Desires, differ from those of the preceding class in many well-marked characters. The appetites, so far as we are able to judge from experience, observation, and analogy, form an essential part of the constitution of every animal. The desires, on the other hand, are not so universally distributed. They constitute rather the distinguishing characters of individuals or species, than of classes, as no one desire (except perhaps that of warmth) appears common to all. Besides, they differ from the appetites, in the uniformity of their excitements. They do not return after certain intervals: and they do not cease to act, upon the attainment of their object.

1. *Desire of Warmth.*—Whether the blood be hot or cold,—whether the animal resides on the land or in the waters, there is a particular degree of heat congenial to the feelings, and necessary to the functions of every animal. There is a propensity, in animals at least, to secure this requisite degree of warmth, and its operations, more, perhaps, than any other circumstance, influence their physical distribution. This instinct is destined to guard animals equally against the extremes of heat and cold, and to keep them within that range of temperature most conducive to their health and comfort. In the fixed animals, this desire, if it exists, cannot excite to action, as locomotion is denied.

In the execution of its important ends, this instinct guides animals to particular places, to the open sunshine, or the refreshing shade; to the heated sand-bank, or the cooling water; to the exposed mountains, or the sheltered plains.

The standard of temperature differs according to the species, and even according to the individuals, and the district of the globe they are destined to dwell in. Hence we find individuals of a species in a variety of situations with regard to temperature, in the same district, and animals thriving in every latitude.

This desire, like all our powers, is greatly influenced by

habit. The standard of temperature may be altered a few degrees, provided the change be slowly brought about. If it is attempted rapidly, a diseased state of the system, in general, is the consequence. In some cases, the sensibility of the nerves is, to a certain extent, destroyed by the change, so that the exposed parts cease to experience those painful impressions which, under other circumstances, would have been produced.

Man is the only being on this globe who, in the gratification of this desire, kindles a fire wherewith to warm himself. He makes use of friction or percussion to excite the heat, in the first instance, and employs vegetables, or bituminous minerals, to increase and continue it. It is impossible to ascertain, whether this was originally the discovery of an individual, and subsequently communicated by example or tradition, or whether it be really a character of this active power as it exists in man. It is at least certain, that the use of artificial heat is coeval and coexistent with our race. The monkey may approach the fire which the savages have left, and warm himself at its glowing embers, but he is never prompted to secure a continuance of the comfort, by the addition of fresh fuel, or by setting fire to combustible matter in another situation.

2. *Clothing.*—Many animals are produced at their birth, in such circumstances, that they stand in no need of any other clothing than the skin, and its natural appendices, and continue through life independent of the protection of foreign bodies. Such animals are, of course, unprovided with this instinctive principle. But there are other animals which, during a part or the whole of life, stand in need of an artificial covering for their protection.

Of those which cover themselves only during a certain part of life, there are some which derive the materials of their habitation from their own body, as the silk-worm, while in its pupa state; while others, as many caterpillars, employ secreted threads in order to draw around them the leaves of the plants on which they feed.

Other animals which likewise require only a temporary covering, fabricate their garment from the materials which they employ as food. This is the case with the *Tineæ*, or clothes-moths. They feed upon wool or hair; and, with the same materials, they construct a covering to fit their body, with which they move about, increasing its length by adding to its extremity, and its breadth, by slitting up the sides, and inserting new materials in the gap. The Phryganeæ, or *Caddis worms*, on the other hand, while they pass the first portion of their existence in the water, clothe themselves with bits of straw, sand, or shells. These they cut into proper shape, and form into a tube a little larger than the body, in which they dwell, and which they likewise carry about with them. However rough the outside of the covering may appear, the inside is smooth, having a coating of slimy matter plastered round the cavity. These animals do not, in obedience to this instinct, employ the materials within their reach indiscriminately; for the covering might, in that case, become too heavy to be easily carried about with them at the bottom of the pool,—or too light, and, by rising buoyant to the surface, remove them from their sources of nourishment. They select and arrange the materials, so as to avoid both these evils.

But there are many animals which employ an artificial covering during every period of their lives. Some we observe, although incapable of collecting and arranging the materials for a habitation, are still provided with a propensity for a covering, and avail themselves of the remains of other animals. The hermit-crab takes up his abode in a deserted univalve shell; and, when his body enlarges, he shifts to a larger shell, and thus preserves himself in a protected state during all the stages of his existence. Others, (as the *Sabellæ*,) construct a covering of particles of sand cemented together, increasing the diameter of the tube with the corresponding increase of the size of the body. Those animals which construct a permanent covering with which they are connected, in every period of life, are all inhabitants of the water. Man is the only animal endowed with this instinct, that inhabits the land. Parental tenderness supplies his wants in the first instance, and afterwards he is guided by this active power, regulated by habit, and the principle of imitation.

As subordinate to this desire, we may here take notice of the *Habitations of Animals*, into which they retire at particular intervals.

The most common display of this instinctive power in the formation of a dwelling, may be observed in those animals which burrow in the earth. These form their holes with their jaws, as many insects, or with their feet, as quadrupeds and some birds. The retreat sometimes consists of a single apartment, while others excavate several chambers, leaving the walls without a covering, or giving them a coat of plaster to prevent them from crumbling down. In many cases, there is only one entry, while others make use of two or more. In the arrangement of the entry, this instinct displays its extraordinary powers. Sometimes the entrance, as in the case of the otter and pole-cat, opens into a thicket, or under the cover of a hanging bank. In other cases, as that of a spider, termed by LATREILLE *Mygale cæmentaria*, the entrance is closed by a door formed of particles of earth cemented by silken fibres, and closely resembling the surrounding ground. This door, or rather valve, is united by a silken hinge to the entrance, at its upper side, and so balanced, that when pushed up, it shuts again with its own weight.

In digging these subterranean dwellings, insects transport the earth to a distance, carrying it in their jaws. Quadrupeds, during the digging, when incommoded by its accumulation at the entrance, turn round, and, with their fore feet, push to a distance the loose rubbish, as may be seen in the efforts of the rabbit and field-mouse.

In many cases, animals unite and construct a dwelling for the convenience and protection of the colony. It is in the architectural displays of this power, exhibited by animals that live in society, that the most wonderful attributes of this instinct have been ascertained. The masonic labours of the bee and the ant are well known; and they produce a habitation regular in its structure, substantial in its materials, and commodious in its apartments. What man considers as a high effort of his understanding, may be here witnessed as the result of an instinct, unaided by experience, uniform in its results, and successful in its plans.

3. *Desire of Place.*—Every one knows, that different animals prefer different situations. The common pigeon, in a wild state, makes choice of the caverns of the rock, in which to sleep, and a smooth grassy bank on which to bask. The wood pigeon, on the other hand, nestles on trees, and spends all its hours of repose or relaxation on their branches. The sparrow prefers, at all seasons, the haunts of men; while the robin flees from their abodes during the breeding season, but delights to spend with them the dreary months of winter. The jack-daw delights to live in the ruined tower, and the rook in the aged wood. In man, this instinctive power exhibits its controul in the love of country, and the pleasure associated with the idea of home. In many cases, the operations of this desire are modified by circumstances connected with food and temperature; but, in a variety of instances, it is impossible to assign any physiological reason for the choice.

Perhaps the most singular attribute of this desire is exhibited in the facility with which many animals, when transported from their usual haunts, can find out the proper direction by which to return. The horse, the ass, and the cat, among quadrupeds, have been known to exhibit very remarkable instances of this power of discovering home *. Among birds, the pigeon has long been celebrated for this quality, and frequent and successful attempts have been made to render it useful to man. A pigeon has been frequently taken to a distance; and, after a letter has been tied under its wing, and the bird then let loose, it has rapidly returned with the despatches. Attention to all the qualities

---

* In the "Introduction to Entomology," by KIRBY and SPENCE, vol. ii. p. 502. the following curious example of this facility of finding home, is communicated on the authority of Lieutenant Alderman, Royal Engineers, who was personally acquainted with the facts.

"In March 1816, an ass, the property of Captain Dundas, R. N., then at Malta, was shipped on board the Ister frigate, Captain Forrest, bound from Gibraltar for that island. The vessel having struck on some sands off the Point de Gat, at some distance from the shore, the ass was thrown overboard to give it a chance of swimming to land,—a poor one, for the sea was running so high, that a boat which left the ship was lost. A few days afterwards, however, when the gates of Gibraltar were opened in the morning, the ass presented himself for admittance, and proceeded to the stable of Mr Weeks, a merchant, which he had formerly occupied, to the no small surprise of this gentleman, who imagined that, from some accident, the animal had never been shipped on board the Ister. On the return of this vessel to repair, the mystery was explained; and it turned out, that Valiante, (as the ass was called,) had not only swam safely to shore, but, without guide, compass or travelling-map, had found his way from Point de Gat to Gibraltar,—a distance of more than two hundred miles, through a mountainous and intricate country, intersected by streams, which he had never traversed before, and in so short a period, that he could not have made one false turn. His not having been stopped on the road, was attributed to the circumstance of his having been formerly used to whip criminals upon, which was indicated to the peasants, who have a superstitious horror of such asses, by the holes in his ears, to which the persons flogged were tied."

of this desire, will often enable the naturalist to find out and to distinguish particular species; and the extent of this kind of knowledge distinguishes the practical from the closet naturalist.

4. *Curiosity.*—We observe in children, a disposition to examine the objects which surround them, and to ascertain their properties. Nor is this tendency confined to the early period. It continues to exercise its controul in manhood, and even in old age, prompting us to attend to new objects, to study the changes which present themselves, and to maintain our acquaintance with the world in which we live.

This principle appears to exercise a greater influence upon the human species than upon the individuals of any other, and urges us to examine a greater variety of objects, in every possible state of combination. It is to this feature of our character, that we can refer our superior capability of receiving instruction. But the same principle operates, although in an inferior degree, among some of the lower animals. We have thus limited our assertion, because, among many animals, we do not perceive even a trace of this desire, the attention being exclusively confined to the supply of the bodily wants. The monkey tribe are fond of examining new objects, and will subject themselves to a considerable degree of trouble in the attempt. The dog is seldom disposed to suffer a stranger to remain in his company, unmolested by his attempts to become acquainted with his appearance and smell.

This instinct, in the lower animals, is confined to the appearances and present properties of objects; while, with man, the case is different. He does not rest satisfied with a knowledge of appearances, but he is disposed to examine the agents which have produced them. The lower animals have their curiosity confined to effects; man alone attempts to investigate causes. He builds, and he pulls down, he combines and divides, in order to satisfy his curiosity; and seldom relinquishes the pursuit willingly, until he has gained the information which he desires.

5. *Society.*—This instinct, like curiosity, is not common to all animals. There are many species, as the oyster and bernacle, which appear to be incapable, from their fixed station, of acquiring any knowledge of the existence of individuals of their species. Others, on the contrary, are social from necessity, as the Mollusca Tunicata, and many zoophytes, constantly growing in groups, or inseparably connected by bodily organization. Even among the free animals, many seem to prefer a life of solitude, and exhibit no desire, unless perhaps during the season of love, of associating with other animals, or even with the individuals of their species.

But in those animals in which this instinct prevails, it exhibits its controul, independent of any feelings which the appetites excite. Among gregarious animals, we observe, with some, the social union so intimate, that this instinct appears to be essentially necessary to their existence, or, at least, to the continuance of their race. Thus, bees congregate by an irresistible impulse of this instinct, and it is only when united that they can perform all the functions, and enjoy all the comforts of their existence. In those animals, where this union is necessary, we perceive such admirable arrangements exhibited by this instinct, such subordination of purpose, and such co-operation of means, that we may consider it one of the most curious, wonderful, and complicated of the active powers.

In many cases this desire operates, although its action is not connected with the immediate wants of the individual. Thus we observe rooks congregating, although, as far as we are able to judge, each pair could subsist, though unconnected with a flock, and even obtain a more copious supply of food than when surrounded with numerous companions, each intent upon the same purpose, and ready to anticipate his neighbour in seizing the enviable morsel. There are advantages, however, which result from this union, and which counterbalance the accompanying evils. There are common interests which are secured by it,—warning is given of danger,—notice communicated of any new store of food, and exertion prompted by example.

Although this desire, in general, brings together the individuals of the same species, and unites them in one flock, yet, in the absence of other individuals of the same species, a social animal will unite itself with the individuals of other species, and frequent the places to which they resort, and follow their movements. Thus, man is often gratified with the company of a dog, and a sheep will associate with a cow. When prevented from indulging this desire, a considerable degree of impatience is exhibited, which not unfrequently terminates in languor, sickness and death.

6. *Imitation.*—We have already taken notice of *sympathy* as affecting the nervous system, and inducing action in parts remote from the place where the first impression was actually produced. We may now consider this desire of imitation, as having its foundation in this organic tendency, and prompting to actions more extensive and complicated. When we see a person yawn, we are disposed to perform the same action; when we see them in motion we have likewise a tendency to follow. The same feeling operates whether it relates to the production of motion, of rest, or of sound.

This desire contributes greatly to the progress of the improvement of the human species. The child imitates the actions of its parents, and acquires a dexterity in performing them. In manhood we repeat the actions of others, and make their acquirements our own. We even attempt to imitate the actions of other species, as in flying, swimming, diving, and a variety of other efforts.

To this principle of imitation can be readily traced, national character, and family resemblance. It is that instinct on which, more than any other, our susceptibility of education depends.

In the lower animals, this instinct displays itself, chiefly in those of social habits. It is proverbially known that the startling of one individual will set a whole herd of cattle in motion, and if one sheep leaps the fence, the others will speedily follow. Even in the lower animals, imitative actions are not always copied from the individuals of the same species. The monkey will imitate the motions of a man, and the parrot his voice.

7. *Approbation.*—The desire of approbation exercises a powerful influence on the human character. It prompts us to the greatest exertions, and it restrains our most violent passions. Long before we are aware of the personal or public advantages which result from its guidance, we enjoy with rapture the applause of others. In order to secure it to our memory, we are disposed, in many cases, to part with life itself.

As this principle could not be called into exercise, but among those animals which are likewise possessed of the social desire, we do not observe this instinct in any others; and even in social animals, we are only aware of its existence in those which we have domesticated. The dog appears delighted with the approbation of his master, and quickly discovers and seems mortified with his displeasure. The cow and the horse exhibit this instinct, but in an inferior degree, arising, perhaps, from our being less intimately acquainted with their manners. Among birds or animals belonging to any of the lower class, we do not perceive very distinct marks of the existence of this instinct. It probably influences the exertions of the social tribes of insects, although their condition precludes the possibility of observing the proofs of its operation.

8. *Power.*—We are fond of exercising controul over all the objects which surround us, checking or accelerating their motions, modifying their properties, and rendering them subservient to our purposes. In accomplishing these ends we have recourse to our bodily strength, and to our acquaintance with the properties of matter, and place them equally under the guidance of this instinct. We train our body by exercise, store our mind with knowledge, and strive to heap up riches, with no other view than to encrease our power over natural objects, or the members of society. When resisted in our obedience to the impulses of this instinct we become impatient, and frequently waste ourselves in fruitless efforts, or sink in despondency. The love of liberty is the offspring of this desire, as well as the horror of degradation or slavery.

Many of the lower animals exhibit this instinct, in their hatred of confinement, but more particularly in the resistance which they offer when any individual, even of their own species, attempts to impose any restraint, or even to exhibit any superiority. What is emulation but our aversion to feel ourselves inferior to others, or to be regarded as such by our neighbours? Horses, in a race, contend with much keenness to excel in speed, and chanticleer, in his seraglio, abhors a rival. The dog delights to receive marks of the approbation of his master, but appears unwilling that his companion should enjoy any share of the favour.

In man this instinct is cultivated with care, and much pains bestowed to extend its influence, in order to secure bodily or mental superiority. But among the brutes, it is confined to bodily superiority, and appears little, if at all, influenced by education or habit.

The desires of imitation, approbation and power, appear chiefly to operate on those animals which live in society, and perhaps may be regarded as attributes of the social instinct, rather than distinct principles of the constitution. Before attempting to simplify the desires in this manner, however, it would be necessary to examine, with greater care than any one appears ever to have done, all the active principles of those animals which are social from instinct, (not necessarily from bodily organization), in order to trace the unity of action in their supposed secondary desires.

9. *Life.*—The existence of this instinctive principle, as a part of our own constitution, is universally recognised. It prompts us to a variety of actions, through every period of our life, retaining its influence even when in old age our field of enjoyment has been very limited. The medical practitioner avails himself of its influence, in order to compel his patient to swallow the most nauseous draughts, or to submit to the most painful and mutilating operations. It is not equally powerful in every individual. Its deficiency contributes to the formation of the hero, its excess constitutes a coward.

The love of life displays itself in the active exertions which we make to avoid pain, as the prelude to death; or any direct attack. If about to fall, we stretch forth our hands, so as to restore our controul over the centre of gravity, or weaken the shock when approaching the ground. That this is not the result of experience, instigating us to avoid pain, is obvious from the circumstance of the universality of the operation, and the similarity of the manner of performing it. Besides, we have seen the most palpable indications of its existence in an infant of a month old.

When any body is falling upon us, we spontaneously hold up our hands, as an effort to defend ourselves.

These instinctive actions, are best calculated for protecting us from attacks which would cause pain, and which come upon us suddenly. In other cases, where the danger is greater, we are guided to oppose it by other kinds of physical force, or seek for safety in flight. In the one case when we fight for safety, we are guided by *anger ;* in the other, when we flee from danger, we are regulated by *fear*.

The inferior animals, in obedience to the same instinct, protect themselves from danger by similar means. Although they do not possess members so well adapted for sudden movement as our hands, yet, by various other expedients, they accomplish the same end. The hedgehog rolls up its body, and presents a surface of prickles to the foe. Birds in general, when surprised, raise and bend back their heads. Some animals, as the hare, uniformly strive to flee from danger ; while the badger, not only flees, when practicable, but fights obstinately when compelled. Many aquatic birds avoid being shot, by diving upon seeing the flash, and getting under water before the shot reaches the place. Upon being surprised, they have recourse to the same expedient. The cuttle-fish conceals himself in his inky fluid, and the torpedo benumbs his foes by an electrical discharge. Others endeavour to strike terror, by their sounds, or gestures, or odours.

There is, however, one attribute of this instinct in the lower animals, which man does not possess, although we can perceive an approach to it. Man conceals himself from the approaching danger, and ceases to move or speak, lest he should discover his retreat, and expose himself to danger. The hare, in like manner, will often remain in her form until the hunters are past, and then steal away unperceived. But the provision to which I chiefly refer, is the propensity, in scenes of danger, to *feign death.* This is chiefly displayed in various insects, and may readily be perceived in the common dorbeetle, (*Scarabæus stercorarius*, Lin.) This insect when seized, will stretch out its legs, rendering them stiff, and will remain motionless, until the danger seems to be over. In this state the limbs may be broken, without any action being excited in the animal.

III. Affections.

The immediate object of the two preceding classes of our active powers, is to secure to the individual the comfortable continuance of existence. In the case of the instincts which are termed Affections, the object is to communicate *pleasure* or *pain* to others. Those of the first class centre in ourselves; the last have a reference to others, binding us in a variety of ways, to encrease their enjoyments or to repress their faults. They have been divided, according to their object, into Benevolent and Malevolent affections.

Benevolent Affections.

1. *Parental Affection.*—We have placed this instinct first in order, because it is the most powerful in its impulses, secures for us the greatest quantity of enjoyment, prompts to the execution of the most complicated movements, and is essentially necessary to the continuation of life. This affection displays its energies,

a. In each species providing a suitable place for the birth of its offspring.—This end is accomplished with the same degree of certainty in those animals which produce their young at first in the form of eggs, as in those which bring them forth alive. Among the animals of the former class, denominated oviparous, we witness fishes approach the shore to deposit their spawn in the crevices of the rocks, on the leaves of sea-weeds, or in the sand; but in all these cases, each species has a peculiar manner of arranging its eggs, as well as a choice of situation. In every case, however, they are placed within the vivifying influence of the solar rays, and are hatched at the season of the year most advantageous for the growth and the comfort of the fry.

In other oviparous animals, such as birds, before the eggs can be deposited, a house or nest must be constructed, often consisting of various materials, collected with great labour, and formed with exquisite neatness, which in a few species is lined with the down which they pull from their bodies. In all these cases, obedience to this instinct is cheerfully complied with, however difficult, and any obstacle to prevent the execution of its purposes occasions pain.

b. In each species securing a supply of suitable food for its offspring.—The simplest form in which this law is observed, consists in the parent depositing its eggs on those substances which are to serve as food for the young when hatched. This is familiarly displayed in the case of the cabbage-butterfly, which deposites its eggs on the leaves which are afterwards to serve as the food of its caterpillar. In the case of the Oestrus equi, the eggs are deposited in such a situation, that circumstances are likely to occur, by which they shall be conveyed to a proper place for the issuing forth of the larva, and for its obtaining a suitable supply of food. The female insect attaches her eggs to those parts only of the horse which are most liable to be licked by the tongue, by which process they are conveyed into the stomach. There, they are almost instantaneously hatched, the larva, known by the name of *bots*, adhere to the coats of the stomach by hooks with which they are provided, obtain food from the juices by which the horse is nourished, and when mature pass out with the dung, to undergo the future changes of life.

In many species this active principle not only prompts the parent to prepare a suitable receptacle, and deposit its egg therein; but, as in the case of many hymenopterous insects, to collect a quantity of food and deposite it in each cell, for the support of the larva when excluded.

The last and most complicated effort of this instinct, consists in providing a regular supply of food for the young animals, while they are incapable of feeding themselves. In the mammiferous animals, the young are supplied with milk at the first, and are accustomed by degrees to partake of that food on which they are afterwards to subsist. In birds, however, the food must be collected with much industry, and, when brought to the nest, distributed in just proportions among the callow young.

In the execution of these various tasks, we never witness any symptoms of reluctance or murmuring, the labour being performed with chearfulness. Neither can we discover any awkwardness, the mark of inexperience. The young and the old, guided by this instinct, perform their duty with equal alacrity and precision.

c. *In each species employing means to secure a suitable temperature.*—In the case of oviparous animals, care must be bestowed to provide for the egg a suitable temperature to ensure its hatching. The crocodile prepares a small hillock of sand at some distance from the water, with a hollow in the middle, which she lines with leaves and other vegetable matter; then deposites her eggs, and covers them over with leaves. The heat from the fermenting leaves, joined with that of the atmosphere, soon hatches them. Birds remain upon their eggs for weeks together, covering them carefully with their bodies, denying themselves, during this period, all the enjoyments of liberty, and scarcely taking enough of food and exercise to keep them in a healthy state. Even in the interval of absence to obtain food, some birds, as the common duck, cover the eggs carefully with down and straw to preserve their warmth, and probably likewise to conceal them from foes. When the eggs are hatched, the old birds for some time continue to sit at intervals on the young brood, to preserve their temperature.

Among the mammiferous animals, the same instinctive carefulness, when requisite to keep their young offspring warm, is equally apparent. Some make a common bed, as the sow, and permit the young ones to lie in her bosom. The rabbit, on the other hand, covers her young ones at the first with hair, and closes up the entrance to her nest, to prevent the circulation of cold air.

d. In each species keeping its own offspring in a suitable state of cleanliness.—The circumstances attending the birth of many animals, call for the immediate exertion of the parent to remove those things which at the time or afterwards would injure or incommode. Thus, in the nests of birds the fragments of the eggshells, if permitted to remain, would bruise and otherwise injure the young. These, however, the parent birds take up in their bill, and remove them to a distance. In the case of young quadrupeds, the rapidity of evaporation from their moist surface, immediately after birth, would prove injurious to them. But the mother, as may be seen in the case of the sheep, cow, or mare, forgetting the pains of parturition, begins to lick the hair and make it dry.

But these are not all the evils which the instinctive power we are now considering prompts the parents of animals to remove. Before the young birds are capable of voiding their excrement over the margin of the nest, the old ones convey away the mutings, which are at first covered with a pellicle, in their bills, and drop them at a distance from their nest. With rabbits and other quadrupeds, whose young dwell in holes, and are born blind, without leaving the nest for many days, the mothers apparently keep them clean, by licking off all moisture and fecal matter, so that you shall find the hair of their bed always dry, as well as the straw beneath and around. When attempted to be reared from birth without a mother, it is extremely difficult, if not impossible, to keep them dry.

e. In each species protecting its own offspring from danger.—This is chiefly accomplished by fighting in their defence against the threatening foe. Even those animals, as the common hen and sheep, which, in general, protect their own lives by flight, will, in defence of their young, brave every danger, and exhibit a degree of courage, amounting to a total disregard to their own safety. Sometimes, however, they employ stratagems to lead the foe to a distance. We have seen the common partridge actually strike at a pointer, who had, by accident, rushed unperceived by the old birds, into the middle of a covey. The ordinary device which she employs is to run off, with her wings hanging down, as if she had been wounded, to entice the dog to follow her, and leave her young in safety. We have seen equally interesting examples of the display of the same instinct of feigning lameness, to lead an obtruder from the young, in the common wild duck, ringed plover, golden plover, and arctic gull.

When accidents prevent the young from reaching the period when they can provide for their own wants, we observe the parents in a state of painful uneasiness, and expressing their grief in sounds which they seldom utter on other occasions.

In the human species, this instinctive affection continues to operate during the whole of life. In the inferior animals, on the other hand, the feeling which binds the parent to the offspring, ceases, when the latter have become capable of supplying their own wants, and securing their own enjoyments. With many animals, indeed, this instinct does not extend to the young, but is limited in its operations to the deposition of the egg in a suitable situation. The parent, in this case, never sees its offspring; the offspring is equally ignorant of the characters of a parent.

2. *Filial Affection.*—This instinct is obviously confined to the young of those animals which are nurtured at the commencement of life by a parent's care. Under its guidance, young animals are restrained from wandering, and induced to remain under the protection of their parents. When forcibly or accidentally removed, they utter sounds peculiarly expressive of their sorrow. In obedience to the same instinct, young animals observe the signals which are made to them by their parents. The chicken quickly obeys the clucking of the hen, the lamb the bleating of the ewe. While young animals are thus under the protection, and obedient to the signals of their parents, they learn to imitate their actions, and by degrees become acquainted with the places which they are afterwards to frequent.

This instinct continues to operate during the period of parental affection. In the human species, both these active powers continue for life; while, among the lower animals, the reciprocal feelings of parental and filial affection cease, when the immediate objects for which they were exercised have been accomplished.

3. *Social Affection.*—We have adopted the epithet Social, in preference to Patriotic, the term by which this affection is generally distinguished; because, under this last, are frequently included the desires of place and society. This is more properly the love which certain animals bear to the individuals of their own species, in preference to those of any other. It is always accompanied with the de-

sire of society, and preceded by the parental and filial affections. This affection exists in its purest form in the human species, in the bosom of a family, in the superiority of the parents, the obedience of the children, and the desire glowing in every breast, to guard from danger, and increase the comfort of all. When several families unite their interests, and form a tribe, we observe, in this patriarchal government, the same subordination prevailing, the same desires and affections operating. Beyond this range of political government, there is unquestionably a greater display of the energies of our nature; but they are accompanied with less disinterestedness and integrity.

In the inferior animals, where the ties of parental and filial affection soon lose their influence, there are few instances of the family union lasting beyond a single season. But the patriarchal form of government prevails in almost all those animals possessed of the desire of society, and appears to be regulated by the same principles which operate in similar establishments of the human race. We observe,

a. A Ruler.—In the lower animals, we do not discover any examples of deliberate choice, in the appointment of a governor; but in all cases, the preference is given to the largest, the strongest, and the most courageous of the tribe. Between what may be termed rival chiefs, there are frequently furious combats; and the victor, without further resistance, assumes the command. This leader is, in general, a male: although, in the common bee, it is a female who holds the reins of government, limited, however, in her power, and subject to the controul of her constituents.

There is in these statements the free use of terms borrowed from human forms; but as we regard these as merely the displays of this instinct in a particular species, the well known phraseology in which their characters have been detailed, is peculiarly adapted to our general view of the subject.

It appears to be the duty of this ruler, to exercise dominion over the individuals of his tribe, to be first at the post of danger, and to regulate the places for the safety of all. Accordingly, in the exercise of his office, a bull, a ram, or a stallion, may easily be detected in a flock, by superior boldness and freedom of action,—by keeping on the outside of the group, and being foremost to hazard the chances of a battle. The herds of wild elephants are likewise governed by a leader; and when these are enclosed in a snare, he is the first that enters the palisades, the first to become captive. When, instead of fighting, the duty of the leader is to retreat, we always observe him first in motion. Thus, in approaching a flock of seals on a sandbank, it is the largest which we observe moving first towards the water; and in springing a covey of partridges, the sire is usually first on the wing.

b. Mutual Support.—It is obvious, that without obedience to the ruler, the tribe would lose the benefit of his protection; so, without a union of strength or stratagem in the hour of danger, the foe would be able to make more extensive havoc. If a dog enter a park among cattle, a general movement of the herd towards him, indicates the sense of common danger. When a wolf approaches the flock, they form a hollow-square, and placing the young ones in the centre, are in this manner prepared for the attack. In such a situation, the bull as the ruler, usually steps forward and chases the foe to a distance. Perhaps, however, the most complicated part of this system of mutual defence, consists in the appointment of sentinels to give warning of danger; and, in the mean time, to permit the herd to eat or rest in safety. The monkeys, when engaged in their predatory excursions, always have one of their number at a suitable place, to give warning of danger, if necessary. In looking at a flock of geese feeding in a field, some of them may always be perceived on the watch, while the others are at work; and these, upon the near approach of a man or a dog, instantly give the signal to their obedient companions.

In many cases, this instinct is not confined to giving warning of danger, but prompts to the communication of news of food. This is familiarly illustrated in the disinterested conduct of the cock, who, upon finding a store of food, immediately calls the members of his family to the feast. It is, perhaps, by a prostituted use of this instinct, that the *decoy ducks* seduce the unsuspecting flocks into the netted ponds in the fenny districts of Lincolnshire, communicating to them some prospect of food or shelter, which, when embraced, leads to capture and death.

In some cases, this affection for the species, while it is generally in exercise towards the tribe, is more especially directed towards a particular individual. In the human race, where this selection is termed *friendship*, the choice is influenced by similarity of pursuits or principles, sometimes by the habits of acquaintance. Among the lower animals, we frequently observe similar instances of the partialities of friendship, without being able to trace them with any certainty to their source.

In some cases, the social affection extends even beyond the individuals of the species to those of other species, whose organization and habits are widely different. Thus, man often forms a very strong attachment to the horse; and the dog, by habit, prefers our society to that of the individuals of his own kind*.

* The late Mr MONTAGU, in the Supplement to his Ornithological Dictionary, article "Grey Lag Goose," relates the following singular attachment which subsisted between a female China goose and a pointer, who had

Before leaving this branch of the subject, we may advert to a very remarkable perversion of this instinct, (if we may venture to call it so,) in the case of several social quadrupeds. If a deer is wounded by a shot, the herd will refuse to admit it again into their number, but will persecute to death the unfortunate individual. The same instinct is said to prevail with wild cattle and elephants. How different this treatment from that disinterested support of each other, which in all other cases prevails in the flock?

As nearly connected with this active power which we are now considering, we may enumerate the feelings expressed by the terms *gratitude* and *pity*. In one case we feel thankful for the kind offices of others, and express our feelings by a disposition to make a similar return. In the other, we sympathise with the afflicted, and endeavour to remove their distress. These feelings are necessary, to prompt to those reciprocal acts of kindness and protection, which we witness to be performed by all animals possessed of the congregating instinct. They are displayed, in the first instance, to the individuals of their own species; and when there is a transference of affection to an individual of another kind, there is a corresponding change in the object of their gratitude and pity.

---

killed the male. "Ponto (for that was the dog's name) was most severely punished for the misdemeanour, and had the dead bird tied to his neck. The solitary goose became extremely distressed for the loss of her partner and only companion; and probably having been attracted to the dog's kennel by the sight of her dead mate, she seemed determined to persecute Ponto by her constant attendance and continual vociferations; and after a little time, a strict amnity and friendship subsisted between these incongruous animals; they fed out of the same trough,—lived under the same roof,—and in the same straw-bed kept each other warm; and when the dog was taken to the field, the inharmonious lamentations of the goose for the absence of her friend, were incessant."

Malevolent Affections.

The benevolent affections have, for their object, the communication of pleasure or protection to others,—those which we are now to consider, are regarded as destined to excite to the commission of actions calculated to inflict pain. The very existence of such a class of principles may, with great propriety, be doubted

In regard to the malevolent affection termed *anger* or *resentment*, it is merely a display of the *love of life*; a desire which we have already considered. And as a proof of the truth of this opinion, we find, that it is usually regulated by the extent of the injury, and the consequent risks of life which it occasions. In like manner when we are united in social affection with our kind, *indignation* is excited when we see others injured; because, without being checked, the evil may extend to ourselves. The same determination to support our neighbours is evinced, when our attachment is fixed on individuals of other species, as a horse or a bird. It displays itself in the dog, who will resent a blow given to his master, with as much keenness as one inflicted on himself. The *hatred* which we observe subsisting between different species, uniformly arises from our appetites or desires. A cat will kill a shrew, but will not eat it. This, however, does not arise from any hatred to the shrew, but in having, from the hurry of capture, mistaken it for a mouse. Our other malevolent affections, as jealousy and envy, may be traced to the selfishness of our appetites and desires. The existence, indeed, of a set of principles in the constitution, whose sole object was to inflict pain, would be worse than useless in the economy of animals. Pain is, without doubt, necessarily inflicted, in the gratification of the appetites and desires of nearly all animals. But although pain or suffering be correlative with our instincts, there are none of these which appear to be exclusively appointed for its production. Besides, the malevolent affections, if they do exist, must be destined to inflict pain on the individual exercising them; since they are always accompanied with emotions in him who indulges them, equally unpleasant as those likely to be excited in the objects of his displeasure. Such a view of the principles which regulate animated nature, has never been warranted by observation, nor contemplated by reflection.

Before closing this account of the instinctive or active powers, it may be thought necessary that we should offer a few observations on the *temperament* of animals, or the relative facility with which the instinctive powers of individuals or species can be excited. On this subject, however, little precise information has been obtained. The temperaments are usually divided into four kinds, the *sanguineous*, excited readily, slightly and transiently; the *choleric*, excited readily, violently and transiently; the *melancholic*, excited slowly, but more permanently; and the *phlegmatic*, excited with difficulty. There are various modifications of these, which do not, however, merit a particular enumeration. The laws which regulate the temperaments of animals have never been developed; and the subject seems scarcely to admit of illustration. We observe such variety in the individuals of the same species, and even in the same individual at different times, as to baffle all attempts to generalise. Carnivorous animals are, in general, the most easily excited, and the most violent; piscivorous animals are less violent, while herbivorous animals, on the other hand, are possessed of gentler dispositions. But these remarks apply to particular instincts merely, and do not embrace all the active powers. Even the same individual differs with regard to particular instincts. How phlegmatic is the hen, in general, in her appetites and desires; but while rearing her young, how choleric? Even in the case of herbivorous animals, there are very marked differences in the species of the same genus, as between the ox and the buffalo,—the horse and the zebra.

IV. On the mutual Communication of Feeling.

It is necessary, for the exercise of the various intellectual, but especially of the active powers, that animals should be able to make known to others, the impressions which they receive, or the propensities which they feel. In this kind of mutual communication, there are various ways in which the object is accomplished. The ear is subservient to the purpose in some cases, the eye, and even the touch, in others. The method of holding intercourse by means of sounds, being the most general, first merits our attention.

a. The communication of Feeling by means of Sounds addressed to the Ear.—Among animals which employ sounds as the medium of expressing their thoughts, those only possess what is termed a *voice*, which breathe by means of lungs. The voice is generated, by the air expelled from the lungs passing through the wind-pipe and mouth, and exciting those vibrations in these parts which produce sound. The variety of structure exhibited by the wind-pipe and mouth, on which the different conditions of the sounds depend, will afterwards come under our more particular examination. We are at present considering the voice as an auxiliary power of the mind. Regarded in this light, the sounds which are uttered in the communication of feeling, are either natural or acquired.

Every animal possessed of a voice, has the power of emitting those sounds which may be termed its *Natural Language*. These sounds, immediately after birth, are exclusively occupied in expressing the presence of pain, or the conditions of the instincts of food and temperature.

By degrees, they are directed to express the wants or the enjoyments of the other active powers. In all the individuals of a species, the same sounds are uttered in expressing the same feelings. But among individuals belonging to different species, the greatest diversity prevails in the sounds of their natural language, as may be observed in the dissimilar cries of the lamb, the foal, and the calf.

In many cases, this language appears exclusively to be a bond of union between the young animals and their parents, by which the former can express their wants, and guide the motions of the latter in supplying them. In proof of the truth of this remark, we need only observe, that this language of infancy is gradually neglected, as the protection of the parent ceases to be necessary; and other sounds are employed, in the independence of maturity, to express the same feelings, and others belonging to their new condition. Thus, the cries of ducklings, when under the guidance of their mother, are different from those which they utter when able to provide for themselves; and when roaming about in the corn-yard, or swimming on the pond. That these established sounds, are equally natural with those uttered in infancy, is demonstrated by this circumstance, that ducklings hatched under a hen, and brought up remote from any other individuals of their own species, utter the sounds common to their kind.

Some animals are destitute of any language at birth, and do not utter sounds until they have arrived at maturity. In those cases, which occur among oviparous animals, there is no connection between the parent and the offspring, the latter having no wants but those which its own instincts can supply. Sounds, therefore, or cries, would be uttered in vain. But in approaching maturity, when it is necessary that an interchange of feeling should take place with others of its kind, a language is provided suited to the occasion.

Thus, the frog, when in a tadpole state, is silent; nor does it ever utter its croakings, until it has reached maturity.

The sounds by which we express pain, are universally disagreeable to those of our own species; while those which indicate happiness are pleasing. There is even a corresponding sympathy excited by the sounds which the inferior animals utter, expressive of pleasure or pain.

The possession of this natural language is peculiar to those animals which live in society, either uniformly or occasionally. In the former case, the faculty of language is generally co-eval with life, while in the latter, it is only possessed during the period of the social union, and disappears when the temporary purposes for which it was necessary have been accomplished.

In this natural language, the vowel sounds are chiefly observable in the child, although the sounds of consonants, particularly K and R, may often be distinguished. Many of the sounds of the lower animals consist of monosyllables, as those of the lamb, while those of the cuckoo or partridge consist of two syllables, and even three may be perceived in the mournful cry of the kitteewake, when disturbed during the breeding season.

In Man, the natural language consists in weeping and sometimes laughing, and to these are added, in after life, other sounds which express anger, joy, or pain. These actions are not confined to a particular tribe, but are common to the race, and universally understood.

In the examination of *acquired language*, we are frequently at a loss to draw the line of distinction between the sounds of nature and those of imitation. Before we can judge with certainty on the subject, we must take the particular animals whose language we are examining under our care, in order to become acquainted with their natural sounds, and the education of which they are susceptible. It is necessary, likewise, that we attend to the conditions requisite for receiving this kind of education. It is obvious, that there is required a peculiar organization, in order, not merely to utter the natural sounds, but to modify in tone, intensity, and expression, the sounds which can be produced, so as to resemble the notes which are to be acquired. This capability of the organs to modify themselves to acquire particular sounds to any extent, only lasts during the period of youth. Afterwards, the muscles of the voice are not so obedient to the will, and it becomes difficult to learn to pronounce readily, sounds to which we have not been previously accustomed. To this cause may be referred the extreme difficulty of ceasing to use the accent of one's native tongue, and of acquiring that of a foreign language. There is, however, a second condition, on which this susceptibility of acquiring language greatly depends, namely, quickness of hearing. In order to learn to imitate sounds, it is necessary that we be able to discriminate their qualities accurately, to know what we are to imitate, and determine on our degree of success. A delicate ear, therefore, is essentially requisite, and as it chiefly exists in early life, when the organ of hearing is in the soundest state, we may discover one reason, at least, why the young learn to pronounce new sounds more readily than the old. But, even with the existence of organs capable of articulating an acquired language, and a delicate ear to discriminate sounds, there is required the desire of imitation to induce us to exert these organs in producing a resemblance to the sounds we hear. It often happens, particularly among birds, that the power of imitation is scarcely exerted in any other manner.

By attending to the conditions requisite to acquire language, we perceive the reason why persons born deaf are at the same time dumb. They do utter, indeed, the sounds of natural language, expressive of their feelings, but as they do not hear the sounds which others utter, they have consequently no guide to their power of imitation. By operating upon their natural cries, and by the help of signs, they may be taught to speak, but their voice is unequal and harsh. It is equally possible to be able to distinguish sounds, and, to a limited extent, the things which they are intended to represent, without possessing the capability of voice necessary for imitating them. This is conspicuously the case in the dog and horse, which understand the language of command, encouragement, and approbation, and yet are incapable of producing any analogous sounds. The parrot, mock-bird, and many others, can learn to imitate many kinds of sound, without comprehending the things of which they are the signs. By much trouble, however, they may be made to perceive the connection, and even to avail themselves of their knowledge, to express their wants or desires. It is necessary, however, to state this important limitation, that these sounds communicate information to *man* alone, their instructor, and do not constitute any channel of communication between individuals of their own species.

The most obviously acquired sounds in the inferior animals, may be observed in the *Singing of Birds*. If a young linnet, for example, be taken from the nest, and brought up in the company of the nightingale or lark, instead of learning to utter the musical notes of its own species, it imitates the song of that bird which it has been accustomed to hear, more or less perfectly, according to the state of its organs. In this mode of educating singing birds, there is frequently displayed a power of selection on the part of the young scholar. "I educated, (says the Honourable Daines Barrington *,) a nestling robin under

* Experiments and Observations on the Singing Birds. Phil. Trans. vol. lxiii. p. 258.

a wood-lark linnet, (a linnet with the song of the wood-lark,) which was full in song, and hung very near to him for a month together; after which, the robin was removed to another house, where he could only hear a sky-lark linnet. The consequence was, that the nestling did not sing a note of woodlark, (though I afterwards hung him again just above the wood-lark linnet,) but adhered entirely to the song of the sky-lark linnet."

If birds thus acquire so easily, in a state of confinement, the song of others, how comes it to pass that, in a wild state, each individual acquires only the notes of its own species? Even in a state of confinement, young birds imitate the notes of those of their own species more readily than those of any other bird. The same observer, to whom we have already alluded, says, "Young Canary birds are frequently reared in a room where there are many other sorts; and yet I have been informed that they only learn the song of the parent cock." Even in a wild state, although the twite and linnet fly in company, "yet these two species of birds never learn each other's notes." The same may be said of many other birds which live in the same place, and nestle in the same hedge. These circumstances probably arise from the structure of the organs of each species enabling them more easily to produce the notes of their own species than those of any other, and from the notes of their own species being more agreeable to their ears. These conditions, joined to the facility of hearing the song of their own species, in consequence of frequenting the same places, determine the character of the acquired language of the feathered tribes. We are even disposed to conclude, that an individual untutored, and without an example to imitate, would, if associated with a mate in the breeding season, acquire, by its own efforts, notes nearly similar to those of its parent. This is, indeed, partly proven by the song of solitary birds, in certain cases, which often approaches the natural, and the presumption is strengthened by the circumstance, that the first attempts at song, termed *recording*, are merely repetitions of the natural cry of the bird.

There is another feature in the song of birds which merits observation. They sing chiefly during the season of love, and in confinement, when in full health. Hence we may regard this language as connected with appetite and expressive of enjoyment or pleasure.

All the individuals of a species acquire the same song in whatever country they have been hatched. Slight differences have indeed been observed; and hence, as we are informed by BARRINGTON, the London bird-catchers prefer the song of the Kentish gold-finches, but Essex chaffinches, and the nightingale fanciers, a Surrey bird to those of Middlesex. These variations may be expected according to the constitution of individuals, affected by the food and temperature of the places in which they have been reared. They are, however, confined to narrow bounds, and require a very delicate and experienced observer to detect their existence.

We have thought it expedient to offer these observations on the acquired language of birds, in order to enable us to form a more correct idea of the language of Man. In treating of this branch of the subject, we shall confine ourselves to what may be termed the natural characters of human speech, leaving the details of its artificial arrangement to the rhetorician and grammarian.

Independent of the natural cries by which, in infancy, we express our wants, we hear sounds uttered by the individuals of our species around us, which we are disposed to imitate, and soon find ourselves equal to the task. These words or sounds we soon perceive to be the names employed to designate particular objects; we learn to use them for the same purpose; and are encouraged in the attempt, by the facility with which we can thus express to others the ideas which external objects have excited in our own minds. By attending to the words themselves by which objects are distinguished, we perceive manifest differences in the nature of the sounds, and in the manner of producing them. We annex to these differences particular names, and employ particular symbols addressed to the eye to distinguish them, and thus form what is termed an Alphabet. The letters are divided, by grammarians, into vowels and consonants. This arrangement is likewise suited to the physiologist.

In uttering the vowel sounds, the mouth is open, and the differences are produced by the position of the tongue, and the form which we give to the opening by the lips. The consonants, on the other hand, are formed by the almost total interruption of the expelled air, for a time, by the tongue, lips, palate, or teeth. The *labial* consonants, (of the English language,) are formed by the contact of the lips, as in *M*, *B*, *P*. The sound of *W* is intermediate between that of a vowel and consonant, as the lips are never so completely closed as in the latter, nor so distant as in the former. The *dentolabial* are produced by the union or separation of the upper front teeth with the under lip, as F, V. The *palatine* consonants are formed by the application of the tongue to the palate, as H, L, N, R, S, X, or as C, D, G, I, T, Z, together with K and Q.

By means of these various sounds, either separately or variously combined, man is able to form symbols by which to designate the objects of nature and their conditions, and to reveal the secret workings of his soul. But in this power of communicating his thoughts, he is limited to the family or tribe in which he has been reared, and whose arbitrary sounds he has learned to imitate and comprehend. Beyond the limits of their territory, the voice of strangers would be unintelligible, and his own sounds unmeaning. The acquired language of man therefore differs, in this respect, from the natural sounds of the inferior animals. In each species, unless controlled by human agency, all the individuals acquire the same notes, and utter the same sounds. But the acquired language of man differs with the country in which he lives, and even in the same country it is ever varying. To what causes are we to refer these differences and changes in human speech?

It is generally supposed, that the volume of Inspiration gives a solution to the question, in the history given by Moses, of the confusion of tongues, in the eleventh chapter of Genesis. But amidst some variety of interpretation of which the passage is susceptible, there is reason to conclude, that the building of Babel was the undertaking of the descendants of Ham only, and that the confusion of languages which there took place, was by the special interposition of Heaven, for the purpose of frustrating the ambitious schemes of Nimrod and his followers, and of preventing them from congregating in excessive numbers in the plains of Shinar.

Before attempting to enumerate those circumstances which have operated in producing the great diversity of speech among the different nations of the earth, it may be necessary to state, that the sounds expressed by the consonants, are more permanent than those indicated by the vowels. We perceive more clearly the mechanical action requisite in the production of the former, while the changes on which the different sounds of the latter depend, are too minute for our comprehension, and their effects frequently too obscure to be properly appreciated by the ear. Hence there is frequently, even in the same country, a substitution of these vowel sounds. The sounds of the consonants being more definite and a substitution of one of these for another does not so frequently take place. When these two kinds of letters are conjoined to form a word, the same changeableness of character in the vowels may likewise be perceived. But in attending to the language of different nations, more remarkable differences than these to which we have now alluded appear to exist. We may detect the same sound of vowels and consonants, but these sounds are the symbols of other objects, and express different ideas. What, then, are those causes which operate in the production of such diversity of speech in the same species? The following sources of change appear to include all the circumstances which exercise any remarkable influence,—the structure of the organs,—the variety of situation, and the progress of civilization.

When we consider the complicated structure of those organs destined to produce voice, the varied movements they execute in articulating the different sounds, added to the influence exercised by age, sex, constitution and habit on the various muscles, we may perhaps be disposed to conclude, that we have discovered causes adequate to account for all the phenomena. It cannot indeed be denied, that these circumstances operate in the production of that peculiar mode of speaking, by which an individual may be distinguished from all his acquaintance. It may even be granted, that the sounds produced by one individual, and which were at first specific, may, by being imitated by the young, become general in the district. Differences may therefore arise in the manner of pronouncing the vowels, and perhaps one or two of the consonants, in the mode of accenting the syllables, and in the tone and energy of expression. A difference of organization, therefore, may account for the existence of provincial sounds or dialects, but it offers no explanation of the fact, that the same syllables among different tribes do not express the same ideas nor appear in words in the same relative position. But as a proof that other causes operate much more powerfully in the production of different languages, than varieties in the structure of the organs employed, or rather that organical differences exercise but feeble influence, we may state, that our capability of learning to speak a strange language, does not depend on any peculiar provincial or national structure of the organs, originating in a peculiar state of the larynx, and strengthened by the custom and habits of generations. There is not one organical conformation qualifying one to speak German, another to speak French, and a third to speak English. The organs of these nations are the same, and their capabilities the same. There is no predisposition to speak one language more than another. Hence, although we admit the complicated nature of the vocal organs, and the constitutional differences which they exhibit, we perceive that these offer no obstacle to the acquisition of any language, since, in the words of one of the most celebrated anatomists of the age, who nevertheless is disposed to refer the variety of languages chiefly to circumstances connected with the vocal organs, "all children acquire the tones, accents, and articulations of those countries in which they are educated; an evident proof, that, prior to the formation of habits, the vocal muscles may be brought to act in any one of the numerous millions of combinations that have ever been adopted by any tribe, family, or nation of the human race, and be made to acquire the habit of pronouncing, with readiness and ease, any one of the almost infinite variety of languages that have been, that are, or that ever shall be on the face of the globe *." Since, then, the condition of the organs exercises but a feeble influence in the

* A New Anatomical Nomenclature by JOHN BARCLAY, M. D. Edin. 1803, p. 79.

production of a diversity of tongues, let us now attend to the influence of situation.

If we suppose a family or tribe dwelling in a district of small dimensions, we can easily imagine, that the same signs would be employed by all. Every one would be familiar with the same natural objects, and be engaged in the performance of the same actions. A sign therefore once adopted and understood, would continue to be employed; and as new objects would seldom present themselves of sufficient importance to call for a new sign, those few which circumstances might render necessary, would speedily be made known to all. But if we suppose a separation to take place, and a branch of this tribe induced to emigrate to a new district, What effect would this change of place produce upon their language? New objects would present themselves, requiring new signs by which to express them; new movements would be exhibited by these objects, and new operations performed upon them, all giving rise to new sounds or signs. In the mean time, the old objects, no longer recurring, would be forgotten, and the signs by which they were expressed, either neglected or annexed to new objects, with which they might be but obscurely connected. In this manner, in the course of a few years, the emigrants would have added many words to their language, expressive of objects, qualities, and actions, unknown to the tribe from which they had separated, and a generation would scarcely have elapsed, before the two tribes spoke different languages, and the sound of the one had become strange to the other. If we conceive a tribe living in a mountainous district, and familiar with glens and precipices, and cataracts, to descend into the plains, how many of their signs would cease to be employed, and how many new ones would be requisite to express the character of the rivers and their inundations, the meadows and pools? If they journey onward to the sea shore, the ebbing and flowing of the water, the waves and their murmurings and roarings, would attract notice and receive particular designations. In each district, therefore, there will be a particular language, expressive of the objects which are peculiar to it, and of their relation to the wants of its inhabitants. We have likewise to bear in mind, that every country furnishes man with a supply of animals and vegetables which are, in a great measure, peculiar to itself,—peculiar modes of collecting and storing fruit or grain must, therefore, be resorted to,—a peculiar mode of hunting and fishing practised,—a peculiar mode of clothing and shelter adopted, all giving rise to new names, expressive of objects, qualities and actions. In the construction of these new words, it is obvious, that, in this stage of society, there are no rules but the caprice of individuals, influenced by the sounds of nature and the resemblance which may be traced between the old and new objects. There is always an aversion to the formation of new sounds; and this aversion has given rise to attempts at generalization, or the introduction of genera and species. By degrees, however, the terms in which they are expressed, coalesce, and the resulting sound, modified by the ear, loses its compound nature, and passes as the simple sign of a particular object.

According to this view of the matter, were two families, of the same speech, to settle in different countries, they would soon assume different languages, although the original connection might be traced in some of those words which express objects occurring every where, or actions which all are compelled to perform.

These circumstances, connected with situation, account for the existence of different tongues prevailing among the inhabitants of different regions; but they offer no explanation of the fact which we have admitted, that even among the same people, the language is perpetually undergoing change. We must, however, place some restriction upon the admission. Among an uncivilised people, new objects, new wants, new employments, seldom arise, so that the necessity of adding to the number of established terms, or enlarging or fixing their meaning, seldom occurs. Hence we may expect the language of a barbarous people to continue stationary, or to alter merely with the progress of improvements. Is it civilization, therefore, that we must look to for those circumstances, by which words are fabricated and language altered.

When man, instead of resting satisfied with the fruits which the earth spontaneously furnishes, or with the beasts of chase, begins to cultivate the fields, to domesticate animals, and to rear them, to build houses, and to engage in trade, he is compelled to invent a host of names to designate his new possessions, and the means by which he has acquired them. With the progress of the arts, therefore, there will be a corresponding enlargement of the number of signs, by which the objects produced, and the instruments used, may be known. When the acquisition of wealth has given to man an opportunity to rest from bodily labour, and occupy his thoughts in reflection; when he begins to examine the laws of nature, and the constitution of his body and mind, he acquires new ideas, adopts new modes of action, creates to himself new scenes, and thus calls for an addition to the signs already in use, to enable him to express the variety of knowledge which he has obtained. In proportion as the number of those who cultivate the arts and sciences increases, so will the words of language multiply, until different terms shall be employed to express even the same object or action. If we suppose each particular tribe thus proceeding on the march of improvement, each inventing new terms, influenced by all the localities of their condition, we shall arrive at the conclusion, that their language, though originally derived from the same stock, will lose the resemblance, as those who use it recede from barbarism; and after the arts of life have been introduced and science cultivated, we shall find scarcely the remnants of a common origin.

If these remarks are founded in truth, we may expect to find the language of a country exhibiting peculiar features, marking the different stages of the civilization of its inhabitants. In the ruder periods of society, the sounds which are employed will be scanty, like the ideas they are intended to represent. Those objects only will receive names which are immediately subservent to the purposes of existence; the words expressive of action will indicate only sudden transitions; and, in addition to their literal import, the various terms will be employed in a figurative sense, to mark the conditions of an event which the language is yet unable to describe. Again, in the more advanced periods of society, we may expect to find the signs by which natural objects are designated, extended in consequence of enlarged knowledge of their number and qualities. The words expressive of action will not only be more numerous, but capable of marking all its conditions with a greater degree of precision. The increase of the signs indicative of the arts of life, government, and social pleasure, will denote the march of the tribe to the attainment of knowledge, wealth, and power. Deficiency of expression, therefore, will mark the infant state of society, and copiousness, the advances of civilization. That the progress of language actually exhibits such a gradation of character, does not appear in this place to require demonstration.

In a civilized country, language is exposed to the influence of many subordinate causes of change. It has already been stated, that the number of individuals engaging in the improvement of the same arts, in the enlargement of the same sciences, and conducting their plans and making known their successes, each in his own city or village, will necessarily enlarge the number of new terms, and even create that ambiguity occasioned by dissimilar sounds being used in synonimous expression. Fashion alone, in these cases, assigns the preference to that of one individual rather than to another, guided by circumstances which follow no law. Again, inventions and discoveries demand new names to express the new combinations, objects, qualities, or actions to which they refer. These new names are speedily employed in a figurative as well as a literal sense, and, by degrees, assume the station which others had formerly occupied.

There is added to all these, another powerful excitement to change. When distinction of rank begins to prevail, there is an attempt to mark the limits by a distinction of sounds. The great avoid the use of many expressions, because indicative of *vulgarity*—the common people, on the other hand, strive to imitate the language of their superiors, considering it as *genteel*. There is thus a constant change induced in the language of a country, by the attempts at novelty on the part of nobles, the eager imitation of the lower ranks, and the departure of both from the ordinary practice.

These seeds of change, then, are coeval with our race, and must always continue to spring up and flourish wherever new objects or new actions present themselves to man. All the expectations, therefore, of rendering a language stationary, by enumerating the terms of its vocabulary, or establishing standards for their pronunciation, are vain, and founded in ignorance of its origin and its relation to the wants of man. In spite of all the labours of the lexicographer, all the rules of the grammarian, or the combined influence of writers of taste, the language which we now use, will, in the course of a few generations, become antiquated, and our phraseology obsolete; even our present arts will be considered rude, and our science infantile *.

The sounds which are addressed to the ear for the purpose of communicating thought, are not exclusively produced by the organs of respiration. Thus, when a rabbit perceives danger, and wishes to give warning to others, it does not utter a sound; but, beating the earth with its feet, produces a noise, whose meaning its neighbours find no difficulty to comprehend. In the insect well known by the name of the Death-watch (Anobium), a sound is produced by striking its mandibles upon wood, and a similar sound is produced in return by another individual when within hearing. In many other insects, the noise is produced by the friction of the wings against each other, the air, or the abdomen. In the Death's-head Hawk-moth (Sphinx atropos), Reaumur found that the noise which it emits when confined, proceeds from the mouth, and is produced by the friction of the palpi against the tongue. In the Tettigoniæ, on the other hand, there is an organ seated in the abdomen, and opening on its under surface, containing cells, elastic plates, and muscles, by whose motions, sounds, loud and disagreeable, are produced †. In all these instances, the sounds are expressive of feelings, and are intelligible

* Dr Barclay, in reference to this subject, states the truth with painful plainness: "Writers of taste, who value themselves on the beauty and elegance of their diction, must often reflect with painful apprehension, on the instability and transient nature of the perishing sounds with which their literary fame is connected," p. 62. Again, "It seems to be owing to the constant operations of such causes, whose influence can neither be checked nor prevented, that no accident ever has occurred, no art ever been discovered, to preserve the stability of vocal language, to call on the forebodings of literary geniuses, and remove the apprehensions, that their laboured eloquence, in a few centuries, must require an interpreter, and the beauties of their diction pass unnoticed without a commentator." Ib. p. 83.

† See Kirby and Spence's Introduction to Entomology, ii. p. 105.

to those to whom they are addressed, although, in the case of insects, the organs which receive the impression have hitherto eluded the researches of the anatomist.

b. *On the mutual communication of Feeling, by means of Signs addressed to the Eye.*—In attending to such signs, it is convenient to divide them into Natural, and Artificial, or Acquired.

The *natural* signs are common to many animals, and exhibit constant and uniform characters in the individuals of each species. They have obtained the name of *gestures,* as they require for their expression the movement of certain parts of the body.

The gestures by which man expresses his feelings, are executed by the face and hands. By the former, he indicates the state of his heart, in a manner perfectly intelligible to all his race, expressing, by the position of his lips, and motion of his eyes and eye-brows, fear, anger, joy, grief or pain. Each sense has an expression peculiar to itself, intimating, whether the sensations which it experiences are pleasant or disagreeable. The hands are chiefly employed in expressing the conditions of distance or place. These different signs, in man, are carefully studied by actors of pantomime; and it is astonishing with what distinctness they are able, by a judicious use of them, to express a great variety of feeling and passion.

In the lower animals, the gestures are not so much confined to the countenance as in man. In the dog, however, we can read in the face, chiefly from the position and motion of the eye-brows, whether he is pleased or offended. The expression of the former feeling is accompanied by a motion of the tail from side to side. The bull expresses his displeasure, by bending his head to the ground, and throwing up earth with his forefeet and horns. The horse testifies surprise or anger by pointing his ears forwards or backwards. It would, however, be a hopeless task, to enumerate the endless variety of gesture employed by other quadrupeds and by birds, since each species uses gestures which are peculiar to itself, and suited to the organs with which it is furnished. The practical naturalist, indeed, studies these with care, as constituting one of the most delightful employments of his science.

In using these gestures, animals, in general, employ sounds at the same time. If a crow is alarmed, and wishes to communicate its fears to the flock, it not only assumes a particular flight, but utters a cry which renders its feelings intelligible, not only to those of its own kind, but even to other birds and to quadrupeds. It may often be observed that ducks in a pond, upon a particular motion of diving being performed as a signal, and a cry uttered by one, begin to dive and swim about in the water as if urged by one common impulse, and execute the most complicated and rapid evolutions in all the apparent giddiness of joy. But the motions of the flocks of Purrs (Tringa alpina), indicate a still more perfect system of signals. When flying, they keep their backs for a time all in one direction; and as suddenly, by a change of position, exhibit their bellies to the spectator. The change of position is very obvious, as their dusky backs, when directed to the observer, exhibit the flock as a dark cloud, which is changed to white, by the exhibition of the light-coloured feathers of the under side. Linnets, and several congregating birds, act in a similar manner.

I have said, that man comprehends the natural expressions of the lower animals, and even hinted that they in part understand the gestures of one another. The dog and cat, however, are, perhaps, the only animals which can discern in the face of their master, the expressions of approbation or displeasure; and, by their behaviour, testify that they feel their influence.

There is one signal addressed to the eye, apparently employed by many to express certain feelings; but with whose characters we are imperfectly acquainted. I here allude to *luminousness* or *phosphorescence*. It has been observed of the common glow-worm, that the females, which possess this luminous power in a much greater degree than the males, are destitute of wings; and, therefore, incapable of flying about in search of a mate; but that during the season of love, when the light is emitted in greatest brilliancy, she is able to guide her vagrant lover to her presence. By some it has been supposed, that this luminous property is destined to direct the animals possessed of it, with greater certainty to their prey. But it is obvious that it would warn the objects of their pursuit, if animated, of their approach, and facilitate the efforts of their foes to discover their retreat.

We come now to consider the *artificial* or *acquired signs* addressed to the eye. This method of holding intercourse between the individuals of a species, appears to be peculiar to the human race, and to be the instrument by which man can gain possession of the information of past generations, and be able to record the transactions of the present time, for the benefit of future ages.

If we consider a person in a barbarous state of society, anxious to communicate intelligence to his tribe, respecting an animal which he had observed in the woods, unlike to any in size and shape which he had ever seen, we are not aware of any other means which he could employ to accomplish his object, but to draw its figure upon the sand, or scratch its outline upon the rock. This figure being difficult or tedious to execute, would be abridged in the course of frequent repetition, until it became a symbol merely, instead of a picture, to represent the animal in question. If a particular name was bestowed upon the animal, then the symbol would either represent the animal itself, or the particular sound by which it was designated. With the same facility with which sounds could be united, symbols could likewise be combined; so that in the same picture might be represented a variety of objects or actions. Such appears to be the origin and structure of much of the written language of China and Japan at the present day.

When the ear began to discriminate the different kinds of sounds used in speeeh, it would not be a difficult operation to devise symbols to express the elementary parts, so that words could be formed by symbols with as much facility as by sounds. The obvious utility of these symbols, in communicating intelligence to the absent, would recommend their general adoption. But it is obvious, that, in the invention of letters, each tribe would devise symbols peculiar to itself; so that a great dissimilarity of form would prevail among the different alphabets. Even some of the circumstances which operated in the introduction of change into speech, would likewise operate in varying the form and import of letters. But, while the principle of change in the former still continues to exercise its influence, in the latter it has been effectually overcome, by the mechanical arts of engraving and printing. But, even with all the advantages of artificial sounds and signs, we feel disposed, when deeply interested in communicating the knowledge of an event to others, to use the language and the gestures of nature; and, by the movement of our body, to draw as it were, a picture of that which description is inadequate to represent. —To prosecute this subject farther, would obviously be foreign to the purpose of the present work.

c. *On the mutual communication of Feeling, by means of Signs addressed to the Touch.*—The more perfect animals employ the sense of touch in a very limited degree, in the communication of their feelings. But among insects it appears to be very generally used. The antennæ perform the office; and in the manner in which they are applied to the antennæ of another insect, and the motions which are executed by them, these organs are capable of communicating the state of feeling with great precision. When a queen bee, for instance, is removed from a hive, those bees which first perceive the loss, when they meet with others, mutually cross their antennæ and strike them lightly. The anxiety and disturbed state of those that have received the blow, indicate plainly the nature of the intelligence which has been communicated. It is probable, that molluscous animals, furnished with tentacula, employ, in like manner, these organs in the communication of their feelings.

We might here take notice of the mutual communication of feeling, by means of smell, were the facts which are known, in connection with the subject, sufficient to illustrate this department of the animal economy. It is, however, well known, that, at the particular season when the procreative appetite is in exercise, smells are emitted which enable the sexes to discover each other. The emission of these effluvia cease with the season of love.

However various these different means employed by animals to communicate their feelings appear to be, it is generally believed, that man alone possesses the power of expressing distinctly *past* events to others. The lower animals are almost exclusively occupied with their present impressions; and their language appears suited only to express these. I have said almost, because there is evidence of the power possessed by various animals, of communicating intelligence to others of a supply of food which has been discovered, and of guiding them to the spot. But still this is in a great measure the business of the present time. Man, on the other hand, by means of his acquired language, can express the feelings which, on any former period, he has experienced; and, by means of artificial signs, he can convey the expression of these feelings even to those who are at a distance. Although in possession of these peculiar powers, he can still use the language and the gestures of nature.

5. *On Restraint.*—When treating of the characters by which the intellectual principles are distinguished, we took notice of the faculty of *Attention*, by whose assistance we can withdraw our various powers of perception and reflection from the contemplation of one object or idea, and fix them on the examination of others. When, from the notions which we have acquired from experience or reflection, we are disposed to action, there is another term employed to designate that principle, by which the kind of action, and the manner of performing it, are determined. This principle, which is identical with attention, is termed the *Will;* and the liberty which it exercises in the choice, is denominated *Free Agency.*

There are some who deny the existence of a liberty of selection, and affirm that all the actions which we perform "are the necessary result of the constitutions of our minds, operated on by the circumstances of our external situation." If the existence and influence of the faculty of attention, such as we have described them, be admitted, and we can appeal to the experience of all, for proof of the truth of our statements on the subject, the doctrine of free agency must likewise be embraced: For, granting even to the Necessitarian, the necessity of the *action*, upon the combination of the particular circumstances which constitute the *motive*, we have still the choice of creating these circumstances. We have the power of directing the eye or the ear to receive particular impressions; we have the power of giving to any particular perception its suitable degree of force to act as a motive, by the progress of reflection; and all this by the aid of a faculty, the existence of which we presume none will deny. According to this view of the subject, we can account for the great diversity of action among individuals, placed in similar external circumstances. The external impressions may be the same; but these, being subjected to different treatment during the process of reflection, by minds whose notions of truth or duty, (acquired under the influence of attention,) are not in unison, the resulting motives and consequent actions will be dissimilar. We urge, likewise, in support of this view of the subject, the pleasure which the mind feels upon acting agreeably to its *rules* of truth or duty, and the pain which accompanies their violation;—the former being the proof that the action was agreeable to the constitution of our minds; the latter, that it was repugnant. Yet such transgressions with regard to our intellectual rules, are but too frequent; while similar violations of the rules of our constitution, in the case of sleeping, eating, or any of our instinctive powers, are seldom observed. Is there a difference in the degree of restraint which we can exercise over the instinctive, when compared with the intellectual powers?

It appears probable, that, in reasoning upon this subject, moral philosophers have seldom drawn a line of distinction between the exercise of the will, when directed to the intellectual, and when directed to the instinctive powers. The supporters of the doctrine of Free Agency usually derive all their arguments from the characters of the former; while the Necessitarians rest their proofs on the powerful influence of the latter. It is to a want of attention to the bearings of the question, joined to the sophistry of words, that we can trace much of that difference of opinion which prevails on this subject.

From the preceding observations, the reader may perceive, that while we contend for the existence of free agency in all the processes about which our intellectual faculties are employed, we are disposed to admit that its influence over our instinctive propensities is feeble and circumscribed. In the intellectual process, there is interposed between the impression (whether external or internal) and the action, a certain degree of deliberation or reflection, after which our free agency exercises its controul. In the case of our instinctive powers, action follows the impression almost instantaneously; so that there is no time to deliberate,—no opportunity to choose. This is the natural condition of our instinctive powers. But, by the force of habit, the influence of example, and the regulations of civilized life, we acquire the power of interposing between the impression and the action, a greater or less degree of deliberation, in consequence of which we have an opportunity of exercising restraint, and of regulating or preventing the action, which the impression, without such controul, would have produced. Frequently, however, the original tendency of the impression triumphs, or gives such a bias to our decision, that it appears rather a necessary result of the original impression, than a choice founded on deliberation. Still, however, there is something confined and unnatural in such interference of the reasoning powers with our instincts; so that we are ever in danger of breaking through the rules which they impose, whether in regard to our appetites, desires, or affections. An appeal to experience, for the justness of this remark, will not here be made in vain! Its truth is likewise established by the conduct of those who, while they prescribe excellent rules for the regulation of life, paint, in the most vivid colours, the dignity of virtue, and the meanness of vice, continue enslaved by their passions, sensual, avaricious, and implacable. The former part of their conduct is an intellectual process, in the performance of which, they meet with no resistance;—in action, they have to contend " with the law in their members waring against the law of their mind."

The acquisition of this restraint over the instinctive powers,—this victory over self, is the most difficult, but the most valuable of all our conquests. It is the ultimate object of a rational and an immortal being.

An eloquent writer, whose authority in the science of mind is deservedly respected, but who appears but imperfectly acquainted with the principles of action in the lower animals, has given it as his opinion, that the reason of man, in reference to his active powers, renders his nature and condition essentially different from that of brutes; and adds, "They are incapable of looking forward to consequences, or of comparing together the different gratifications of which they are susceptible; and, accordingly, as far as we are able to perceive, they yield to every present impulse. But man is able to take a comprehensive survey of his various principles of action, and to form a plan of conduct for the attainment of his favourite objects. Every such plan implies a power of refusing occasionally, to particular active principles, the gratification which they demand *." The merits of this distinction between man and the lower animals will best appear, by comparing both as placed in nearly similar circumstances. The cultivation of the power of restraining the instinctive or active principles, does not appear to be necessaay in order to form the character of a savage. He has appetites, desires, and affections whose gratification is necessary to his welfare, and few opportunities occur where indulgence to excess is either practicable or desirable. It is only in the progress of civilization, when property has been acquired, and personal rights recognized, that man feels the necessity of imposing upon himself *laws* or restraints upon his active principles. It is difficult to find an opportunity of examining man in a very low state of intellectual improvement. The intercourse of Europeans has

* "Outlines of Moral Philosophy," by DUGALD STEWART, p. 109. Edin. 1808.

chiefly been with nations who have in part emerged from barbarism. In order to observe the power of restraint which they are capable of exercising over their propensities, we must retire to a factory established on the confines of a hunting tribe. There we shall find the appetite for spirits and tobacco, or the desire for ornament or dress, overcome every dictate of prudence or experience. A few glasses of brandy will often purchase the whole produce of a summer's labour.

Even in a civilized state, this ability to control the active powers, if it really exists to the extent here supposed, is never exerted by all the individuals of society, nor by any one individual over all his propensities. Neither rank nor learning, youth nor age, can plead exemption from the charge. The reason of this, according to our view of the matter, is obvious. The power of controlling the intellectual powers is *natural*; of regulating our active powers, *acquired*. In the acquisition, so many difficulties present themselves, so many privations must be endured, so much energy of mind must be exerted, that we need not be surprised that the object is not generally sought after, and only gained by few. Indeed, mere worldly considerations hold out no suitable motives for encouraging us to gain the victory over *all* our instincts. It is enough if we restrain them from being externally inconvenient to others. Christianity alone furnishes us with sufficient motives, and urges us to subdue *every lust*.

We have thus endeavoured to mark the difference of free agency, as exerted upon our intellectual and active powers; and we are now prepared to examine how far, in reference to the brutes, their nature and condition are essentially different from our own. Do they always yield to every impulse? And are they incapable of acquiring the power of occasionally restraining their propensities?

In a wild state, the inferior animals, like man in his savage state, stand in no need of restraining, to any great extent, their instinctive powers, indulgence being necessary for their own comfort, and the safety of the race. In the case, however, of social animals, (including man), where every individual contributes to the welfare of the group, it is obvious, that, in many cases, some degree of self-denial must be practised, and a self-denial, which, so far as we are able to judge, is not instinctive, but acquired or voluntary. In illustration of this opinion, we may quote the case of congregating animals, who, while feeding, have a sentinel to give them warning of danger, as apes and geese. The sentinel, in this case, may look forward to be released from duty; but, in the mean time, he must feel the cravings of an empty stomach, and witness his acquaintances enjoying their repast. In all this he yields not to present impulses, but restrains his appetite for food, in order to comply with the arrangements of the social affection. In the case of animals which have escaped from a snare, and which refuse to be again enticed, there is a still more decided example of self-denial. The bait still allures; but the temptation is overcome through the sense of danger.

As it is in man, when civilized, that we meet with the most unequivocal proofs of controul exercised over the instinctive powers, so, among domesticated animals, we may expect to find its existence most distinctly exhibited. We have seen a dog enter a larder, even when hungry, and smell at the cold meat and bread, without presuming to touch them. That he had an inclination to eat, could not be doubted; but he had acquired the power of controlling it. The same animal exhibits, in many cases, great sagacity in the exercise of his controul over his feelings. Thus, if you conduct an experienced spaniel to a place from whence he has seen a covey of patridges spring, he will pass on, indifferent to the scent which they have left behind them; but, if he did not observe their flight, his actions are widely different;—" He treads with caution, and he points with fear." But it is needless to multiply examples: for all our domesticated animals exhibit the power of restraining their instincts; and the extent of this power is in the ratio of their obedience. We shall not here inquire into the motives which regulate the obedience, knowing that the moralist is aware that compliance with the laws of society, in regard to man, is often disagreeable, and even forced.

"There is another" (says the same philosopher *) "and very important respect, in which the nature of man differs from that of the brutes. He is able to avail himself of his past experience, in avoiding those enjoyments which he knows will be succeeded by suffering; and in submitting to lesser evils, which he knows are to be instrumental in procuring him a greater accession of good. He is able, in a word, to form the general notion of happiness, and to deliberate about the most effectual means of attaining it." We are compelled, however reluctantly, again to differ from this celebrated moralist, and to advance the opinion that the brutes do control their instinctive powers under the guidance of experience; avoid enjoyments which are succeeded by sufferings; and submit to lesser evils, to avoid greater ones. We by no means venture to state, that the lower animals are *always* so prudent, and we presume that none will contend for the universality of such discretion in the human species. But that they are guided in their attempts to avoid evils and secure happiness, by the experience of the past, cannot admit of a doubt. A horse will submit to the lesser evil of mending his pace, rather than to the greater evil of being spurred. Dogs

* Outlines, p. 112.

will often submit to the evil of continuing for a time in a constrained position, with a piece of bread upon their nose, until the signal for taking it be given, and exhibit unequivocal symptoms of satisfaction at obtaining happiness at so easy a rate. A goldfinch in confinement, will submit to the evil of drawing up a small bucket by its chain, for the sake of the enjoyment of a draught of the water which it contains. Those who are conversant with the history of animals, must be acquainted with many other proofs of a similar kind.

In the course of the preceding remarks on the phenomena of Mind, we have frequently had occasion to take notice of the force of HABIT, or the effects of custom, upon our functions. When external circumstances continue, for any length of time, to operate upon our feelings, in a particular manner, our attention is directed towards them, in order to accommodate ourselves to their influence. In many cases, where they are disagreeable at first, they afterwards become inoffensive and ultimately objects of desire. This is equally descriptive of our intellectual and active powers. The facility with which we yield to circumstances, differs greatly in different individuals, and appears to depend upon constitutional organization. Even in the same individual, habits are more easily acquired in youth than in old age. Man acquires habits, or conforms himself to a variety of circumstances, with greater facility than any of the inferior animals. He lives in all climates; subsists on all sorts of food; and performs all kinds of labour. Many of the lower animals acquire habits with considerable facility; and these we have inlisted into our service. They almost exclusively belong to those which are united in groups, by the social affection, as the ox, horse, and dog. Solitary animals, as the cat, are reclaimed with greater difficulty, and the education of which they are susceptible is much circumscribed. The social animals, on the other hand, easily submit to restraint; and, under our guidance, acquire habits essentially necessary to our comfort.

4. *On the difference between Reason and Instinct, and on Man's superiority over the Brutes.*—After the details into which we have entered respecting the organs of perception, and their functions, and the intellectual and instinctive powers, we consider ourselves prepared to give a satisfactory solution to those questions, which have been so often agitated, respecting the claims of our race to be considered as the lords of the earth. We profess neither to be influenced by any desire to conceal the defects, or to magnify the qualities of our species, nor to elevate the brutes to a higher station than they were destined to occupy. We admit neither pride nor malevolence to be judges in the case.

The mysterious and unsettled opinions which prevail upon this subject, probably proceed from the ambiguity of the phrases employed, arising from our confounding the characters of the Intellectual and Active powers, and making Instinct represent the operations sometimes of the one, and sometimes of the other. In speaking of the faculties of man, the term Reason is generally confined to express the efforts of his intellectual powers; and, with regard to what are termed Instincts, many authors may be pointed out, who consider him as nearly destitute of these, the qualities of his reason supplying the defect. In speaking of the faculties of brutes, the term Instinct is sometimes restricted to the efforts of their active powers; sometimes it also includes those of the intellect. If an observer, therefore, compares the instincts of brutes, or the efforts of their *active powers*, with the reason of man, as consisting in the operations of his *intellectual powers*, a difference of mental constitution will be observed, not in degree merely, but in kind. Again, if we compare the instincts of brutes, considered as consisting of the efforts both of their *active and intellectual powers*, with the reason of man, in his savage state, in which the operations of his active powers are chiefly conspicuous, we shall be led to conclude, that there is no difference in kind between reason and instinct, and if we form our conception of this instinct, from its collective prominent qualities, derived from all the different species of animals, we shall be tempted to conclude, that the power which regulates the manners of the inferior animals, is even superior to the boasted reason of man *.

It would serve no good end to produce particular examples from the writings of the most eminent cultivators of moral and natural science, to prove that the terms Reason and Instinct have had assigned to them such varied and indefinite significations; for there is no author whose observations on this subject we have had an opportunity of examining, who has been sufficiently careful to avoid the confusion.

If we restrict the term *Reason*, then, to express the operations of our intellectual powers, and inquire, is there any thing analogous in the constitution of brutes; and restrict the term *Instinct* to express the operations of our active powers, and make a similar comparison, we shall be able to point out, not only wherein the difference between reason and instinct consists, but between man and the lower animals, in reference to mental attainments. The solution, indeed, of the whole difficulty, appears to us to be obtained by this simple limitation of the terms.

* There is evidently a much greater difference between a citizen and a savage, than between a savage and many of the lower animals. But still we are to bear in mind, the unlimited capacity of the one for improvement, and the limited capacity of the other.

In our review of the intellectual powers, we have demonstrated, that man and the brutes possess, in common, attention, memory, and imagination; and that, in their ideas of reflection, they have a knowledge of personality, time, power, truth, and duty. But, in all these, man has a decided superiority. This arises chiefly from the strength of his power of Attention, by which he can direct his mind to any subject, until he is satisfied with regard to its qualities or connections; in other words, he can practise abstraction more successfully. Hence all his ideas of reflection are more perfect and extensive, and have given rise to the belief in God, and the corresponding feelings of a piety, of a future state, and the preparation necessary for a blessed immortality. If the facts which I have adduced in support of the statements be admitted, it follows, that the intellectual powers of man differ, not in kind, but merely in degree, from those of brutes.

In our examination of the active powers or instincts, we have likewise demonstrated, that man and the brutes possess appetites, desires, and affections, regulated by the same laws, and destined to accomplish the same objects, in the animal economy, exhibiting, however, slight shades of difference, according to the species. The superiority of man over the brutes, in reference to the active powers, (except perhaps some of the desires,) is so small, that doubts may be entertained respecting his claims of supremacy. The comparison, indeed, cannot well be made, as every species has peculiarities of condition to contend with, for which peculiar active powers or instincts are required. The number of qualities of each instinct, therefore, is in proportion to the wants of each species.

We have likewise seen, that man and the inferior animals naturally utter particular sounds, and execute particular gestures, as the means of communicating thought or feeling, and that this natural language is chiefly occupied in expressing the wants or the enjoyments of the active powers. In addition to this natural language, man has contrived artificial sounds, and artificial signs, to express all his perceptions, reflections, and feelings, to others of his race, absent as well as present,—attributes which give him a decided superiority over the brutes. It is by means of these, that the improvements or the experience of one generation are transmitted to another; so that, while the brutes must acquire all their knowledge by experience, man derives his most valuable acquisitions from tradition.

We have likewise seen, that the power of restraining the impulses of the active powers is weak in the lower animals; but, in man, it is capable of exerting itself in a greater degree, thus giving him a more complete command over himself, and consequently over the objects around him.

These various instances of superiority, unfold to us the steps by which man is destined to rise in the scale of civilized life, and acquire that dominion over nature which he upholds by his wisdom. Man, therefore, is far exalted above the brutes, by a superior degree of perfection in his intellectual faculties; by a greater power of restraint over his instincts; and by readier methods of communicating his ideas and feelings,—rather than by a difference in the nature of his mental constitution.

The importance of the nervous system to the existence and wellbeing of animals, naturally excited inquiries concerning its structure and modifications in the different races of animals, and led to the adoption of the characters which these furnished in their classification. We have already presented a view of the animal kingdom, at the conclusion of our remarks on the structure of the nervous system, in which the classes are arranged according to differences in the organization of this system.

When we consider the difficulty attending the examination of the functions of the nervous system, our ignorance of the circumstances in the organ which determine its action, the numerous instances which are known, where the function is performed, without the organ existing which was supposed peculiarly destined to its execution, we could scarcely have expected an arrangement of animals to have been proposed, depending on the characters furnished by the nervous system in action. The celebrated M. LAMARK unappalled by these difficulties, has adopted a method, founded on such principles, in his " Extract du Cours de Zoologié," 1812, and since in his " Histoire Naturelle des Animaux sans vertebres," 1815 *.

The proofs of the existence of sensation, as an attribute common to all animals, have been given in an early part of

---

* The following synoptical view of this arrangement will enable the reader to understand its peculiar features :

Animaux sans Vertebres.

* *Animaux Apathiques.*—Ils ne sentent point, et ne se meuvent que par	
1. Les Infusoires.	leur irritabilitie exciteé. *Caract.* Point de cer-
2. Les Polypes.	veau, ni de masse medullaire allongeé ; point de
3. Les Radiaires.	sens ; formes variées ; rarement des articulations.
4. Les Vers.	
(Epizoaires.)	

** *Animaux Sensibles.*—Ils sentent, mais n'obtiennent de leurs sensations	
5. Les Insectes.	que de *perceptions* des objects, especes d' idées sim-
6. Les Arachnides.	ple qu'ils ne peuvent combiner entrelles pour en
7. Les Crustacés.	obtenir des complexes. *Caract.* Point de colonne
8. Les Annelides.	vertebrale ; un cerveau et le plus souvent une
9. Les Cirrhipèdes.	masse medullaire allongée ; quelques sens dis-
10. Les Mollusques.	tincts ; les organes du movement attachés sous la
	peau ; forme symetrique par des parties paires.

Animaux Vertebrés.

*** *Animaux Intelligens.*—Ils sentent, acquierent des idées conservables ;	
11. Les Poissons.	executent des operations entre ces idées, qui leur
12. Les Reptiles.	en fournissent d'autres ; et sont intelligens dans
13. Les Oiseaux.	differens degres. *Caract.* Une colonne vertebrale, un
14. Les Mammiferes.	cerveau et un moelle epiniere ; des sens distincts ;
	les organes du movement fixes sur les parties d'un
	squelette intérieur ; forme symetrique par des par-
	ties paires. Tom. i. p. 381.

this volume, and, if satisfactory, demonstrate that *Apathique* animals do not exist. With regard to the distinction between *Sensible* and *Intelligent* animals, founded on the circumstance of the ideas of the former being all simple, and in the latter both simple and complex, we can regard it in no other light than an arbitrary assumption of a character which has no existence in nature. Among his *Sensible* animals, the sense of hearing, taste, smell, and touch, are well known to exist; and if an animal derives ideas from all these sources, is it conceivable that each class shall be preserved distinct, and no combination take place where the ingredients are already in contact. When a bee departs from its hive to collect food at the place where, on the preceding day, it obtained a bountiful repast, it is obvious that both the distance and direction must previously be contemplated, intimating the existence of complex ideas both of time and space. But it is not our intention to occupy the time of the reader in a refutation of the theoretical opinions of an author, who, in his delineation of the mental powers of animals, substitutes conjectures for facts, and speculation for philosophical induction. Fortunately for his reputation, he possesses much real merit as a systematical naturalist.

In conclusion we may observe, that the different members of the body are equally subservient to the purposes of intellectual and the instinctive powers. We open our mouths by instinct to eat, and we execute, by means of the same muscles, a similar movement as an act of the will, in consequence of reflection. The impression made upon the senses by external objects, excite the movement of the intellectual powers, and they operate equally on our instincts. The instinctive powers may be said to comprehend the relations of our impressions almost intuitively. The will can excite the senses to action, and the instincts can do the same. It is impossible, therefore, in treating of the origin of the motions of animals, to separate the volition of intellect from instinct, because few actions can be excited or continued by the latter, without being perceived by the former. It was in consequence of this intimate connection, that we treated of the instinct of animals, along with the functions of mind which depend on the nervous system. We by no means, however, give it as our opinion, that instinct and intelligence are the same, either in kind or in degree. The former we know to be common to every organised being, the latter to be correlative with the nervous system. Their intimate union in animals, therefore, is no proof of their identity. Vitality, in its simplest state, possesses irritability and instinct *; in its most complicated, irritability, instinct, sensation, perception, reflection, and volition.

The instinctive powers are essential to the existence and comfort of the individual; the intellectual powers increase the resources of instinct, and in a savage state, are exclusively subservient to its purpose. It is only when man advances in civilization, that his intellectual powers assume their preeminent station, control his instincts, and exhibit the emblems of his greatness.

Under the first class of functions, we considered those organs which are subservient to the purposes of protection, stability, and action. Under the second, we have investigated the nervous system,—the control which it exercises over the body,—and the intercourse which it permits us to hold with the world around us. In the third class of functions, to which we are now to advert, the organs are employed in obtaining nourishment for the system,—prepar-

* In such beings, instinct cannot excite to action by volition, which is coexistent with sensation, but must operate exclusively by means of irritability.

ing it for the particular parts,—depositing it where wanted,—and removing those portions which have become superfluous.

CHAP. XII.

Digestive System.

In treating of the appetites, as occupying a conspicuous place among the active powers, we had occasion to delineate the characters of the instinct for food. In this place, we have to investigate those organs over which this instinct presides, and the important purposes which they serve in the animal economy.

Every one knows, that the sensation of hunger is felt when the stomach is empty, but the nature of the connection between these circumstances has not been satisfactorily explained. Some have ascribed the feeling to the fatigue of the muscular fibre from the stomach being contracted, and others to the friction of the rugæ of its internal covering, while there have not been wanting many who refer it to the stimulus of the accumulated gastric juice, or to the acrimony of that fluid, arising from its unusual detention. But the total absence of proof in support of any of these opinions, renders it unnecessary to enter upon their examination.

Many circumstances, however, are known, which promote hunger, especially exercise, cold air applied to the skin, and cold, acid, or astringent fluids introduced into the stomach. Inactivity, warm covering, the attention diverted, and warm fluids, have a tendency to allay the sensation *. But

* These facts serve to explain the circumstance of the inhabitants of warm countries making use of food highly seasoned with hot spices, while

whether these effects indicate the local state of the stomach as the cause of the sensation, or merely its connection with other circumstances, which give rise to the feeling, can scarcely be determined. Indeed, it is not proved that the sensation of hunger arises from the state of the stomach, in preference to the condition of the duodenum, or the orifices of the lacteals.

Thirst is accompanied with a sensation of dryness in the mouth. This dryness may be occasioned by excessive expenditure of the fluids, in consequence of the dryness or saltness of the food which has been swallowed; or to their deficiency, from the state of the organs.

Both hunger and thirst, besides being greatly influenced by habit, exhibit very remarkable peculiarities according to the species and tribes of animals. In each, the wants of the system are provided by instincts, which, in the attainment of their object, exhibit movements the most complicated, and employ means the most various.

Animals are frequently characterised by the kind of food on which they subsist. Those which live on the spoils of the animal kingdom are said to be *carnivorous*, when they feed on flesh,—*piscivorous* when they subsist on fishes,—and *insectivorous* when they prey on insects. Again, those animals which are *phytivorous*, or subsist on the products of the vegetable kingdom, are either *granivorous*, and feed on seeds,—*graminivorous*, pasturing on grass,—or *herbivorous*, browsing on twigs and shrubs.

Besides those substances which animals make use of as food, water is likewise employed as drink, and as the vehicle of nutritious matter. Salt is necessarily mixed with the drink of the inhabitants of the ocean, and is relished by

---

those of colder districts are satisfied with a less stimulating diet. The vigour of the digestive organs of the latter is fit for the accomplishment of its object, but the languid condition of the former calls for assistance.

man and many other animals *. Other inorganic substances are likewise employed for a variety of purposes. Many savages make use of steatite and clay along with their food. The common earth-worm swallows the soil, from which, in its passage through the intestines, it extracts its nourishment.

In some cases, substances are swallowed for other purposes than nourishment. Stones are retained in the stomach of birds, to assist in triturating the grain. The wolf is said to satisfy his hunger by filling his stomach with mud. We have found in the stomach of the eared grebe (Podiceps auritus), in the month of January, a large ball of its own breast feathers, probably pulled off and swallowed for the same purpose.

When we consider the number of elementary substances which enter into the composition of the bodies of animals, and the varieties of combination which they form in the individual parts, we may perceive how vain it is to assign to any one substance the exclusive property of nourishing, and to consider the value of the food to depend on its presence. In the ordinary secreted fluids, the number of elementary bodies of which they consist, is greater than one would be led to expect, without some deliberation. The spittle, for example, is, in man, necessary to mastication, and, therefore, during the continuance of health, must be constantly produced. Yet we find the following elements entering into its composition: Oxygen, hydrogen, azote, carbon, phosphorus, muriatic acid, lime, and soda. We

* The eagerness with which many quadrupeds and birds press towards salt-springs and lakes, situated in inland districts, for the purpose of tasting their contents, indicates a constitutional fondness for salt. The saline mineral spring of Dunblane, discovered a few years ago, first attracted notice by its being the constant resort of pigeons, which flocked to it from great distances at all times of the day.

have no proof that these substances are derived from any other source than the food. Hence, for the supply of this single secreted fluid, substances containing, at least, these eight elementary bodies, must be obtained for the use of the system. It would be difficult to determine which of these substances could be spared, so as to produce the least injurious alteration in the secretion; but it is obvious, that, as they all occur constantly, they are all necessary to the perfection of the secretion and the healthy state of the system.

As animal matter contains a great deal of azote, physiologists have differed in their opinion with respect to its origin, some attributing its source to the air in respiration, others to the food. M. F. MAGENDIE, instituted a series of experiments, for the purpose of determining this question. He selected a dog in good condition, three years old, and fed him with substances which contained no azote. These consisted exclusively of white sugar and distilled water. During the first week he appeared to enjoy his new regimen, but in the second week he began to grow lean, his eyes became ulcerated, and gradually losing strength, he died on the thirty-second day. The urine was found to be alkaline, and to present no trace of uric or phosphoric acid. The bile resembled that of herbivorous animals, and the excrements contained but a small proportion of azote. In the body the fat had disappeared, the muscles were reduced in size, and the volume of the intestines diminished. Similar results were obtained, when the experiment was repeated upon two other dogs; and likewise in two cases when olive oil, and in others, when gum was substituted in place of sugar *.

---

* Precis Elementaire de Physiologie, Paris 1817, 2 vols. 8vo. ii. p. 390,—394.

The value of these experiments is reduced almost to nothing, when we consider, that an animal, naturally carnivorous, and, in a domestic state, omnivorous, is, all at once, restricted to diet and drink foreign to its nature, and with which it was never accustomed. In such circumstances, we may expect a derangement of the system to take place from a variety of causes.

In the first experiment, where the dog was fed with white sugar and distilled water, the elementary substances which the system could obtain from these, were oxygen, carbon and hydrogen, together with a little azote, contained in the air within the pores of the sugar, or mixed with both the sugar and distilled water in the process of mastication. But we have seen, that in the saliva alone, a fluid essentially necessary to the very commencement of the process of digestion, phosphorus, muriatic acid, lime and soda, enter as constituent ingredients. Azote may have been obtained from the air in respiration. We may, therefore, with great propriety, regard the decay and death of the animal, as originating in the want of these other elementary substances, rather than to the azote, the one to which they are referred by the author. If saliva could not be produced from such food, much less could blood, which, in addition to the elementary bodies now mentioned, contains magnesia, potash and iron. It is obvious, therefore, that in these experiments there were too many deranging causes operating, to permit a safe conclusion to be drawn from them.

In attending to the animal economy, in reference to the digestive system, it is interesting to observe the various means which are employed, in order to bring the food within the reach of the organs employed in deglutition. In man, the hand is extensively used for this purpose, and for which it is admirably adapted by the pliability of the fingers, and the opposition of the thumb in the act of grasping.

In some cases, the claws in which the toes terminate, by penetrating the substance of the object, retain it more securely. These claws are sometimes double on the same finger, and have their motions so arranged, that they seize the object like a pair of forceps.

In place of arms, protected or supported by hard articulated substances, many animals possess soft flexible threads of various thickness, termed *tentacula*. These encompass the objects used as food, and convey them to the mouth. Sometimes, for the more certain retention of the prey, these tentacula are armed with suckers, which adhere readily to any surface to which they are applied.

Some animals are furnished with a lengthened *snout*, for the purpose of turning up the ground in search of food, as the sow and mole. In the case of the elephant, the snout is so constructed at the extremity, as to grasp an object, and so flexible throughout, as to serve the purpose of tentacula.

These organs for seizing which have been enumerated, though, in general, situated in the anterior part of the body, in the neighbourhood of the mouth, do not enter into the composition of that opening. Those which follow, however, form constituent parts of that organ.

The margin of the mouth itself, termed the *lips*, is, in many cases, well adapted for the purpose of seizing objects. In some cases, they are capable of being protruded to some distance, and form a tubular proboscis; and being beset with teeth or prickles, retain possession more secure of the objects of their pursuit.

In some instances, the tongue is used as an instrument of seizure. In which case it is capable of being extended considerably beyond the margin of the mouth.

In many aquatic animals, a variety of means are employed, in order to produce currents in the water, to bring the food which it contains more immediately in contact with the mouth. Tubes are used for this purpose among the mollusca, and circular moveable ciliæ in some of the infusory animalcula.

Where the food is soft, and easily broken or torn, and where the lips have sufficient strength, the mouth is easily filled with food. But where the animals are weak, or the portion of the food for mastication difficult to be separated, other means are provided than those to which we have already referred. The lips, as in the case of the mandibles of birds, are covered with a horny shell, admirably adapted for cutting off portions of food. Teeth placed in front, with sharp edges, are given to many quadrupeds for the same end; and are termed, from their use, *cutting teeth.*

By these various contrivances, the food is brought in contact with the mouth; and the requisite quantity separated from the mass. The cavity of the mouth, in the most perfect animals, is formed by the lips in front, the hard palate above, the soft palate behind, the tongue below, and the cheeks on each side. All these parts contribute to modify the motions of the food, while retained in the cavity. In some cases, the food, upon entering the mouth, is conveyed immediately into the gullet; while in many animals it is subjected to the process of *mastication* or chewing. For this purpose, the under jaw is indispensably necessary; and the motions of which it is susceptible indicate the kind of action which is performed. When the jaw is articulated in such a manner, that a vertical motion only is permitted, we may infer, that the food consists chiefly, if not exclusively, of flesh. If, on the other hand, the joint admits both of a vertical and horizontal motion, the food consists chiefly of vegetables. Between these two extremes there are many intermediate modifications, enabling certain animals to feed on both kinds of food. With each condition of motion in the jaw, there are, likewise, in many cases, co-existing forms of teeth. In the jaw, with the vertical motion, the teeth are sharp or pointed on the summits, and they divide the food by the edges passing each other like scissars, or by the mutual opposition of the unequal surfaces. Where the jaw has an extensive motion, the surfaces of the teeth are flattened and grooved, and well adapted for triturating grain or herbs.

While the food is thus undergoing the process of mastication, it is necessary to change its position, in order to place all the parts under the action of the cutting surfaces of the teeth. In the accomplishment of this purpose, the cheeks render essential service, by bringing back the food which falls on the outside of the teeth. The tongue, however, is the instrument which chiefly serves to regulate the motions of the food in the mouth. Its delicate sense of touch and taste, fit it for ascertaining the changes which are taking place; while its flexibility enables it to turn the substances in every direction, and give to the whole the requisite preparation. Here we can discern three coexisting organs, where the food requires to be chewed: the jaws must perform certain motions; teeth are requisite; and a flexible tongue. Where chewing is not performed, the tongue neither possesses the same flexibility of motion, nor the same delicacy of feeling.

The food, while subjected to the process of mastication, is, at the same time, mixed with a peculiar fluid, termed *saliva* or *spittle*. This fluid is secreted by almost the whole internal surface of the mouth, and from the parotid, submaxillary and sublingual glands. Its chemical nature, as a whole, has been already noticed; but it remains to be ascertained, whether all these glands secrete the same kind of fluid, or each a liquor peculiar to itself. On this subject, however, there are no satisfactory observations, neither is it determined, whether the saliva acts mechanically, merely, by assisting the teeth to reduce the food to a pulp, or whether it acts chemically, by aiding the conversion of the food into chyle.

After the food has thus been reduced to a pulpy mass, by the attrition of the teeth, and sufficiently mixed with saliva, it is fit to be conveyed to the stomach. Between the mouth and the stomach, however, in the more perfect animals, there intervene two cavities, the *pharynx* and the *gullet*. The former occupies the back part of the mouth, behind the base of the tongue and the soft palate, and the posterior openings of the nostrils. The windpipe and gullet open into it. By means of the tongue and cheeks, the food is thrust into this cavity, and at this time, its entrance into the nostrils is protected by the soft palate, and its suspended uvula. The opening into the windpipe is also closed by a valvular arrangement, and concealed by the base of the tongue and a cartilaginous valve, termed the *epiglottis*. The muscular structure of the walls of the pharynx propels its contents into the gullet or œsophagus.

In many animals, the mouth may be said to open immediately into the stomach. In the more perfect kinds, however, the gullet forms the intermediate connection, and consists of a tube, varying greatly in its dimensions, according to the species. Its external coat consists of longitudinal or transverse muscular fibres, which assist in propelling the food to the stomach; and within this is a layer of cellular substance, lined, on the interior, with a mucous membrane. The food is here still farther softened, by being mixed with the secretions of the gullet, and becomes prepared for entering the stomach, to experience still farther changes.

The stomach exhibits such remarkable variations of form

in the different tribes of animals, as to render a general description of its appearances and structure impracticable. In the human subject, and many other animals, it is in the form of a bag; simple in the interior, and furnished with two openings; the one, by which the contents of the gullet enter, being termed the *cardia*, and the other, which conveys these out of the stomach, the *pylorus*. The portion of the stomach next the cardia, is termed the cardiac portion, and the other the pyloric.

The food is mixed in the stomach with another animal fluid, termed the *gastric juice*. The glands in which this juice is prepared, vary greatly in form as well as position. In some cases they are simple; while in others they are botryoidal, or divided into lobes. They appear in some animals at the base of the gullet, in others at the cardia; while in many they appear to be confined to the pyloric portion. The nature of the fluid which is secreted, and its use in the process of digestion, have given rise to great differences of opinion. The impossibility of obtaining it in a pure state, has hitherto prevented any confidence being reposed in those experiments which have been attempted to determine its nature. Indeed, if we reflect on the different kinds of food employed by animals, we may expect to find the gastric juice varying in its nature, according to the substances with which it is destined to be mixed. The experiments of Reaumeur, Spallanzani, and others, demonstrate its resolvent power. If grains of corn be put into a perforated tube, and forced into the stomach of a granivorous bird, no alteration will take place, though suffered to remain a considerable time; but if the husks be removed, and the grains replaced in the tube, they will soon be reduced to a fluid mass. Bones are dissolved by the stomach of a dog; and it is no uncommon thing for sea-fowl to swallow a fish too large for the stomach, so that a part remains in the gullet. That which is retained in the gullet continues unaltered; but the portion which has entered the stomach is very speedily reduced. Again, the gastric juice is well known to possess the power of coagulating milk, and of redissolving the coagulum. An antiseptic power was likewise ascribed to the gastric juice by SPALLANZANI. He states, that pieces of meat may be preserved a long time in this fluid, without undergoing putrefaction; and that putrid meat becomes sweet, after remaining a short time in the stomach of a dog. The existence of this antiseptic power has been recently called in question by M. MONTEGRE *, who found that the liquid, which he considers as gastric juice, is nearly similar to saliva, and putrefies as readily as that secretion, when kept in phials at the temperature of the human body. But it appears probable, that he operated on saliva only, or on saliva mixed with the mucous secretions of the gullet and stomach, as he obtained the subject of his experiments from his own stomach, after fasting, by means of vomiting; since it is not yet demonstrated, nor even rendered probable, that in the living system the gastric juice is secreted, unless when the stomach is distended with food.

In some cases, as among the ruminating quadrupeds, the stomach consists of several divisions. Into the first of these the food is conveyed as into a storehouse, to be afterwards thrown up in small quantities into the mouth, to be remasticated and transmitted to the true digesting stomach, in which only the gastric juice is to be found. The two first stomachs of the ox may be regarded as appendages of the gullet, and more connected with mastication and deglutition, than digestion. In other animals, as birds, the food is exposed in the stomach to a process of trituration

* Report of the French Institute for 1812.

or pounding, performed by the strong muscular gizzard, assisted by stones which have been swallowed, and are retained on purpose. In other animals this trituration is accomplished by hard bony parts in the stomach, as among some of the crustacea. These expedients supply the deficiencies of imperfect mastication, and enable the gastric juice, and the other secretions, to become more intimately mixed with the food. What, then, are the changes which the food undergoes in the stomach?

Without attempting to give a history of the opinions which have been advanced on this subject, and expressed by the terms Concoction, Putrefaction, Fermentation, Dissolution, and Maceration, let us attend to the phenomena. The food, after having been mixed with the saliva and gastric juice, and exposed to the heat and motion of the stomach, in a short time begins to exhibit changes in its mechanical properties. That some changes likewise take place of a chemical nature appears obvious, from the circumstance that an acid is produced. This has been considered by some as the phosphoric, by others as the acetic or carbonic acid. This much seems to be determined that it is a volatile acid; as its effects on litmus paper are temporary. But the mechanical effects are the most remarkable. The appearance of the food by degrees changes, and the whole is reduced to a somewhat uniform and pultaceous mass, termed *chyme.* This mechanical change takes place from the surface to the centre of the contents of the stomach; and by means of the movements of the organ, the reduced portion is withdrawn as it is prepared, and passes out at the pyloric opening of the stomach.

After all this preparation, we might expect that the nutritious portion should now be withdrawn and enter the system. But other changes are still necessary. The pylorus opens into the *duodenum*, which constitutes the commencement of the intestinal canal. In this cavity the chyme is destined to be mixed with other secretions, before it be rendered fit to be taken up for the use of the circulation. These are the pancreatic juice and the bile.

The *pancreas* is a large conglomerate gland, bearing a very close resemblance in its texture and secretions to the salivary glands. In the different vertebral animals, it exhibits remarkable modifications of size, colour, consistence and form. It is destined to prepare the pancreatic juice. This liquor is collected in the gland by small radicles, which gradually unite, to form the excretory ducts. These last sometimes unite with the biliary ducts; in other instances they continue distinct, and pour their contents into the duodenum. In some animals the stomach itself receives the fluid. Sometimes there is but one opening, in other cases the ducts are as numerous as the lobes or the divisions of the gland. The excretion from the ducts is promoted by pressure and stimuli.

The nature of this juice has not been satisfactorily examined. It is, by some, considered as similar to the saliva; by others, as of the nature of the gastric juice. It likewise remains to be ascertained, what effeets are produced on the chyme by its presence, whether it acts chemically as a solvent, or, by assimilating it nearer to the nature of the blood, renders it more fit for being absorbed and mixed with that fluid. The other organ which is concerned in the process of chylification, is the

Liver.—This gland exists in all those animals which are furnished with a heart and circulating system. It varies greatly in form, size and subdivisons; but in all cases there is a very striking analogy in structure, colour and consistence. It receives an artery in the mammalia, termed the *hepatic;* and a vein which likewise terminates in its substance. This last circumstance is peculiar to this gland.

This vein is termed the *vena portarum*. It is formed from the union of the branches of nearly all the veins which arise from the abdominal viscera. Both these vessels spread their minute ramifications throughout the liver, and both their extreme divisions terminate in true veins. Whether, then, is the bile prepared from venous or arterial blood? In the mammalia, it is probable that while the latter serves for the nourishment of the liver, the former furnishes the materials for the formation of the bile. The peculiar termination of the vein, its superior size, when compared with the artery, appear to countenance such an opinion. On the other hand, instances have occurred, in which this vein did not terminate in the liver; consequently, the bile must have been derived solely from the hepatic artery. But these extreme cases do not enable us to determine the nature of the different parts of the organ in a sound state; although they furnish us with interesting displays of the compensating power of nature.

The bile is collected in the liver by very minute delicate vessels, which, by their union, form the *hepatic duct*. Sometimes, instead of forming a single duct, the bile is conveyed by several canals to the stomach or intestine. In some cases, the hepatic duct conveys the bile directly to its destination; in other cases, a portion of it is diverted into a particular receptacle, termed the *gall-bladder*. The structure of the coats of this reservoir is similar to that of the stomach. In shape and size this bladder varies according to the species. The canal by which it is filled, and by which it is likewise emptied, is termed the *cystic duct*. The canal which is formed by the union of the hepatic and cystic ducts, has been denominated *ductus communis choledochus*.

Let us now attend to the nature of the bile itself; a fluid which has long occupied the attention of the physiologist and chemist. THENARD considers bile as consisting, besides the saline ingredients and water, of a yellow matter, of a resin, and a substance which he terms *picromel*, white, solid, and soluble in water and alcohol. BERZELIUS considers bile as consisting of water; a *peculiar biliary matter*, of the nature of albumen, and of mucus of the gall-bladder dissolved in the bile, together with alkalis and salts, common to all the secreted fluids. In this peculiar biliary matter BERZELIUS could not detect any azote.

That different animals should furnish bile with properties depending on the food which they consumed, and consequently dissimilar in its nature, was reasonably to be expected. THENARD found the matter which he describes as picromel, to be present both in some herbivorous and carnivorous animals; while in others he could not detect its presence.

When the chyme has become mixed with the pancreatic juice and the bile, it enters into that state in which a part of it is fit to be absorbed for the nourishment and support of the system, the remaining portion being thrown out as excrement.

The intestines have been divided by anatomists into the smaller and larger. The smaller intestines, in the more perfect animals, consist, first, of the *duodenum*, into which the food passes from the stomach, and receives the two secreted fluids which we have now mentioned, and in which the food is changed into two parts, a nutritive and a useless. This part of the intestine is larger in its dimensions than that which immediately follows, and which is termed the *jejunum*, from the circumstance of its empty collapsed appearance. The remaining portion, termed *ilium*, is the longest and most convoluted. During the passage of the food through the small intestines, the chyle is separated by the mouths of the absorbing lacteals. In order to expose the food to a greater surface, and aid the action of the lacteals, the inner coat of the intestine is villous, and likewise covered with various rugæ, which have been termed *valvulæ conniventes*. The villi possess a vascular structure, and, when destitute of chyle, have been compared to little loose pendulous bags. It is this villous coat, which, when there is a separation of the intestines into large and small, characterise the latter. In the duodenum and jejunum, the chyme acquires the nature of chyle, and in the ilium the separation is completely effected between it and the fæces. These last consist of the useless part of the food, coloured by the union of a part of the bile, which is never reabsorbed. When the lacteals have finished the absorbing process, the fæces now leave the portion of the intestine destined for digestion, and enter the larger intestines, which are storehouses, in which the useless part of the food is placed until it can be conveniently evacuated. The larger intestines are characterised by their superior dimensions and strength, but more particularly by the absence of the villous coat.

The larger intestines are divided into three parts, the cœcum, colon, and rectum. The upper extremity of the *colon* is closed, and as the ilium enters laterally and a little way from the end, the head of the colon consequently forms a blind gut or *cæcum*. Sometimes there is attached to this caput cœcum coli, another tubular appendage closed at its free extremity, termed *appendix vermiformis*. The orifice of the ilium where it enters the colon is furnished with folds, which probably regulate the quantity of matter discharged into the colon, and prevent regurgitation. These folds constitute what is termed *valvula coli*. When the fæces have undergone all the changes which are necessary in the colon, and traversed its tortuous cavity, they enter a portion of the intestine, which, from the straightness of its course, is termed the *rectum*, which opens externally by the anus.

The inner surface of the intestines is lubricated by a mucous secretion, and the progress of the contents is accomplished by the contraction of the muscular coats, occasioning what is termed the *peristaltic motion* of the intestines.

During the passage of the contents of the intestines, various gases may be detected in different parts. Oxygen, azote, and hydrogen appear, together with carbonic acid, and even traces of sulphur. These may in part be derived from the air swallowed along with the food, and from the decomposition of the drink; and it is probable that the action of the bile on the chyme may, in certain cases, aid their production. But neither the experiments of MAGENDIE and CHEVREUL on the intestinal gases of man, nor those of VAUQUELIN on the elephant, lead to any satisfactory results.

The general view of the digestive process, and of the organs which are employed, which has now been given, is applicable only to the higher orders of animals. In many species the stomach is merely a bag with one orifice, serving for the entrance and ejection of the food. Here no gastric, pancreatic, or biliary juices appear necessary. In other species, the alimentary canal is open at both its extremities, but exceeding simple throughout, so that there is some difficulty in distinguishing between the stomach and intestines. In all animals, there must exist digestive organs, but nature here exhibits such an astonishing variety of forms and structures, as to baffle all attempts at generalization. Very remarkable differences prevail in stomachs destined to receive the same kind of food, and a striking similarity sometimes prevails between stomachs destined to receive different kinds of food. These irregularities will again occupy our attention.

Before concluding the remarks on this part of the subject, it is necessary to take notice of some parts which are more or less intimately connected with the digestive organs. The first of these is the

Spleen.—This organ is confined to the vertebral animals. In general it is single. In a few species, however, it is divided into several lobes. Its texture is vascular, consisting of a great assemblage of bloodvessels, united by a small quantity of cellular substance. It is covered externally by the peritoneum, beneath which there is another integument peculiar to it, which penetrates its substance along with the vessels. The arteries by which the spleen is supplied with blood, are, in general, branches from those of the stomach; and the veins in which these terminate, ultimately pour their contents into the liver. In position, the spleen is, in some cases, connected with the stomach; in others, more intimately related to the liver, but united to both by means of its bloodvessels.

The use of this organ in the animal economy remains to be ascertained. Though largely supplied with bloodvessels, it has no excretory duct, nor is its vascular, rather than glandular structure, favourable to the supposition that it is a secreting organ. Its lymphatic vessels are regarded by some as "nowise remarkable for their number or size *;" while others assert, that they "are both larger and more numerous than in any other organ †."

It seems to be generally admitted, that the blood of the splenic vein is more fluid, and coagulates with greater difficulty than the blood in the splenic artery. Therefore, some change must have taken place on the blood in the spleen. By this change, the blood is probably prepared for the use of the liver, to which it flows. When the spleen has been removed from a dog, the cystic bile has been

* Monro on Fishes, p. 37.

† Home's Lect. on Comp. Anat. vol. i. p. 235.

found pale and inert. These facts seem to warrant the conclusion, that the spleen is subservient to the functions of the liver, and probably supplies the deficiency of fluid which the branches of the vena portarum are unable to obtain from the abdominal viscera.

Sir EVERARD HOME at one time considered the spleen as the medium by which fluids, unnecessary for the process of digestion, were conveyed from the cardiac portion of the stomach, into the system, or into the urine to be excreted. But, in subsequent experiments which he performed, coloured fluids injected into the stomach, reached the urine, even when the spleen had been previously extirpated, and even after a ligature prevented the contents of the thoracic duct from being poured into the circulating system. Although he adopted this opinion hastily, with a praiseworthy candour he frankly published the demonstration of his error *. I have said hastily, because the well known circumstances of quadrupeds which do not drink, possessing a spleen equal in size to those which swallow large quantities of fluid, and the attachment of this organ to the first stomach of ruminating quadrupeds, into which the drink does not enter, furnished objections of great weight.

The next organ in connection with the digestive system, is the *Peritoneum*. This membrane lines the walls of the abdomen, and insinuates itself among the intestines, forming various duplicatures, distinguished by appropriate names. It is composed of cellular tissue, and contains numerous bloodvessels and absorbents. Its free surface is kept constantly moist by an aqueous exhalation. It varies in thickness and tenacity in different animals; and although transparent or whitish among quadrupeds and birds, is found variously coloured in reptiles and fishes. Its obvious use is to protect the intestines from rubbing against one an-

* HOME's Lect. on Comp. Anat. vol. i. p. 221,—245.

other, and to retain them in their proper position. For this purpose, its duplicatures enclose all the digestive organs I have already enumerated; and, from their peculiar position and contents, appear to serve some other end in the animal economy.

The *Omentum* is the most remarkable of these duplicatures or processes, and belongs only to the mammalia. It arises from the peritoneal covering of the stomach, and forms a bag which hangs from the stomach and liver, the other end being free in the cavity of the abdomen. Where it is in connection with the liver, it has been termed *omentum parvum* or *hepato-gastricum*. The walls of this sac are remarkable for containing fatty bands, disposed in a reticulated form. The quantity of fat depends on the condition of the animal. It appears to be deposited here by the system, as a storehouse of nourishment, and is quickly absorbed when food is scantily supplied. The omentum likewise protects the intestines from friction against the walls of the abdomen; and, according to some, contributes to retain their heat.

The *Mesentery*, which exists in all the vertebral animals, is the duplicature of peritoneum which invests the intestines, unites them to one another, and to the abdomen, and which serves as a support to the nerves, vessels, and glands which belong to them. That portion which is more intimately connected with the smaller intestines, is considered more particularly as the mesentery, while other portions receive peculiar appellations, as *Mesocolon*, and *Mesorectum*, from the parts of the large guts with which they are united.

Although the digestive organs furnish important and highly useful characters, in the discrimination of the habits of the species of certain genera, they can seldom be employed with advantage to distinguish any divisions of a higher kind.

CHAPTER XIII.

CIRCULATING SYSTEM.

By means of the digestive organs, the food is reduced to a pultaceous mass, and mixed with a variety of secreted fluids. In this state it is denominated *Chyme;* and, whether it owes its origin to vegetable or animal substances, exhibits, in its last condition, a chemical constitution nearly approaching that of blood, into which it is destined to be converted. In this stage of the process, however, it is necessary to effect a separation between that portion of the chyme which is fitted for the use of the system, and the other part, which is either superfluous or useless.

In giving an account of the circulating system, it is necessary that we attend to the state of the blood, and the fluids which enter into its formation, previous to its *aeration*; that we investigate the changes which it undergoes during this important process; and, lastly, consider the uses to which it is applied in what is considered its most perfect state.

The vessels in which the blood is collected previous to aeration, are with great propriety denominated *Pulmonic*, by Dr Barclay *. They have their origin in every part of the body, and terminate in the lungs, or other organs, in which respiration is performed.

The vessels which are destined to absorb the nutritious portion of the chyme, in the higher orders of animals, are termed *Lacteals*, from the milky appearance of the fluid which they contain, and which is denominated *Chyle*. These vessels take their rise in the villous coat of the intes-

* Anatomical Nomenclature, p. 176.

tines, in the form of extremely minute radicles, which uniting, form the trunks in the coats of the intestines. These trunks afterwards pass through the mesentery, occasionally become much convoluted, and form glands, liberally supplied with bloodvessels, and ultimately terminate in the *Thoracic Duct*. This duct, the commencement of which is usually termed *Receptaculum Chyli*, has slender walls, and internal valves, and empties its contents into the subclavian vein. In many quadrupeds, birds, and reptiles, there are two ducts, one on each side; but neither the size nor proportion of the parts appear uniform, especially in domesticated animals.

With regard to the nature of the chyle itself, little satisfactory information has been obtained. When procured from the thoracic duct of quadrupeds, and allowed to rest, a coagulum speedily separates, which has been compared to fibrin, and the serous fluid which remains contains a considerable proportion of coagulable albumen. When the number of secreted fluids which are mixed with the food previous to the separation of the chyle is considered, we need not be surprised at the near approach which it makes in composition to the blood, into which it is about to be changed. But does it undergo no change from its entrance into the lacteals, at the villous surface of the intestine, until it is poured into the thoracic duct?

Many physiologists are disposed to consider the mesenteric glands as destined to effect some change in the nature of the chyle, during the passage of the lacteals through them. This opinion has been supposed to receive considerable support from the observations of Mr ABERNETHY "on some particulars in the Anatomy of a Whale *." He found the mesenteric glands forming bags, on the inner

* Phil. Trans. vol. lxxxvi, p. 27.

surface of which, numerous arteries terminated, and veins originated. The orifices of these last were open. While some of the lacteals terminated in these bags by numerous orifices; others formed merely a plexus on their walls, and passed on with their contents to the thoracic duct. What becomes of the chyle after it enters the bag? To this question there is no satisfactory answer given, although it appears probable that it passes directly into the surrounding open mouths of the veins.

Besides the lacteals, there is another class of absorbents, termed *Lymphatics*, whose contained fluid is a transparent lymph. The office of these vessels is to collect the superfluous fluids from every part of the body, and to bring them back again to the general circulation, either to be renovated, or thrown out of the system as useless. Hence it is that the lymphatics take their rise in the skin, in the mucous web immediately underneath the cuticle, in the integuments of all the viscera, or wherever there is a part subject to increase and diminution during life. The radicles of the lymphatics unite into trunks, which are furnished with valves. They frequently anastomose, and enter conglobate glands. They either empty their contents into the thoracic duct, or pass on to one of the subclavian veins, in which they terminate.

Mr ABERNETHY, in the paper quoted above, has found reason to conclude, that some of the conglobate glands of the lymphatics of the horse are mere bags; while others appear to have a cellular structure. These glands are liberally supplied with arteries and veins. The effects which they produce upon the lymph have not been determined, although it is probable they are similar to those of the mesenteric glands. Mr BRACY CLARK has found the trunk of the lymphatic system of the horse to have several openings into the lumbar veins.

The structure of the lacteal and lymphatic vessels is similar. They are composed of two coats, an internal and external, the former being thin and smooth, the latter fibrous. The extent of the absorbing power, however, is not the same in both. The lacteals appear to be incapable of absorbing any thing else but chyle. M. MAGENDI *, administered to a dog diluted alcohol during digestion; and, although he was able to detect its presence in the blood, yet in the chyle no traces of it could be perceived. The experiments of Sir E. HOME, already quoted, and which have been repeated by M. MAGENDI †, (in reference to the use of the spleen,) lead to the same results. The absorbing power of the lymphatics is not so limited. Dispersed throughout the whole body, and destined to absorb occasionally every substance of which it consists,—fat, bone, the watery part of the bile, urine, and fæces, when too long retained,—they are likewise capable of taking up a variety of foreign substances, when brought into contact with the orifices of their roots. It is owing to the action of the lymphatics, that alcohol, camphor, rhubarb, and a variety of other substances thrown into the stomach, which the lacteals do not absorb, and different substances rubbed on the skin, are conveyed into the system. Attempts have been made to establish the opinion, that absorption of foreign matter takes place independent of either lacteals or lymphatics, since the blood exhibits proofs of the entrance of such into the system, which are not afforded by the contents of the thoracic duct. But the connection which has been pointed out between the lacteal and lymphatic vessels and the veins, independent of the thoracic duct, throws a considerable degree of doubt over the supposition.

The lacteals and lymphatics have been long known as the absorbents in the mammalia. HUNTER, HEWSON, and

* Précis Elementaire de Physiologie, vol. ii. p. 168. † Ibid. p. 182.

Monro, succeeded in demonstrating their existence in birds, reptiles, and fishes. Previous to these discoveries, the veins were supposed to perform the office of absorbents. In birds the chyle is transparent; so that the lacteals can only be distinguished from the lymphatics by their origin and office. There are no mesenteric glands. The lymphatics, however, possess numerous conglobate glands. The thoracic duct terminates in two branches. In reptiles, the thoracic duct is double as in birds, but neither the lacteals nor lymphatics appear to possess any glands. In the turtle these systems frequently unite, by anastomosing branches. In the crocodile, according to Hewson, the chyle is white. In fishes, the absorbing system is equally simple as in reptiles; but the vessels have no valves. In the lower animals, as the mollusca and annulosa, the existence of lacteals or lymphatics has not been demonstrated. The veins are considered as officiating in their stead.

Besides the lacteals and lymphatics which we have now been considering, there is another class of vessels, termed Veins, by means of which the pulmonic system completes its operations. These veins are connected by their radicles, with the extremities of the arteries, and probably also with the lymphatics. These radicles, by their union, form branches, which frequently anastomose. In the mammalia these branches unite into two trunks, termed *venæ cavæ*, and are either *superior* or *inferior*, according as they collect the blood from the *atlantal* or *sacral* extremities.

The structure of the veins has been investigated chiefly in the larger stems. The walls consist of three coats. The external one is composed of a dense cellular substance. The middle one is muscular, and consists of fibres, interwoven in all directions. The internal covering is very thin and smooth on its central surface. It assumes, in many places, the form of loose folds, which act the part of valves, in preventing the retrograde motion of the blood. This inner coat is found in all the veins; but the external ones cannot be detected, where the vessels penetrate bone, or are otherwise protected.

In many animals, the blood appears to be conveyed directly to the aerating organs, by means of the veins, without the intervention of any other apparatus. In other cases, especially among the higher classes of animals, it has to pass through a muscular enlargement or bag, termed the HEART, by which its motion is accelerated.

The heart is inclosed in a membranaceous sac, termed the *Pericardium*, of dimensions somewhat larger than its greatest expansions. In the walls of this sac, fibres, passing in different directions, may be readily perceived. It is kept moist internally, by a serous fluid, consisting chiefly of water, with a little albumen, mucus, and muriate of soda.

In the mammalia and birds, the heart consists of four cavities. Two of these are connected with the pulmonic, and two with the systemic vessels. With the former our attention is at present occupied. The venæ cavæ empty their contents into what is termed the *Pulmonic Auricle*; or, in reference to its position in man, the *Right Auricle*. The blood passes directly from the auricle, by a valvular orifice, into the pulmonic or right ventricle, and from thence by a vessel termed the *Pulmonary Artery*, to the aerating organs. The structure of the auricle and ventricle exhibits a dense muscular substance, consisting of fibres intimately interwoven, and liberally supplied by bloodvessels and nerves, lined with a smooth membrane. Hence we may regard the heart merely as an enlargement of the veins, and its walls endowed with corresponding strength by the increase of the thickness of the muscular coat. The walls of the auricle are much thinner than those of the ventricle. The former bag has merely to convey the blood to the ventricle; while this last, by a contractile movement, propels it at intervals through the pulmonic artery, to its remotest subdivisions. When the auricle relaxes, it receives the blood from the venæ cavæ; and, when it contracts, the ventricle relaxes to receive its contents. When the ventricle contracts, the blood is thrown into the artery, and suddenly enlarges its dimensions. These motions, which are sensible to the eye and touch, have been denominated the Pulsation, or beating of the heart and arteries; those vessels being termed Arteries into which the ventricles discharge themselves.

The number of pulsations of the heart, in a given time, varies according to the species, and in the same individual, according to circumstances connected with food, exercise, or the state of the mind. The same individual, likewise, exhibits remarkable variations in this respect, according to age*.

Climate, likewise, is supposed to exercise a considerable influence on the frequency of the pulsations, their number being greatest in warm climates.

Before proceeding to consider the aeration of the blood, it is necessary to make some inquiry into its *mechanical structure*, and chemical constitution. When blood, newly taken from an animal, is sufficiently diluted, and placed on the stage of the microscope, it is observed to consist of a

* BLUMENBACH found the pulsations of the heart of a new born infant, while placidly sleeping, amount to 140 in a minute.

Towards the end of the first year, about	124
———————— second ————	110
———————— third and fourth,	96
When the first teeth begin to drop out,	86
At puberty, - - -	80
At manhood, - - -	75
About sixty, - - -	60

In those more advanced, scarcely two were found alike.—The Institutions of Physiology, p. 58.

transparent, nearly colourless fluid, surrounding particles of a denser substance, and highly coloured. The enveloping fluid is called *Serum*, and the bodies which are suspended in it the *Particles*, or *Globules* of the blood.

A great difference of opinion has prevailed among naturalists, with regard to the form and structure of the coloured particles of the blood. Although making use of the same kind of instrument, and professing to abide by the evidence of the senses, their statements exhibit very remarkable differences. ADAMS considered them as ramified, and resembling the branches of a tree *. Father DI TORRE, with a spherical lens, which magnified 512 diameters, regarded them as oblate spheroids, much compressed, with the middle part much darker than the margin. With a magnifying power of 1280 diameters, he considered them as perforated in the centre, and the surrounding ring as composed of joints from two to seven in number †. This jointed annular structure had been remarked half a century before, by LEEUENHOEK, and a figure of the appearance given ‡. HEWSON, who appears to have conducted his observations with caution, and used lenses which magnified the diameter of objects from 184 to 1280 times, considered the particles of blood to be " as flat as a guinea." The dark spot in the middle, which other observers had perceived, he considered " as a solid particle contained in a flat vesicle, whose mid-

* " Human blood is so far from shewing any Red Globules swimming in Serum, that immediately after its Emission, it appears to be a Body of infinite Branches, running in no certain order, variously coloured : where it lies thickest on the glass, it's of a Red, where thin, inclining to Yellow ; but the whole so blended, as to represent, very near, the top of a Yew-tree in a very fine landskip, having its supposed Branches of a Red and Yellow confusedly intermixt."— Phil. Trans. vol. xxvii. p. 26.

† Phil. Trans. vol. lv. p. 255.

‡ Opera Omnia, vol. iii. p. 221. fig. 4. I. K.

dle only it fills, and whose edges are hollow, and either empty, or filled with a subtile fluid *." The form of the particles in some animals, he found to be oval, as LEEUENHOEK had previously observed, and in others circular. YOUNG admits, that " their axis is sometimes not more than one-third or one-fourth of their greatest diameter," and that they resemble a soft substance with a denser nucleus, not altogether unlike the crystalline lens, together with the vitreous humour, as seen from behind †." AMICI has found them of two kinds, both with angular margins; but, in the one, the centre is depressed on both sides; while, in the other, it is elevated ‡.

Independent of these discrepancies in the results of different observers, it appears to be the general opinion at present, that the particles of the blood consist of a dense nucleus, surrounded by a film of matter of less consistence; and that the particles of the blood of some animals, as man, are circular; while, in others, as the skate, they are oval ||.

The *size* of these particles of the blood has not been satisfactorily determined, as considerable difference prevails in the result of their measurement, even when the blood of the same animal has been subjected to the micrometer by different observers. The measurements of KATER, WOLLASTON, YOUNG, and BAUER, give the diameter of each particle a range from $\frac{1}{1700}$th part of an inch, to $\frac{1}{8000}$th. From the testimony of several observers, it appears, that, in the same blood, the particles are not all of equal size, and that the particles of the blood of different animals exhibit different magnitudes. In this last case, there is no proportion

* Phil. Trans. vol. lxiii. p. 310.

† Annals of Philosophy, vol. ii. p. 116.

‡ Edin. Med. and Surg. Journ. vol. xv. p. 120.

|| Mr HEWSON has given figures of the particles of the blood in a variety of different animals, in Phil. Trans. lxiii. tab. xii.

observed between the size of these particles, and that of the body of the animal. The particles of human blood are larger than those of the ox; while those of the ox are only equal to those of the mouse. Age appears to exercise some influence on their size. Mr HEWSON found in a chicken, on the sixth day of incubation, the particles larger than in a full grown hen; and also larger in the blood of a very young viper than in that of its mother, out of whose belly it was taken. He could not, however, detect any difference in size between those of a child at its birth, and those of an adult person *.

Every one is acquainted with the red *colour* of the blood of the vertebral animals. The colouring matter is not equally diffused, but adheres to the particles, which it surrounds with a thin film. Mr BAUER found an entire particle to be $\frac{1}{1700}$th part of an inch in diameter, but when deprived of its colouring matter, only $\frac{1}{2000}$th part. While the blood of some animals, as all the vertebrosa, is uniformly of a red colour; in many other animals it exhibits a different shade. Among the crustaceous animals, as the lobster and shrimp, it is *white*; while, among some insects, as the grasshopper and white caterpillar, it is green †.

When blood, newly drawn from an animal, is allowed to rest, it thickens, and spontaneously separates into two parts, a pale coloured fluid termed *Serum*, and a denser coagulum, termed *Clot*, *Cruor*, or *Crassamentum*. The serum is the fluid basis of the blood; the *clot* is formed from the floating particles; and hence the difference of colour and consistence exhibited by the two portions.

During the act of coagulation, a sensible quantity of heat is evolved, amounting to about 6° Fahr. This appears

---

* Phil. Trans. vol. lxiii. p. 321.

† HEWSON, Phil. Trans. vol. lxiii. p. 321.

to have been first stated by FOURCROY, and afterwards established by the author of the article Blood, in REE's Cyclopædia, and by the late Dr GORDON of Edinburgh *.

Previous to coagulation, if the atmospheric pressure be removed from the blood in an air-pump, a considerable quantity of air bubbles is disengaged, as was first clearly established by the experiment of DARWIN †. VOGEL determined this air to be carbonic acid ‡; and BRANDE found that two cubic inches were disengaged from every ounce of blood, and that the quantity was the same in arterial and venous blood ||. Even during spontaneous coagulation, Mr BAUER found the blood, when confined in a glass tube, to give out a considerable quantity of carbonic acid §. It appears, likewise, from his observations, in company with Sir EVERARD HOME, that when a drop of blood is placed on a watch-glass, in the field of a microscope, the following changes may be perceived. "The first thing," (he says,) "that happened, was the formation of a film on the surface, that part beginning to coagulate sooner than the rest. In about five minutes, something was seen to be disengaged in different parts of the coagulum, beginning to shew itself where the greatest number of globules were collected; and, from thence, passing in every direction, with considerable rapidity, through the serum, but not at all interfering with the globules themselves, which had all discharged their colouring matter; wherever this extricated matter was carried, a net-work immediately formed, anastomosing with itself, on every side, through every part of the coagulum ¶."

Are we to conclude from these statements, that carbonic acid exists in a gaseous state in the blood? The experi-

* Annals of Phil. vol. iv. p. 139.

† Phil. Trans. 1774, p. 344.

‡ Annals of Phil. vol. vii. p. 56.

|| Phil. Trans. 1818, p. 181.

§ Ibid.

¶ Ibid. p. 182.

ments of DARWIN, already alluded to, demonstrate, that there is no air in the blood while confined in the proper vessels, and render it probable that the air which appears, when the blood is placed under the exhausted receiver of an air-pump, has been mixed with the blood upon leaving the vessels and passing through the atmosphere *.

Chemical Constitution.—The serum from which the clot has been removed, is more fluid than the blood itself, but retains its taste and smell. It is slightly alkaline, and consequently renders vegetable blues green. When heated to the temperature of about 160°, it coagulates, and if this coagulum be cut into thin slices, a fluid oozes out, which is termed *Serosity*. The coagulum has all the properties of coagulated albumen. The serosity still contains a considerable portion of albumen, whose properties are much modified by the salts with which it is combined. The following are the contents of the serum, as stated by different chemists:—

	MARCET.	BERZELIUS.
Water, - - -	900.0	905.0
Albumen, - -	86.8	80.0
Muriate of potash and soda, - -	6.6	6.0
Subcarbonate of soda, - - -	1.65	
Sulphate of potash, - - -	0.35	
Earthy phosphate, - -	0.60	
Lactate of soda united with animal matter,		4.0
Soda, phosphate of soda, and a little animal matter,		4.1

* The following, among many of his experiments, may be quoted:—"A part of the jugular vein of a sheep, with the blood in it, was included between two strict ligatures, during the animal's being alive; and, being cut out with the ligatures, was immediately put into a glass of warm water, and placed in the receiver of an air-pump. It sunk to the bottom of the water, and would not rise when the air was diligently exhausted. It was then wiped dry, and laid on the brass floor of the receiver, and the air again exhausted; but there was not the least visible expansion of the vein, or its contents."—Phil. Trans. 1774, p. 345. Similar experiments, with the same results, were repeated on the blood in the vena cava inferior of a large swine.

When the coagulated serum is heated in a silver vessel, the surface of the metal is blackened, being converted into sulphuret. This is the proof brought forward, that the serum contains sulphur; but the previous demonstration of the existence of albumen rendered it unnecessary, since into that substance sulphur appears to enter as an ingredient of its constitution.

Clot.—The clot, or crassamentum, as we have already stated, consists of the coloured particles which were suspended in the serum, and with which a portion is still mechanically mixed. If upon the clot, (in a great measure freed from the serum, by cutting it into thin slices, and pressing it upon blotting paper,) cold water be poured, the colouring matter is removed, and the portion which remains exhibits the properties of fibrin.

The colouring matter is not dissolved by the water, but merely mechanically suspended, and gradually subsides when allowed to rest. When thus mechanically diffused, the colouring matter coagulates by heat, and may readily be obtained upon a filter. This matter appears to be a substance intermediate between fibrin and coagulated albumen, but distinguishable from both, by the colour which it retains, and the salts which it yields by incineration *.

LEMERY, and several of the earlier chemists, discovered iron in the blood. SAGE and GREEN considered that it existed in combination with phosphoric acid. FOURCROY

* BERZELIUS found that 400 grains of colouring matter, when burnt, yielded 5 grains of ashes, composed of

Oxide of Iron, - - -	50.0
Sub-phosphate of iron, - - -	7.5
Phosphate of lime, with a small quantity of magnesia,	6.0
Pure lime, - - - -	20.0
Carbonic acid, and loss, - -	16.5
	100.0.

Annals of Phil. vol. ii. p. 197.

and VAUQUELIN regarded it in the state of a sub-phosphate, and as giving colour to red blood. This opinion was, for a considerable time, servilely embraced by chemists, but is now as generally abandoned. It was first successfully attacked by Dr WELLS, in his "Observations and Experiments on the Colour of Blood *." Subsequent experimenters have fully confirmed his views, particularly BERZELIUS †, BRANDE, and latterly VAUQUELIN ‡, and have greatly extended our knowledge of the habitudes of this substance with acids and alkalis. Mr BRANDE has even succeeded in dying cloth with it, but he found considerable difficulty in fixing the colour. The most effectual mordants he discovered, were the nitrate and oxymuriate of mercury §. No experiments have hitherto been performed on the colouring matter, or mechanical or chemical

---

* As his remarks have not attracted the attention they deserve, I shall here state his arguments in his own words :—

"1. I know of no colour, arising from a metal, which can be permanently destroyed by exposing its subject, in a close vessel, to a heat less than boiling water. But this happens with respect to the colour of the blood.

"2. If the colour from a metal, in any substance, be destroyed by an alkali, it may be restored by the immediate addition of an acid; and the like will happen from the addition of a proper quantity of alkali, if the colour has been destroyed by an acid. The colour of the blood, on the contrary, when once destroyed, either by an acid or an alkali, can never be brought back.

"3. If iron be the cause of the red colour in blood, it must exist there in a saline state, since the red matter is soluble in water. The substances, therefore, which detect almost the smallest quantity of iron in such a state, ought likewise to demonstrate its presence in blood; but, upon adding Prussian alkali, and an infusion of galls, to a very saturate solution of the red matter, I could not observe, in the former case, the slightest blue precipitate; or, in the latter, that the mixture had acquired the least blue or purple tint."—Phil. Trans. vol. lxxxvii, p. 429.

† Annals of Phil. vol. ii. p. 24. ‡ Ib. vol. vii. p. 230.

§ Phil. Trans. 102. p. 110.

constitution of the white blood of the inferior animals. In nearly all the experiments which have been performed on the blood, that fluid has been obtained from man, or the more common domestic quadrupeds.

We come now to consider the changes produced on the blood by the *aerating organs.*

We have already stated the necessity of a constant supply of atmospheric air to the continuance of the life of organised beings. We are here to consider this supply in connection with the circulation of the vital fluid.

The aerating organs of animals may be divided into two kinds, *Lungs* (Pulmones), and *Gills* (Bronchiæ), both destined to accomplish the same end. The lungs are suited for bringing free air into contact with the blood, and therefore belong to those animals which have their residence on the land. The gills are calculated to separate air from water, with which it is always united, and bring it in contact with the blood, and belong therefore to those animals which reside in the sea or in fresh water. It is to be observed, however, that many animals which reside in the water, breathe by means of lungs, and are obliged, at intervals, to come to the surface to respire, such as whales; but there are no animals which reside on the land and are furnished with gills which are obliged to return to the water to respire.

Whether the aerating organs be lungs or gills, it appears to be the object of nature in their construction to expose a large surface to the contact of the air. This object is accomplished by their division into numerous cells and leaf-like processes, or by their extension on the walls of cavities, or the surface of pectinated ridges. The blood brought to these organs by the pulmonic vessels, is there distributed by their terminating branches. Although still retained in vessels, it can nevertheless be easily acted upon by the air on the exterior. PRIESTLEY found the colour of blood changed by the air, when enclosed in a moistened bladder, and the same effect was observed by HUNTER, when it was covered with goldbeaters' skin. Need we be surprised, then, at the air having the necessary access to the blood in the lungs or gills, since these organs have been constructed for the particular purpose?

In those animals which possess cellular lungs, and which belong chiefly to quadrupeds and birds, the air is conveyed to them by means of a tube, termed the Windpipe, Trachea or Aspera arteria. This tube is composed of annular cartilages, united by a ligamentous elastic substance. On its peripheral surface, it is invested with a strong membrane, consisting of very distinct longitudinal fibres. The central surface is covered with a thin, delicate, extremely irritable membrane, which is kept continually moist by a mucous liquor which exhales from it.

Where the windpipe terminates in the lungs, it subdivides into two or more branches, termed *Bronchiæ*, which, by farther subdivision, at last terminate in the larger, and these again in the smaller cells of the lungs.

The upper extremity of the windpipe, or glottis, terminates in the pharynx by a peculiar arrangement of cartilages, denominated the *Larynx*. These are moveable, and connected together by membranes, which suffer them to vary their position. The summit of the windpipe ends in a broad annular cartilage, termed the *Cricoid*, on which the others rest. Immediately above this, on the lateral and sternal sides, is the broad angular cartilage, termed the *Thyroid*, or *Pomum Adami*, Adam's Apple, from an absurd allusion to the first transgression of our first parent. Two processes, termed horns, connect this cartilage with the bones of the tongue.

On the dorsal side of the thyroid cartilage, are two other cartilages, termed *Arytenoid.* To the side of each of these cartilages, and joined to the internal angle of the thyroid, there is attached a broad fibrous band, which forms the margin of the opening into the windpipe or *rima glottidis*.

This singular structure of the larynx is not necessary to respiration, but is subservient to the purposes of voice or speech. The air that is expelled from the lungs being acted upon by the variations in the form and tension of the rima glottidis, produces those various sounds by which different species are distinguished. The precise manner in which these sounds are formed, does not appear to be established. By different authors it has been compared to a drum, a violin, a flute, and an eolian harp.

Several glandular bodies are connected with the larynx, whose functions have not yet been satisfactorily explained. The largest of these is the thyroid gland, so called, because it is in part situated on the thyroid cartilage. It consists of two lobes, which descend on each side the windpipe. Two glands are likewise seated at the base of the arytenoid cartilages, which appear to secrete the fluid with which the inner surface of the larynx is moistened. Another glandular body at the root of the epiglottis, appears to secrete a fatty substance.

In consequence of the liberal supply of serous fluid to the surface of the larynx and windpipe, and, we may add, the cells of the lungs, the air which is returned from the lungs is always found to be loaded with moisture. In man, the quantity given out during twenty-four hours has been rated by some as low as six ounces, and by others as high as twenty.

In order to ascertain the changes which the blood undergoes when thus exposed to the influence of the air, it will be necessary to attend, in the first place, to the changes produced in the air itself.

In man, regarding whose respiration the greatest number of experiments have been performed, it has been observed, that the air which is alternately inspired and ejected, becomes unfit for future use; and is likewise rendered incapable of supporting combustion. The analysis of this altered air indicates the change to have taken place in its oxygenous portion *. A part thereof has disappeared, and an equal bulk of carbonic acid is found occupying its place. The quantity of oxygen in this carbonic acid is equal to that which has been abstracted from the air. In this case, either carbonic acid escapes from the blood, and an equivalent bulk of oxygen is absorbed; or, the blood furnishes the carbon only, with which the oxygen of the air unites. The former supposition was long countenanced by chemists; the latter is at present the prevailing opinion. In the adoption of the former, many difficulties present themselves. There is no apparent cause to produce the expulsion of the carbonic acid, as there is no substance in the blood with which it could be combined, and from which it could so easily escape; neither is there the slightest reason to suppose, that it could be displaced by the same bulk of oxygen, since these two gases have very different combining values. But when it is considered, that, in all cases where carbon unites with oxygen, the carbonic acid produced is equal in bulk to the oxygen consumed, we are

* There is no reason to believe that any change is produced on the azote. In some experiments on the respiration of fishes, by HUMBOLDT and PROVENÇAL, a loss of azote was indicated. But the sources of error in performing such experiments are so many, and the results which were obtained differ so much from one another, that no confidence can be placed in the conclusions which have been drawn from them.

led to adopt the second opinion, and to conclude, that, in the process of aeration, the blood parts with a portion of carbon, and that its carbon combines with the oxygen of the air inspired, and passes into the state of carbonic acid. The changes which the blood itself undergoes, support the conclusion.

In man and the red-blooded animals, the pulmonic or venous blood changes its *colour* in the lungs, and passes from a dark to a florid red. That this effect is produced by the formation of the carbonic acid, is demonstrated from the circumstance, that pulmonic blood exposed to the air of the atmosphere acquires the florid colour, that part of its oxygen disappears, and an equivalent bulk of carbonic acid occupies its place. When the blood is exposed to azotic gas, no change takes place, but when oxygen is substituted, the red colour appears, a portion of the gas is consumed, and a similar quantity of carbonic acid formed.

But not only is the colour of the blood changed, but likewise its *density*. The specific gravity of venous blood obtained from a sheep, was found, by Dr John Davy, to be 1051, while the aerated or arterial blood was 1049. It would perhaps be rash to found any reasoning with regard to the proof that carbon is emitted, from this change of density in the blood, owing to the insufficiency of the experiments which have been performed, to serve as standards of comparison; at the same time, it would be easy to determine the difference in density between the venous and aerated blood, if the change consists merely in the emission of carbonic acid, and the absorption of a corresponding bulk of oxygen.

Besides the change of colour and density experienced by the blood in the process of aeration, an alteration likewise takes place it its *specific heat*. Dr Crawford found the specific heat of venous blood (water 1.000) to be 0.8928, while aerated blood was 1.0300. This circumstance was considered as furnishing a strong support to a theory of animal heat, long in vogue, and which we shall notice afterwards. Dr John Davy, however, in the course of his experiments, obtained results which differed from the foregoing. He found the specific heat of venous blood to be 0.903, while that of aerated blood was 0.913. Both these results indicate a diminution in the specific heat to have taken place. It may be proper, however, to state, that it is very difficult, if not impracticable, to obtain accurate results in such attempts. Whether the blood is poured into an equal bulk of another fluid, whose specific heat is known, for the purpose of ascertaining the resulting temperature of the mixture, or its rate of cooling be employed, there are changes which begin to take place in its constitution from the moment it leaves the bloodvessels, which must affect the accuracy of the comparison. The experiments of Darwin, already quoted, prove, that the blood is altered in the course of flowing through the air from the veins into the cup in which it is received.

The *quantity* of oxygen consumed by animals in a given time, is variable, not only in regard to species and individuals, but in the same individuals in different circumstances. In man, the quantity of oxygen consumed in a minute, has been differently rated by chemists. Allen and Pepys found it to be 26.6 cubic inches in a minute, Davy 31.6, and Murray 36. The quantity, however, is found to vary under the following states of the system.

Muscular exertion appears to increase the consumption of oxygen, according to Seguin, nearly fourfold beyond the usual quantity. Dr Prout *, who has examined this

---

* Annals of Phil. ii. p. 328—343, and iv. p. 331—337.

subject with great attention, was led from his experiments to conclude, that moderate exercise increased the consumption of oxygen, but if continued so as to induce fatigue, it occasioned diminution. The exhilarating passions appear likewise to increase the quantity, probably by exciting muscular action. On the other hand, the depressing passions and sleep, alcohol *, and tea, diminish the quantity.

Dr PROUT likewise found, that the quantity of oxygen consumed, is not uniformly the same during the twenty-four hours; but is always greater at one and the same part of the day than at any other. He determined, that its maximum occurs between 10. a. m. and 2. p. m., or generally between 11. a. m. and 1. p. m., and that its minimum commences about 8^h $30'$ p. m., and continues nearly uniform till about 3^h $30'$ a. m. He is inclined "to believe, that the presence and absence of the sun alone regulate these variations." Here it may be observed, that in diurnal animals, the forenoon, during which it appears that the consumption of oxygen is at its maximum, is the period of activity or muscular exertion; while in the afternoon, in which the con-

* Dr PROUT found that "alcohol, in every state and in every quantity, uniformly lessens, in a greater or less degree, the quantity of carbonic acid elicited, according to the quantity and circumstances under which it is taken." This result is certainly the reverse of what might have been expected, considering the temporary exhilarating effects of spirituous liquors. That the quantity should be found diminished, after the exhilarating effects of the alcohol had ceased, and the consequent depression had taken place was precisely what might have been expected, and which the experiments of Dr FIFE realize. "On the 8th of June, (says Dr FIFE) a much greater quantity of wine than usual was taken, and the next day the quantity of carbonic acid was repeatedly found as low as above stated, (5.75 *per cent.*) On making the experiment again with less wine, the quantity of carbonic acid was considerably reduced, though not so much as before," (only between 2 and 3 *per cent.*) ib. iv. 335. This diminished state certainly indicates a previous excess. The apparent discordance of the results of these experimenters probably arises from peculiarities of habit or constitution.

sumption of oxygen diminishes, the body begins to experience fatigue. With the nocturnal animals, this arrangement is probably reversed, to correspond with their season of activity.

The influence of temperature in increasing or diminishing the consumption of oxygen is likewise considerable. CRAWFORD found that a Guinea-pig confined in air at the temperature of 55°, consumed double the quantity which it did, when placed in air at 104°. He likewise found, in such cases, that the venous blood, when the body was exposed to a high temperature, had not its usual dark colour; but, by its florid hue, indicated that no change had taken place in its constitution, in the course of circulation. When the temperature of warm-blooded animals is greatly augmented, exertion becomes laborious, a great degree of lassitude is speedily induced, and a condition of the system is produced, similar to that which follows great muscular efforts. In the cold-blooded animals, on the other hand, whose exertions are so much under the influence of temperature, that they become torpid when cooled below a certain degree, the effect of a moderate degree of heat will be to increase muscular action, and a corresponding consumption of oxygen. The experiments of SPALLANZANI and others, are in conformity with such suppositions.

Upon a review of the different circumstances which have been stated, as influencing the consumption of oxygen, it appears obvious, that it keeps pace with the degree of muscular action, and is dependent upon it, consequently, a state of increased consumption is always followed by an equally great decrease, in the same manner as activity is followed by fatigue. Yawning and drowsiness indicate muscular exhaustion, and they likewise indicate a decreased consumption of oxygen.

This consumption of oxygen is the index of the quantity of carbon which is thrown out by the system. In man, it amounts nearly to half an ounce every hour. Experiments are still wanting to determine the relation of the quantity to the bulk of the animal, or the amount of the food which is consumed.

It now remains that we trace the progress of the aerated blood in the *Systemic vessels* to its final destination. The radicles of the systemic vessels take their rise in the aerating organs, and receive from the terminating twigs of the pulmonic vessels, the blood which has parted with its carbon, and suffered the changes consequent upon the separation of the superfluous portion of that ingredient. These radicles unite into branches, and either transmit the blood directly to the different parts of the body, or suffer muscular enlargements, forming a *systemic heart.* In quadrupeds, the aerated blood is collected from the lungs by the radicles of the systemic veins: these, by their union, form four trunks, which proceed to their common sinus, and, through it, pour their contents into the systemic or left auricle. The blood now enters the systemic ventricle, and, by means of its contractions, is sent into the systemic artery or *Aorta,* of which it may be considered as an expansion. Through the subdivisions of the aorta, the blood is conveyed to all the parts of the body.

The structure of the systemic heart is similar to that of the pulmonic, with this difference, that its walls are much stronger, and, consequently, are fitted for exerting the requisite force in order to propel the blood through the arteries to the remotest parts of the system *.

* It appears probable, that the arteries themselves, by the contraction of their coats, serve to promote the circulation of the blood, and that these contractions depend on the nerves with which they are supplied. See a paper by Sir E. HOME, "*On the Influence of the Nerves upon the Action of the Arteries.*" Phil. Trans. 1814, p. 583.

Although these two hearts, the pulmonic and systemic, are united in quadrupeds and birds, it is otherwise with many genera of molluscous animals, in which they are removed to a considerable distance from each other, and exhibit apart the functions of the two systems of circulating vessels.

The arteries observe usually a very tortuous course. In dividing into branches, there may, in general, be observed, the continuation of a principal stem. These branches frequently communicate with each other, or anastomose. The number of branches, or the quantity of blood which they convey to any particular part, may be considered as proportional to the quantity of action which it performs, whatever be the structure of the organ. Mr CARLISLE* has observed, that in the limbs of slow moving quadrupeds, the trunk of the artery subdivides into a number of parallel branches, the sum of whose areas is much greater than the trunk from which they have proceeded. By this arrangement, the rapidity of the circulation must be diminished, probably to make it correspond with the slow but continued exertion of the muscles, to which such parallel arteries are distributed.

The arteries terminate in those minute twigs which have been termed *capillary vessels*. In these, the blood flows in such minute quantity, and their coats are so thin and transparent, that they appear colourless. These capillary vessels open into the extremities or radicles of the vein, so as to form a continuous circulation. This, however, is not universally the case. For these capillaries, in some cases, open into cavities, into which they pour a watery liquor. When exercising this function, of whose mechanical or chemical arrangements we know nothing, they have been

---

* Phil. Trans. 1800, p. 98, and 1804, p. 17.

termed *exhaling* or *secerning vessels.* In other cases, they terminate in glands, whose office it is to separate some ingredients from the blood, for the benefit of the system, or to be thrown out as useless. It is in these terminating twigs of the arteries or the capillary vessels, that the changes take place in the blood, by which its qualities are altered, and it passes into venous blood.

In the process of digestion, the food is mixed with a variety of secreted fluids, by which it is gradually prepared for the action of the absorbing vessels or lacteals. These separate the nutritious portion, and convey it to a particular receptacle. Another set of absorbents, the lymphatics, take up all the substances which have been ejected from the circulation, and which are no longer necessary in the particular organs, and communicate their contents to the store already provided by the lacteals. The veins receive the altered blood from the extremities of the arteries, or the glands in which they terminate and proceed with it towards the lungs to be again aerated. In their progress, they obtain the collected fluid of the other absorbents, and, in the lungs, again prepare the whole for the use of the system. Thus, during the continuance of life, the arteries supply the materials by which the system is invigorated and enlarged, and oppose that tendency to decay, produced by the influence of external objects. The process continues during the whole of life, new matter is daily added, while part of the old and useless is abstracted. The addition is greatest in early life, the abstraction is greatest in old age.

This continued system of addition and abstraction has led some to conclude, that a change in the corporeal identity of the body takes place repeatedly during the continuance of life, that none of the particles of which it consisted in youth, remain in its composition in old age. Some have considered the change effected every three, others every seven years. This opinion, however, is rendered doubtful by many well known facts. Letters marked on the skin by a variety of substances, frequently last for life. There are some diseases, such as small-pox and measles, of which the constitution is only *once* susceptible; but it is observed to be liable to the attack of these diseases, at every period of human life.

Urinary System.

We have been induced to add this section as an appendix to our observations on circulation, from the persuasion, that the urine is an excrementitious fluid separated from the blood. The glands which are employed for this purpose, are termed *kidneys*.

These organs, as they exist in quadrupeds, are two in number, one on each side of the spine, at the upper part of the loins. Each kidney is made up of numerous lobes, which are more or less intimately united according to the species. They are situated behind the peritonæum, and surrounded by a peculiar vascular membrane. They consist of two parts, an exterior, termed *cortex*, and an interior *medulla*. In the cortical part, the urine is secreted, it then enters into conical shaped bodies in the medullary part. These cones have their base towards the surface of the kidney, and their apex towards the concavity of its central side. They consist of hollow tubes, which convey the urine into the great cavity of the kidney, termed its *pelvis*. From this pelvis or receptacle, a tubular irregular vessel, termed the *ureter*, conveys the urine to the bladder. The kidneys are supplied with blood from the aorta. The veins return it to the vena cava inferior. Their nervous energy is derived from the great sympathetic nerve.

The bladder consists principally of two coats. The external is muscular, and consists of fibres which are variously decussated. The internal is a serous membrane, which secretes a lubricating fluid, to protect the organ from the action of the contained fluid. The ureters enter the bladder towards its mouth on the dorsal side, passing through its coats in an indirect and tortuous manner. The bladder itself is retained in its place, partly by the folds of the peritoneum, and partly by its own ligaments. These either arise from the neck of the bladder, and are attached to the pubis, or from its fundus, constituting the *urachus*. The urine, at stated intervals, is expelled from the bladder, through a canal termed the *urethra*, which accompanies the reproductive organs.

The modifications of this system in the inferior classes of animals, are numerous. In none of them do the kidneys in their structure appear to consist of two parts. The ureters, while they pour their contents into a bladder of urine, in some reptiles and fishes, do not, in others, terminate in any common re ceptacle. In birds, in general, and many reptiles and fishes, the urine, before expulsion from the body, is mixed with the excrement, while in many fishes, it either passes out by a peculiar opening, or in a common passage with the melt or spawn. Nothing analogous to urinary organs has yet been detected in the mollusca or annulose animals, although in the dung of the caterpillars of several insects, traces of the peculiar principle of urine, or urea, have been detected.

There is a singular compound body adhering to the upper part of the kidney in quadrupeds, termed the *renal gland*, whose use is unknown. It is of a yellowish colour and firm consistence, and frequently contains a cavity filled with a dark serous fluid. In the fœtus, it is equal in size to the kidneys, but in the adult, it is about one-fifth less.

The *urine* itself has been repeatedly examined by different chemists, and a variety of products obtained from its analysis. The colour of urine is usually a pale yellow, and its specific gravity rather higher than water. It has a peculiar odour. When first discharged, it is sensibly acid, readily reddening vegetable blues, but in a short time it suffers decomposition, ammonia is evolved, the excess of acid is neutralised, and alkaline properties predominate. Such is the case with the urine of man and many quadrupeds, but VAUQUELIN found that of the lion and tiger alkaline, owing to the excess of ammonia*. It is not determined which of the acids existing in urine, are in an uncombined state,—the uric, phosphoric, acetic, lactic, and carbonic acids having been successively referred to, as occasioning the change in vegetable blues †.

In the urine of different species of quadrupeds, nearly

---

* BERZELEUS lays it down as a rule, (Annals of Phil. ii. 206,) that all the excreted fluids, such as urine, are acid. The alkaline nature of the urine of the lion and tiger, however, furnish a striking exception.

† The following are the ingredients contained in a 1000 parts of urine:

Water,	933.00
Urea,	30.10
Sulphat of potash,	3.71
Sulphat of soda,	3.16
Phosphat of soda,	2.94
Muriat of soda,	4.45
Phosphat of ammonia,	1.65
Muriat of ammonia,	1.50
Free lactic acid,—lactat of ammonia,—animal matter soluble in alcohol, and usually accompanying the lactat,—animal matter insoluble in alcohol,—urea not separable from the preceding,	17.14
Earthy phosphate, with a trace of fluat of lime,	1.00
Uric acid,	1.00
Mucus of the bladder,	0.32
Silex,	0.03
	1000.00

BERZELIUS,—Annals of Phil. ii. p. 423.

the same ingredients are found as in that of man In all, urea exists, but the nature and proportion of the acids, alkalis, and earths which are present, are somewhat different. Thus, in the gramenivorous animals, there is no uric acid, while the benzoic acid is present in such quantity, as to allow of its being extracted for commercial purposes. The phosphoric acid and its combinations, are likewise absent in the urine of many quadrupeds.

Urea, and uric acid, appear therefore to be substances peculiar to urine, and as they have not been detected ready formed in the blood, it appears reasonable to conclude, that they are produced by the kidneys.

The contents of the urine in the same individual, appear to differ with age, food, or disease. In infants, there is no phosphoric acid, and but feeble traces of uric acid; while in adults, both uric acid and earthy phosphats abound. A variety of substances, such as nitre, alkalis, carbonic acid, rhubarb, or madder, when taken into the stomach, communicate their sensible qualities to the urine in a very remarkable manner, leading some to believe, that there is a passage for fluids from the stomach to the kidneys, independent of the ordinary course of circulation.

But, the most singular changes which take place in urine are effected by disease. In diabetes, the quantity is increased in a great degree, as well as its density. The proportion of urea his diminished to such an extent, that it can with difficulty be detected, while sugar occupies its place in abundance. These two substances contain the same elements, but in different proportions. While the hydrogen of the urea remains the same, its azote disappears, and the quantity of carbon and oxygen being doubled, produces the sugar. Mr Rose has likewise remarked the absence of urea in the urine of patients, labouring under chronic and acute inflammations of the liver.* During jaundice, the presence of bile may be detected in the urine.

In some cases of the disease called suppression of urine, the body is relieved from this excrementitious liquid by means of perspiration. Dr DAWSON mentions the case of a young woman, in which the urine was conveyed out of the system by spontaneous vomitings. "She vomited sometimes every day, and sometimes only every third or fourth day; and though these vomitings usually came on presently after dinner, yet what she vomited seemed to be mere urine, without any thing which she had eaten mixed with it †."

Some notice ought here to be taken of *urinary calculi*. These are found both in the kidneys and in the bladder. They consist of concretions of one or more of the ingredients of urine, and exhibit considerable variety of structure and external appearance. In addition to the ingredients of the urine, oxalic acid, and its alkaline and earthy compounds, have likewise been detected in them. Their origin is obscure. In the kidneys, these morbid concretions usually consist of uric acid, or oxalat of lime. These passing into the bladder, become nuclei, round which layers of the phosphats, urats and carbonats are deposited, and also urea. But these concretions may form likewise in the bladder itself, from the thickening of the mucus, or the deposition of those ingredients of the urine, which, from being in excess, may be only mechanically suspended. In the human species, these concretions produce the most ex-excruciating pain; and it has hitherto baffled the skill of the chemist, to point out a method by which their formation may be prevented, or, when formed, their dissolution effected. But man is not alone subject to this painful disease.

---

* Annals of Phil. v. p. 424.

† Phil. Trans. 1759, p. 216.

Urinary concretions have likewise been detected in the horse, ox, sow, dog, rabbit, cat and rat. In none of these, however, has any trace of uric acid been detected. In all of them the base is lime, or magnesia, united with phosphoric, carbonic, or oxalic acids.

CHAP. XIV.

Peculiar Secretions.

Under this head, it is our intention to consider the circumstances under which light, electricity and caloric, are generated in the animal system. The subject is exceedingly interesting, and it has been enriched by numerous observations, the united efforts of the naturalist and chemist. Still, however, there prevails a considerable diversity of opinion, respecting the manner in which these substances are produced, or the share which the vital principle contributes towards their developement. As these bodies frequently appear to be produced contemporaneously, in the changes to which matter is subjected, they are to be viewed in this chapter in connection, in expectation that the mode of production of one, may help to illustrate the origin of antoher.

I. Luminousness of Animals.

The faculty of emitting light does not appear to be possessed by any individuals of the classes Mammalia, Birds or Reptiles. Several fishes, however, exhibit this remarkable property; particularly the Dorado, Mullet, Herring and Sprat. The *Sparus chrysurus*, an inhabitant of the seas of Brazil, is said to be luminous in such a remarkable manner, that when a few individuals are swimming in company, they emit so much light, that in a dark night a person might see to read by its aid. But considerable suspicions may be entertained on this subject, whether the light is emitted by the bodies of fish, or by the number of minute parasitical animals which adhere to the surface of the skin. It is well known, that when fish begin to putrefy, they appear luminous; but the light is occasioned, not by the flesh of the animal, but by numerous animalcula, whose growth the putrefaction has promoted. To such a cause may be referred the light observed by WILLOUGHBY, in the Sparus pagrus*.

Among the Mollusca conchifera, the Pholades, according to the observations REAUMUR, possess a liquor which is luminous; and the same property is possessed by the Pyrosoma of PERON, inserted among the Mollusca tunicata. The crustaceous animals exhibit several examples of luminous species, such as the Cancer pulex of LINNÆUS, and the Cancer fulgens of Sir JOSEPH BANKS†. The Myriapoda exhibit this property in the species of the genus Scolopendra. The Insecta furnish examples in many genera, as Elater, Lampyris, and others. Among the Vermes, the Neries noctiluca has been long known; and, according to BRUGUIERE, the common earthworm‡. The genera Medusa and Beroe among the Acephala; and the Pennatula and Sertularia among the Zoophyta, are all remarkable examples.

The luminous quality appears to reside in a fluid substance in the Pholas, Pennatula, Medusa and many other animals. In a species of cancer observed by Captain TUCKEY, in the

* Historia Piscium, 312. † Phil. Trans. 1810, p. 262. Tab. xiv. f. 1, 2.

‡ Journ. d'Hist. Nat. ii. p. 267.

Gulf of Guinea, the luminous property resided in the brain; which, when the animal was at rest, resembled a most brilliant amethyst, about the size of a large pin-head; and from this there darted, when the animal moved, flashes of a brilliant silvery light. In the Lampyris, the luminous matter is a soft yellowish substance, of a close texture, and resembling paste. It is not attached to the same part of the body, even in animals of the same class: thus, in the Lampyris, it is spread on the internal surface of the last rings of the abdomen. In the Elater, under the skin of the thorax; in the Fulgora, in the remarkable projections on the head; and in the Pausus, in the club of antennæ*.

Among insects, this luminous quality is peculiar to the season of love, and disappears during the remaining part of the year. The effect of season on the other luminous animals, has not been observed.

In many of the medusæ, particularly *Medusa hemisphærica* of MACCARTNEY †, exposure to light destroys the luminous power; hence these animals are not observed at the surface of the sea in moonlight nights. The light of insects appears to be in a great measure independent of external circumstances. In the *Scolopendra*, on the other hand, the luminousness does not appear, unless the animal has been previously exposed to solar light.

The luminous quality is not destroyed, extinguished, or altered, by immersing the animal in oxygen gas, chlorine, hydrogen, alcohol, or water, unless death be produced, in which case, the luminous power speedily ceases. In some cases, however, this power may be restored. Thus Mr MACCARTNEY, to whom the public is indebted for the most valuable information on the subject which has yet been

---

* Pausus spherocerus, Lin. Trans, vol. iv. p. 261.

† Phil. Trans. 1810, p. 268.

published, having cut from the bellies of living glow-worms the sacs containing the luminous matter, found that it shone uninterruptedly for several hours in the atmosphere; and, after the light became extinct, that it was revived by being moistened with water. Some of the sacs were put into water in the first instance, and they continued to shine in it unremittingly for forty-eight hours.

Whatever excites to muscular action, increases the luminous appearance,—as heat and electricity. In the case of the medusæ, they give out their light upon being disturbed, the emission of the luminous jets following the line of the muscular contractions. The minute species are very common in the sea; and produce those sparks or globes of light, constituting the luminousness or phosphorescence of the sea, so visible in a dark night in the wake of a ship, or when the water is struck by an oar?

Mr MACCARTNEY, in the course of his dissections of luminous insects, did not find that the organs of light were better or differently supplied with either nerves or air-tubes, than the other parts of the body. The emission or suppression of the light, however, appears to be under the influence of the will of the animal *.

II.—ELECTRICITY OF ANIMALS.

THE production of the electrical fluid, one of the most singular secretions of the animal frame, is peculiar to the

* When fish are steeped in water, until the whole fluid becomes luminous, especially the surface, this luminousness is increased by shaking the vessel. A heat of 118° destroys, in a great measure, this property. The luminousness appears to be caused by the infusory animalcules with which such water abounds. See "Experiments to prove that the luminousness of the sea arises from the putrefaction of its animal substances,"—by J. CANTON, Phil. Trans. 1769, p. 446.

following species of fishes. Rhinobatus electricus, Torpedo vulgaris, unimaculata, marmorata and galvani*, Gymnotus electricus, Silurus electricus, Tetraodon electricus, and Trichiurus Indicus. The claims of the first and last of these species to rank as electrical fishes, is doubtful; those of the others have been established.

The electrical organs of the *Torpedo* are double, and occur in the fore-part of the body, one on each side of the cranium; and extend backwards as far as the gill-openings, occupying the whole skin between the upper and under surfaces. In the *Gymnotus electricus*, these peculiar organs occupy nearly one-half of the body of the animal. They are four in number, two on each side; and extend along the sides and belly, from the head to near the tail. They are of unequal size; the superior one on each side being the largest. It is covered on its dorsal aspect by the muscles of the back, and on its ventral aspect by the smaller organ. This last reaches to the middle of the belly, and is scarcely one-fourth of the size of the superior one. In the *Silurus electricus*, the electrical organ commences at the head, where it is thickest, aud extends along the back and sides towards the tail, gradually decreasing in thinness, as it approaches the caudal extremity.

The electrical organs in all the fishes which have been examined, appear to have a reticulated cellular structure. In the Torpedo marmorata, which HUNTER† examined,

* Doubts may be entertained with regard to the propriety of constituting so many species, out of the Raia torpedo of naturalists, as has been done by M. RISSO, in his Ichthyologie de Nice, p. 18., in a great measure from the colour-markings of the body; a character in the torpedo liable to considerable variation, according to Mr TOD. (Phil. Trans. 1816, p. 121.) The T. marmorata, is the species referred to by British writers.

† Phil. Trans. 1773, p. 481.

they consisted of perpendicular hollow columns, reaching from the ventral to the dorsal surface of the animal; and, where the organ is thickest, extending to an inch and a half in length; and where thinnest, to one-fourth of an inch. These columns he found to be four, five, and even six-sided; but Mr Tod is inclined to consider them cylindrical*. The coats of the columns are thin and transparent, closely connected with each other, by a loose network of tendinous fibres, passing transversely and obliquely between the columns. They are, likewise, attached by strong inelastic fibres, which pass directly from one column to another. The cavity of each column is divided into a number of cells, (containing a fluid which M. GEOFFROY found composed of albumen and gelatine), by transverse partitions, which, in a column of one inch, amount to 150. These partitions consist of a very thin translucent membrane; their edges appear to be attached to one another, and they are connected to the inside of the columns by a fine cellular membrane. The whole organ is covered by a thin membrane, composed of fibres running in a longitudinal direction, and united to the skin, or surrounding parts of the body, by a cellular substance. Within this external membrane of the electrical organ, there is another, consisting likewise of fibres, running, however, in a transverse direction. The sides of the columns take their rise from this membrane. The arteries by which the electrical organs are nourished, after penetrating the investing membrane, ramify upon the sides of the columns, and send in, to the partitions, numerous small branches, which anastomose with one another, and with the vessels of the adjacent partitions.

* Phil. Trans. 1816, p. 121.

The nerves of the organs arise from the *medulla oblongata*. Each organ is liberally supplied by three large trunks, which, after communicating a few filaments to the gills, ramify in every direction between the columns, and send in small branches upon each partition.

The electrical organs of the Gymnotus electricus, as examined by HUNTER*, differ considerably from those of the Torpedo. The largest, or superior organs, consist each of a series of thin tender membranes, parallel to one another, extending longitudinally the whole length of the organ; and, in breadth, reaching from their central to the dermal surface. The uppermost membranes are concave dorsad, the middle ones are nearly horizontal; and the inferior ones are concave sternad. Their dermal edges are united with the skin and its muscles, their central edges with the middle partition, the air-bag, and the dorsal muscles. They are farthest distant from each other at their dermal edges, and gradually approach as they proceed to their central attachment. Where the organ becomes narrower towards the tail, two of the membranes sometimes unite, and form into one. In a fish, 2 feet 4 inches in length, they are $\frac{1}{27}$th of an inch distant from one another; and the whole organ, where broadest, is an inch and a quarter in breadth, and contains thirty-four membranes. In the inferior or smaller organ, the membranes are nearer each other, being only about $\frac{1}{56}$th of an inch asunder. The superior ones are the broadest, and they decrease in breadth the nearer they are situated to the ventral line of the animal. In both organs there are numerous partitions, dividing the spaces included between the different layers, into narrow vertical transverse cells. Each cell is in the form of a compressed qua-

* Phil. Trans. 1775, p. 339.

drangular pyramid, the height of which is equal to the breadth of the organ, the apex centrad, and the narrow sides dorsad and sternad. These partitions are so close to one another, as even to appear to touch, making the cells exceedingly narrow. In an inch in length, there are about 240 partitions. These electrical organs have no peculiar covering, being attached directly to the surrounding parts by cellular substance. The two organs on the side, are separated from each other by a thick membranous partition, reaching from the skin to the centre. The organs are liberally supplied by bloodvessels. The nerves are numerous, and arise from the *medulla spinalis* coming from it in pairs, between all the vertebræ of the spine. Previous to reaching the organ, they give out filaments to the muscles of the back, and afterwards to the skin and air-bag. Dr HUNTER was unable to trace their termination in the organ itself.

In the electrical organ of the Silurus electricus, according to GEOFFROY, the structure is more simple than in the Torpedo or Gymnotus. It consists of a bed of filaments which cross each other in every direction, and form meshes of very small dimensions. The whole is covered by a ligamentous membrane, which is itself covered with a layer of fat. The nerves with which the organ is provided, proceed from the eighth pair ; but are not so large in proportion as in the torpedo.

When the hand is applied to the peculiar organs of an electrical fish, the animal is observed to twist its body, as if about to make a vigorous muscular exertion; and a benumbing sensation is instantly felt in the fingers, and even as far as the elbows. This sensation, however, is not always felt, as the animal appears to excite it only when irritated, or otherwise disturbed. It is capable of making this benumbing effort many times in succession in the water, as well as in the air, when arrived at maturity, and even previous to the natural period of exclusion from the uterus of the mother. When caught in the net, it gives a shock to the hands of the incautious fisherman who ventures to seize it. When concealed in the mud, it is capable of making its most violent efforts; and is able to benumb the limbs to such a degree, as to throw down the passenger who inadvertently places his foot upon the body.

Although the benumbing powers of the torpedo were known to PLATO and ARISTOTLE, and had frequently been proclaimed by the verses of the poet, and the exaggerated statements of the fishermen, it was not until the doctrines of electricity had been established, that the circumstances under which these were exerted, and the effects which they produced, were investigated with any degree of success. The first person who turned his attention particularly to this subject, was Mr WALSH, and he succeeded in demonstrating, that the animal could exert its benumbing power at pleasure, and that the shock was regulated by all those circumstances which influence the discharge of the electric fluid from the Leyden phial; in other words, he established the identity between the electric fluid and the benumbing power. His experiments were communicated to the Royal Society, July 1. 1773 *. According to the experiments of this observer, the shock of the torpedo is prevented by all electrical non-conductors, as glass or sealing-wax; while it readily passes along brass-wire, water, or persons whose hands are joined. The torpedinal fluid was unable to force itself across the minutest tract of air, or from one link of a small chain, suspended freely to another; or through an almost invisible separation, made by the edge of a penknife

* Phil. Trans. 1773, p. 461.

in a slip of tinfoil fixed on sealing wax. The animal is capable of giving its shocks in the air, as well as in its natural element the water. Those given in the former are even much stronger than in the latter, owing probably to the fluid being more confined by the surrounding medium. The shock is given with the same degree of force, even when the animal is insulated, and the person receiving it likewise insulated. In such circumstances, fifty shocks have been received in the space of a minute and a half. The most powerful effect is produced by these organs, when the circuit is made between their upper and under surfaces, as these appear to be in opposite states of electricity.

The history of animal electricity received some valuable additions from the experiments of Dr WILLIAMSON in Philadelphia *, on the Gymnotus. He found that this animal could communicate its shock to the hand, in water, even when held at the distance of three feet. It killed small fish by the shock, without coming into actual contact, and stunned larger ones in the same manner.

In both these fishes, the electrical energy is insufficient to produce appearances of attraction and repulsion in the most delicate electrometers, although feeble sparks have been perceived by different observers, in consequence of using the utmost precaution in interrupting the circuit. The animal has the power of making the discharge from any part of the surface of its peculiar organ, or from the whole, at pleasure; but it is incapable of communicating shocks for any length of time, without exhibiting unequivocal symptoms of fatigue. If the electrical energies of the fish are too much excited, it becomes debilitated, and expires.

---

* Phil. Trans. 1775, p. 94.

The influence of the nerves in the production of the electrical shock has been examined with care by Mr Tod, and the results which have been published * throw considerable light on the mysterious process. He made an incision on each side of the cranium and gills of a lively torpedo, and pushed aside the electrical organs so as to expose and divide their nerves. The animal was then placed in a bucket of sea-water. On examining it in about two hours afterwards, he found it impossible to elicit shocks from it by any irritation ; but it seemed to possess as much activity and liveliness as before, and lived as long as those animals from which shocks had not been received, and which had not undergone this operation. Two of these animals being procured, the nerves of the electrical organs of one of them were divided after the manner above described. They were placed each in separate buckets of sea-water, and allowed to remain undisturbed. This was performed in the morning; and, when examined in the evening, it was impossible to distinguish between the liveliness or activity of either. Of other two of these animals, the nerves of the electrical organs of one of them were divided. Being placed each in separate buckets of sea-water, they were both irritated as nearly alike as possible. From the perfect animal, shocks were received: after frequent repetition, it became weak and incapable of discharging shocks, and soon died. The last shocks were not perceptible above the second joint of the thumb, and so weak as to require much attention to observe them. From the other, no shocks could be received : it appeared as vivacious as before, and lived until the second day. This experiment was frequently repeated, with nearly the same results. The nerves of one electric

* Phil. Trans. 1816, p. 120.—Ibid. 1817. p. 32.

organ only being divided in a lively torpedo, from which shocks had been previously received, on irritating the animal, it was still found capable of communicating the shock. Whether there was any difference in the degree of intensity, could not be distinctly observed. One electrical organ being altogether removed, the animal still continued capable of discharging the electric shock; and the same circumstance took place when only one of the nerves of each electrical organ was divided. When a wire was introduced through the cranium of a torpedo, which had been communicating shocks very freely, all motion immediately ceased, and no irritation could excite the electrical shock.

He likewise found that no alteration took place in the electrical condition of the organs, when the muscles of the fins were intersected. When the organs themselves were divided by a longitudinal incision, no change was produced in the electrical energies, but they seemed weakened by the removal of part of the organ. When the surfaces of the electrical organs were denuded, the animal could still give shocks; and the same power remained, when, by incisions, no other attachment existed but by the nerves.

It is obvious, from all the circumstances which have been stated, that the production or condensation of the electric fluid in these animals, is a vital action, dependent on the will of the animal, and acting through the nerves of the peculiar organs. The power appears to be subservient to the continuance of life, in a twofold manner. It affords protection, by enabling the animal to benumb its foes, and it assists in procuring food, by stunning or killing the smaller animals, on which its sustenance depends.

The electricity of animals which do not possess peculiar organs for its condensation, has not been investigated with sufficient care. From the observations of SAUSSURE, HEM-

MER, and others, on the human body, it appears to vary greatly, being, in the same individual, sometimes positive and at other times negative, and liable to sudden change, upon any quick motion being performed. In making these observations, there is some difficulty in avoiding the effects of friction of the cloth, and other circumstances foreign to the natural state of the body.

III.—Animal Heat.

In every animal there is a certain degree of heat necessary to its existence, and the full exercise of its functions. In quadrupeds and birds this heat is considerably greater, in general, than the surrounding atmosphere, while in fishes and the animals of the inferior classes, it is seldom very different from the temperature of the objects with which they are usually in contact. The ordinary temperature of the human body is rated at between 96° and 98° of F. In the common hen, the temperature is between 103° and 104°. Those animals whose temperature is high, and not greatly influenced by the changes in the heat of external objects, are denominated *warm-blooded animals*. Those, on the other hand, whose temperature is greatly influenced by that of surrounding objects, are termed *cold-blooded animals*. In both classes, the temperature is regulated by the vital powers of the animals, and limits are assigned beyond which it is dangerous to pass. The range of warm blooded animals is confined, that of the cold blooded extensive. Both, however, are influenced by the same agents, and appear to be governed by the same general laws.

1. *When an animal is exposed to a change of temperature, some corresponding change likewise takes place in the heat of its body.*—The propriety of establishing this general law, is sanctioned by the experience of our own

feelings, and confirmed by a variety of decisive experiments. Dr CURRIE found the temperature of a man, plunged into cold salt-water at 44°, to sink in the course of a minute and a half, after immersion, from 98° to 87°, and in other experiments, it descended as low as 85°, and even 83° *. In these experiments, the pulse sunk from 70 beats in a minute, its natural state, to 68 and even 65; and after the first irregular action of the diaphragm from the shock of immersion had ceased, the breathing became regular, and unusually slow. When the human body is exposed to a higher temperature than its natural standard, a corresponding change likewise takes place. Dr FORDYCE tried a variety of experiments on this subject. By exposing the body to heated air in a close room to the temperature of 120°, and in some cases as high as 211°, the heat of the body rose to 100°. In this situation the pulse beat 145 times in a minute. No change, however, was produced on respiration, it became neither quick nor laborious †. Similar experiments have likewise been performed on other warm-blooded animals. Dr HUNTER found the temperature of a common mouse to be 99°, when the atmosphere was 60°; but when the same animal was exposed for an hour to a cold atmosphere of 15°, its heat had sunk to 83° ‡. Dr CRAWFORD exposed a dog, whose natural heat is 101° or 102°, to water whose temperature was raised to 112°, and found that his temperature was raised to 108° and 109°. In air at 130°, the temperature of the dog was 106° §.

---

* Phil. Trans. 1792, p. 199. † Ibid. 1775, p. 114.

‡ Ibid. 1778, p. 21.

§ Phil. Trans. 1781, p. 486. The venous blood, in such experiments, appeared of a light florid colour, like the arterial, indicating the diminished action of the capillary vessels upon the blood.

From these experiments it appears, that the temperature of the human body may be raised 4° above its natural standard, and sunk 15° below it, by the operation of external circumstances. In the mouse, the heat was diminished 16° below,—and in the dog, raised 7° above the natural standard.

The same law extends to the cold-blooded animals. Dr HUNTER found the temperature of a healthy viper, when exposed to 108°, rise to 92½°; and the same animal, when placed in a temperature of between 10° and 20°, had its heat sunk to 37°, 35°, and even 31° *.

It is in consequence of the unceasing influence exerted on the animal frame, by the varying temperature of surrounding objects, that the different parts of the body are seldom of the same degree of heat. The stomach, heart, and liver, are always nearest the standard, the other parts, which are more remote, varying one or two degrees in the warm-blooded animals, when not greatly acted upon by deranging causes.

2. *When the body is exposed to a temperature greatly above the ordinary standard of the animal, a counteracting influence is exerted, and cold is generated.*—In all animals, whether hot or cold blooded, the heat may be raised a limited number of degrees. But, even when the external temperature continues high, the heat of the body remains stationary at its maximum of elevation †. Thus, in

---

* Phil. Trans. 1778 p. 25.

† Governor ELLIS, in a letter to the celebrated JOHN ELLIS, dated Georgia, 17th July 1758, makes a slight reference to this law, which was afterwards developed by Dr FORDYCE. " I have frequently (says the Governor) walked an hundred yards under an umbrella, with a thermometer suspended from it by a thread, to the height of my nostrils, when the mercury has rose to 105°; which is prodigious. At the same time, I have confined this

Dr FORDYCE's experiments, the heat of the human body, in a high temperature, speedily reached 100°, but exposure to 211° did not raise it higher *. The temperature of the dog could be raised no higher than 109°, though exposed to 112°. The viper, though exposed to a temperature of 108°, only had its heat raised to 92½°. Here, then, it is obvious, that the body of a living animal is capable of resisting the influence of a high temperature, and remaining comparatively cool, though exposed to a heating cause. That the body is not kept cool by evaporation, is demonstrated by a variety of circumstances. In Dr FORDYCE's experiments, with his body in heated air, water poured down in streams over his whole surface. But that this water was merely the vapour of the room, condensed by the coldness of his skin, appeared clear, from his having placed a Florence flask filled with water of the same temperature with his body, viz. 100°, when he observed the vapour in like manner condense upon its surface, and run down the sides in streams. Besides, the cold is generated

---

instrument close to the hotest part of my body, and have been astonished to observe, that it has subsided several degrees. Indeed, I never could raise the mercury above 97° with the heat of my body." Phil. Trans. 1758, p. 755.

* Dr BLAGDEN, who assisted in performing these experiments, observes, "Being now in a situation in which our bodies bore a very different relation to the surrounding atmosphere, from that to which we had been accustomed, every moment presented a new phenomenon. Whenever we breathed on a thermometer, the quicksilver sunk several degrees. Every expiration, particularly if made with any degree of violence, gave a very pleasant impression of coolness to our nostrils, scorched just before by the hot air rushing against them when we inspired. In the same manner, our now cold breath agreeably cooled our fingers whenever it reached them. Upon touching my side, it felt cold like a corpse." Phil. Trans. 1775, p. 118.

whether the air is moist or dry. And in the case of frogs and dogs, Dr CRAWFORD found that the cold was generated even when the body was immersed in water.

Even when the heating cause is applied only to a particular part of the body, as (in the experiments of Dr HUNTER) to the urethra, its temperature is not increased beyond that degree to which the whole body may be raised.

In all these experiments in which the body is exposed to a high temperature, the generation of cold is effected by a great expenditure of the vital energy. Hence a great degree of weakness is produced by continuing the effort.

As the power of the whole body for the production of cold appears to be limited, so is likewise the power of particular parts, " which (says BLAGDEN *) may be one reason why we can bear for a certain time, and much longer than can be necessary to fully heat the *cuticle*, a degree of heat which will at length prove intolerable." It is probable that this power of the human body is greatly influenced by habit, in the case of washer-women, whose hands are frequently immersed in warm water, and glass-blowers, whose bodies are exposed to a degree of heat, which to others would prove painful.

3. *When the body is exposed to a temperature greatly lower than the ordinary standard, a counteracting influence is exerted, and heat is generated.*—In the experiments already quoted it appears evident, that there is an inferior as well as a superior limit to the change of temperature in living bodies, from the influence of surrounding objects. In the experiments of Dr CURRIE, the heat of the human

* Phil. Trans. 1775, p. 121.

body, in one instance, sunk rapidly from 98° to 87°, when placed in water at 44°, but at the end of twelve minutes it rose to 93½°. In another experiment, in water of the same temperature, the heat of the body fell from 98°, in the course of two minutes to 88°, but, at the end of thirteen minutes, it had risen to 96°. Dr HUNTER found that a dormouse, whose heat, in an atmosphere at 64°, was 81½°, when put into air at 20°, had its temperature raised in the course of half an hour to 93°: an hour after, the air being 30°, it was still 93°: at another hour after, the air being 19°, the heat of the pelvis was as low as 83°, but the animal was now less lively. In this experiment, the dormouse had maintained its temperature about seventy degrees higher than the surrounding medium, and for the space of two hours and a half.

In the cold-blooded animals, heat is generated under similar circumstances. HUNTER found that the heat of a viper, placed in a vessel at 10°, was reduced in ten minutes to 37°, in other ten minutes, the vessel being 13° to 35°, and in the next ten minutes, the vessel at 20° to 31°. In frogs, he was able to lower the temperature likewise to 31°, but beyond this point it was not possible to lessen the heat, without destroying the animal.

When the cooling cause is applied to particular parts of the body, the heat of these parts sinks lower than the minimum of depressed temperature of the body. Although HUNTER was unable to heat the urethra one degree above the maximum of the elevated temperature of the body, yet he succeeded in cooling it twenty-nine degrees lower than the minimum of depressed temperature, viz. to 58°. He succeeded in cooling down the ears of rabbits until they froze, and when thawed, they recovered their natural heat and circulation. The same experiment was performed on the comb and wattles of a cock. Though he found that, in many instances, parts of animals which had been frozen recovered their functions when thawed, yet in no case did recovery take place when the whole body had been frozen.

The efforts of the body to generate heat, when thus exposed to a cold medium, rapidly exhaust the vital energy, and when these cease to be made, or have become too weak to accomplish their object, the temperature sinks, and freezing and death simultaneously take place.

Dr HUNTER found, that even eggs possessed this power of generating heat, when exposed to a cooling medium. An egg which had been frozen and thawed, was put into a cold mixture along with one newly laid. The fresh one was seven minutes and a half longer in freezing than the other. In another experiment, a fresh laid egg, and one which had been frozen and thawed, were put into a cold mixture at 15°; the thawed one soon came to 32°, and began to swell and congeal; the fresh one sunk to 29½°, and in twenty-five minutes after the dead one, it rose to 32°, and began to swell and freeze *. During these twenty-five minutes, it must have generated a very great deal of heat, before it yielded to the influence of the cold mixture.

In what manner, it may now be asked, are these extraordinary capabilities of varying the temperature, maintained and regulated in the animal frame?

The influence of the *cutaneous system* in modifying the heat of the body, is very considerable. The skin is a bad conductor of caloric, and consequently protects the internal organs from the sudden influence of heat or cold. In the performance of this function, the cuticle and its appendices act a conspicuous part. In the warm-blooded animals, we observe their bodies covered with hair or feathers, and these in the greatest abundance, where the parts stand most in need of protection. Among these animals, the

* Phil. Trans. 1778, p. 29, 30.

whales alone are destitute of hair. But, besides the protection yielded by the extraordinary thickness of the skin, and the layer of fat (likewise a bad conductor) with which its interior is lined, these animals reside constantly in the water, and are in a great measure protected by the uniform temperature of that element. The seals, bears, and walruses, on the other hand, which seek their food in the same seas, are, nevertheless, obliged to come to land to sleep and bring forth their young, and consequently have a coating of hair to protect them from the varying temperature of the atmosphere in which they occasionally sojourn. In the mammalia, the quantity of hair on the body is, in general, proportional to the cold of the climate in which they reside. The same law does not seem to prevail to the same extent, in regard to the quantity of feathers on the birds of different climates.

The influence which the muscular system exercises over the heat of the body, is of great extent. When we exert ourselves in speaking, walking, or running, our animal heat is kept at its natural standard, even when the body is exposed to a great degree of cold. On the other hand, if, while the body is exposed to the cooling influence of a low temperature, we remain at rest, our heat is speedily diminished, and we are aroused to action by the painful sensations of cold. During sleep, our temperature sinks a little, obviously, in consequence of our state of rest. But to guard against the prejudicial effects which might follow, we observe all animals, before going to sleep, retire to such places, or assume such positions, as are best calculated to protect them from the cooling influence of external objects.

The nervous system likewise exercises a great controul over animal temperature. As it influences respiration *,

* Phil. Trans. 1811, p. 36;—and 1812, p. 378.

and, by means of respiration, the action of the heart, and consequently the whole circulating system,—as it exercises likewise an unlimited controul over the arteries and glands in the office of secretion, it may be said to regulate all the movements of vital action, and consequently the production of that temperature requisite for the continuance of existence.

The digestive system is destined to furnish the means by which the vital energies are to be recruited, and its indirect influence over the power of an animal to regulate its temperature, must be considerable. In Dr Currie's experiments already quoted, the body, which had been much cooled by exposure to cold, was most speedily restored to its natural warmth, and made comfortable, by the application of a bladder of hot water to the pit of the stomach. In one case, during the application of the cold, the person complained of a coldness and faintness at the stomach. Hence this observer concluded, that there was some peculiar connection of the stomach, or of the diaphragm, or both, with the process of animal heat.

The changes which take place in the circulating system, viewed in connection with animal heat, are not so great as might have been expected. In resisting the application of heating media, Dr Fordyce found the circulation of the blood to proceed very rapidly, as his pulse gave 145 beats in a minute. "The external circulation was greatly increased; the veins had become very large, and an universal redness had diffused itself over the skin." With this increased rapidity of circulation in the blood, there was no change, however, produced in the frequency of respiration. In Dr Currie's experiments, where the body was exposed to a cooling medium, and where a considerable quantity of heat must have been generated, the ordinary velocity of the blood was diminished, and the breathing became unusually slow. In Mr BRODIE's experiments, in which respiration was kept up by artificial means, after the animal (a rabbit) had been killed, he found the heat of the body to diminish as rapidly as in a dead animal of the same kind, in which no attempts were made to keep up the respiration. Yet, in the animal in which artificial respiration was carried on the heart continued to beat for nearly two hours; the blood circulated, and was changed from arterial into venous blood in the capillary vessels; it was aërated in the lungs, and carbon given off equal in quantity to that which is evolved in a natural state; and the aërated blood had the usual florid colour *.

The changes which take place in the temperature of the body, in consequence of morbid states of the organs, are calculated to throw some light on this mysterious secretion. Where the nerves leading to particular members are compressed or injured, these soon become cold, in comparison with the rest of the body, although the circulation of the blood (the only means by which any heat is communicated to them) continues as usual. This is frequently exhibited in the case of paralytic limbs. In a gentleman who was seized with an apoplectic fit, HUNTER says, that "while he lay insensible in bed, and covered with blankets, I found that his whole body would, in an instant, become extremely cold in every part; continue so for some time; and, in as short a time, he would become extremely hot. While this was going on for several hours alternately, there was no sensible alteration in his pulse †." Dr CURRIE states an equally remarkable case: "I have seen a young woman, once of the greatest delicacy of frame, struck with madness, lie all night on a cold floor, with hardly the covering that decency requires, when the water was frozen on the table by her, and the milk that she was to feed on was a mass

* Phil. Trans. 1812, p. 378. † Ibid. 1775, p. 458.

of ice *." The quantity of heat evolved during the inflammation of any particular part, is probably much greater than is generally supposed. Dr THOMSON found that a small inflamed spot in his right groin, gave out, in the course of four days, a quantity of heat sufficient to have heated seven wine pints of water, from 40° to 212°. Yet the temperature was not sensibly less than that of the rest of the body at the end of the experiment, when the inflammation had ceased †.

Two hypotheses have been devised to account for the origin of animal heat. The first is that of the justly celebrated Dr BLACK. He supposed that the specific heat of oxygen gas was greater than that of carbonic acid gas, and consequently, when the former was converted into the latter in the lungs, a quantity of latent caloric would be disengaged, sufficient to heat the parts in contact, more especially the blood, which, by its circulation, would likewise communicate its high temperature to the distant parts of the body. As the lungs, however, are not warmer than the neighbouring viscera, it was supposed that they were kept cool by the evaporation of the pulmonary vapour. When the quantity of oxygen consumed in respiration is compared with the whole of the azote, and the remaining unchanged portion of the oxygen of the air, which have their temperature raised forty or fifty degrees,—when the quantity of heated vapour given off by the lungs and the skin is considered,—when we likewise estimate the portion of heat abstracted from the body by contact and radiation; we clearly perceive, that all the heat which the oxygen consumed can impart, (supposing the inference respecting its specific heat to be just,)

* Phil. Trans. 1792, p. 218. † Annals of Phil. vol. ii. p. 27.

is far from sufficient to supply the quantity which is constantly abstracted. But the hypothesis has fallen, by attacks in another direction. Dr CRAWFORD inferred from his experiments, that the specific heat of oxygen was as 4.7490, carbonic acid 1.0454, azote 0.7936, atmospheric air 1.7900. The more recent and accurate experiments of MM. DELAROCHE and BERARD, however, have established the relative specific caloric of the same gases as follows: Oxygen 0.2361, carbonic acid 0.2210, azote 0.2754, atmospheric air 0.2669. It follows from these experiments, therefore, that the quantity of heat given out by oxygen during its conversion into carbonic acid gas, would be insufficient to heat the residual air that is expelled in breathing to its ordinary elevation, and consequently could contribute nothing towards sustaining the high temperature of the body.

The objection that had been brought against the hypothesis of BLACK, that the lungs were not hotter than the rest of the body, was attempted to be obviated by Dr CRAWFORD, who assumed, from his experiments, that the specific caloric of arterial blood was 1.0300, and that of venous blood only 0.8928, and inferred, that, during the conversion of the former into the latter, in the course of circulation, a quantity of heat must be set free; and as this conversion takes place in every part of the body, heat must consequently be every where disengaged. But the basis of this hypothesis has not been demonstrated to be true. It is next to impossible, by the methods at present known, to determine with any degree of accuracy the relative specific heat of arterial and venous blood; and the conclusions of CRAWFORD on the subject, differ widely from those of Dr JOHN DAVY already quoted *.

* Page 353.

But admitting the principles on which the opinions of Black and Crawford rest, it will be seen, from the conditions of the problem which have been stated, that they offer no explanation of many of the changes which occur. Nay, it may be affirmed with confidence, that the supporters of these opinions have not put themselves to any trouble in order to obtain a knowledge of these conditions, but have suffered themselves to be seduced by the explanation they seemed to afford of some of the phenomena. In consequence of this neglect, a theory which was long the boast of the chemist, has been rejected as visionary; and the disappointed physiologist is now left to re-examine the properties of that vital principle he had inconsiderately abandoned.

In the present state of physiological science, no rational theory can be offered to account for the production and regulation of animal heat. Perhaps the changes which take place in the fluids of animals may occasion a disengagement of heat or cold, in consequence of the difference in their capacities. But a variety of other circumstances appear to operate. The compression of the air in the lungs in the act of breathing, and the compression of the blood, by the muscular power of the heart and arteries, may likewise exercise some influence. We are inclined to believe, however, that the principal source of animal heat may with propriety be referred to the electrical changes which accompany those endless combinations and decompositions which take place in the system through the whole of life, changes which are necessary to the wellbeing of the principle of life, and demonstrated to be subject to its control.

CHAP. XV.

Reproductive System.

The organs which we have hitherto been considering, refer exclusively to the individual, and are necessary to the support of the system throughout the whole of life; those which now claim our attention, do not refer to the wants of the individual, but are subservient to the continuation of the species. The instincts by which the organs of reproduction are governed, have been already enumerated, when treating of the active powers of the mind. At present, it only remains for us to enquire into the different modes by which animals are propagated, and the various organs which are called into exercise in each.

The simplest mode of generation does not require sexual organs for the accomplishment of its purpose. Part of an individual drops off, and speedily exercises the functions of an independent being. This is termed Generation by Spontaneous Division. In other cases, a bud is produced from the surface, which gradually evolves, drops off from the parent when ripe, and begins to exhibit a separate life. This is termed Gemmiparous Generation.

In those animals which possess peculiar organs for the preparation of the germ or ovum, some are Androgynous*, and either have the sexual organs incorporated, and capable of generating without assistance, or the sexual organs are distinct, and the union of two individuals is necessary

* ανδρογυνος, man-woman.

for mutual impregnation: others have the sexual organs separate, and on different individuals. The young of such animals are either nourished at first by the store of food in the egg, or by the circulating juices of the mother. Those species in which the former arrangement prevails are termed Oviparous, while the term Viviparous is restricted to the latter. As the organs of reproduction are displayed in the greatest perfection in those animals with distinct sexes, we shall proceed to consider,

I. Viviparous Animals.

Quadrupeds alone are truly viviparous. The manner in which the fœtus is nourished previous to birth, the peculiar configuration of the sexes, and the nature of the reproductive organs, indicate an arrangement for the generation of quadrupeds to which there is nothing analogous among the rest of living beings.

Before entering upon the consideration of the structure of the peculiar organs employed, and their functions, it is scarcely necessary to remark, that it is the business of the *Female*, in all animals, to prepare the *ovum* or germ, and bring it to maturity. For this purpose, the germ is produced in the *ovarium*, farther perfected in the *uterus* or matrix, and finally expelled from the system through the *vagina*. The office of the *Male* is to impregnate the germ by means of the spermatic fluid. This fluid is secreted in the testicles, transmitted by the spermatic ducts, and finally conveyed by the external organ to its ultimate destination. In proceeding to our account of the reproductive organs in viviparous animals, we shall first consider those which are peculiar to the male sex, and afterwards those by which the female is characterized.

1. *Male Organs.*—The spermatic fluid is secreted in two glandular bodies which are called the *testicles*. These,

when they occur externally, are contained in a *scrotum* formed by the common integuments. They are enveloped by two coats, the first of which, termed the *tunica vaginalis*, is derived from the *peritoneum*, and the second, called, from its white colour, *tunica albuginea*, is merely a reflected duplicature of the former. This second coat adheres closely to the surface of the testicle, and is only in union with the first in a line on the dorsal edge *.

Each testicle is usually of an oval form, and of a firm compact substance, consisting of an infinite number of ramifications of arteries and veins, termed *spermatic*, united together by a cellular substance. The *spermatic tubes* take their rise in different parts of the testicle, and gradually uniting, at last emerge from its body, in the form of a single canal, termed the *spermatic duct*. This duct, which, at its commencement, where it is exceedingly convoluted, adheres to the surface of the testicle by cellular substance, and is in part covered by the *tunica albuginosa*, is termed *Epididymis*. The remaining part of the spermatic duct, termed *Vas deferens*, usually preserves a straight course towards the base of the urethra, where it terminates. In many species, however, it is slightly tortuous, and its walls, in some cases, assume a considerable degree of thickness, and become obviously glandular. In a few instances, it is enlarged in its diameter towards its termination. The spermatic ducts usually terminate at the base of the urethra, but, in some cases, they are continued to the extremity of the penis.

* In the human fœtus, the testes are at first lodged in the belly, and only descend into the scrotum a little before birth. In some instances, the descent never takes place. This, however, is unnatural, but in many animals termed *testicondi*, the testes are always internal, and in a few the descent is periodical.

Besides these organs, there are others which appear to be necessary towards perfecting the seminal fluid. These are the Seminal Vesicles, the Prostate Gland, and Cowper's Glands. The *Seminal vesicles* are two in number, corresponding with the spermatic ducts *. In some cases their cavity is simple, while in others it is divided into numerous sacs, united together by cellular substance. The walls of these vesicles are, in some animals, thin and simple, and lined with a continuation of the mucous membrane of the urethra; while in others they are thick and glandular, and are evidently destined for secretion. These vesicles terminate in the urethra, either by canals which are common to the spermatic ducts, or by orifices peculiar to themselves.

The seminal vesicles are wanting in many quadrupeds. Their presence does not seem to be regulated by any uniform plan, or to be exhibited in any peculiar habits. Their use, where they do exist, has not been satisfactorily determined. By some, they are considered to be receptacles of the spermatic fluid; while by others they are regarded as destined to secrete a fluid peculiar to themselves. This last opinion is rendered probable by the judicious remarks of HUNTER †, and confirmed by the whole history of their structure, contents, and terminating canals. That they are still connected with the reproductive system, is indicated by their increased size during the season of love.

The *Prostate Gland* is situated at the commencement

---

* The *vesiculæ accessoriæ* of CUVIER, are membranous tubes, varying in number, adhering to the sternal side of the vesicles, or situated around the base of the urethra. They are filled with the same fluid as the vesicles, and empty their contents into the urethra by a common duct, or by a separate opening. They are very obvious in the mole and hedgehog.

† "Observations on certain parts of the Animal Economy." London, 1786, p. 27,—42.

of the urethra, and usually on its dorsal aspect. Internally it is cellular, and is destined to secrete a glairy albuminous fluid. In many animals this gland is single, or the lobes of which it consists are attached to one another, and the secretion is poured into the urethra by numerous orifices. In other cases this organ consists of two, or even four separate glands, each having a common cavity, into which the different cells empty their contents, and a common duct opening in the urethra.

The organ in its simple form exists in man, the quadruma, and many other mammalia. It appears in the elephant and ruminating animals, in its most complicated form; while it is wanting in the hedgehog, mole, and many of the glires.

Besides the seminal vesicles and prostate gland, there are two bodies which have been named, in honour of the discoverer, *Cowper's Glands*. These are situated immediately behind the bulb of the urethra, with the walls of which they are intimately united. They differ greatly in texture, and empty their contents into the urethra by a single duct. The fluid which they secrete is bluish-white, semitransparent, and gelatinous. They are surrounded by a muscular covering, which serves to expel their contents. In size and appearance they vary less with the season, than the two organs which have last been noticed. They are wanting in many mammalia, and, where present, exhibit many different combinations with the seminal vesicles, accessory vesicles, and prostate gland.

The seminal fluid prepared in the testes, and conveyed through the spermatic duct, is thus finally mixed with the secretions of these assisting glands, and brought to its destination through the urethra. What are the properties of the seminal fluid itself?

The *Seminal Fluid* has never been examined with care in its pure state, or unmixed with the contents of the vesicles and prostate gland. In its compound state, VAUQUELIN found that of man to be heavier than water, to be alkaline, and to consist, in the hundred parts, of 90 of water, 6 of mucus, 3 of phosphate of lime, and 1 of soda.

The most remarkable character, however, of the seminal fluid, consists in the number of animalcula which are contained in it. These were first observed by one LEWIS HAMME, and afterwards examined with care by LEEUWENHOEK and SPALLANZANI. They are exceedingly minute, and require, for their examination, a practised eye, aided by powerful magnifiers. The want of these qualifications led BOMARE to deny their existence, and LINNÆUS to regard them as inert particles, set in motion by heat. They differ in size and form in different species, and are found in the semen of oviparous, as well as viviparous animals. They move about in the fluid with a progressive motion, in various directions, and are easily killed by cold, or mixing the sperm with other liquors.

These animalcules are not peculiar to the semen. SPALLANZANI found them in the mesenteric blood of frogs, and newts, even females; in the blood of a sucking calf, and of a ram.

It was the opinion of LEEUWENHOEK, that these animated beings were destined to expand into maturity, and ultimately to exhibit the forms of the animal in the semen of which they resided; or that they were the germs of the animal. BUFFON regarded them as organic particles, no longer required by the body for its support, but treasured up in the genital organs, and ready to unite with the organic particles of the female, to give rise to a new individual. They are now generally regarded as exercising no influence whatever on generation, but as analogous, in their station and habits, to the intestinal worms *.

2. *Female Organs.*—The female organs, in viviparous animals, consist of the *ovaria* and *oviducts*, the *uterus* and *vagina*.

The *Ovaria* are two in number, of an oval or globular form, and situated in the cavity of the pelvis. They are covered by a fold of the peritoneum, and likewise possess a covering of a finer texture, peculiar to themselves. The surface of the ovaria is smooth in some animals, but uneven, or tuberculated, in others. The texture of an ovarium is spongy, and it abounds in cells, especially towards the surface. In early life it is small in size, and only reaches its destined dimensions at the age of puberty. About this period, irregular oval bodies, of a glandular structure, begin to form in its substance. These are frequently of a yellow colour, and hence have been termed *corpora lutea*. In the centre of these bodies is a cavity, in which the egg or germ is generated, and brought to a certain degree of maturity. This body at length bursts, and suffers the egg to escape. The *corpus luteum*, after discharging its germ, becomes absorbed gradually, and finally disappears. Others, in the mean time, form, to prepare for successive births. Attempts have been made to examine the structure of the ovum by M. BAUER †. But the smallness of the object, and the softness of its parts, prevent any microscopical observations of a satisfactory kind being made.

---

* The reader who wishes further information on this curious subject, may peruse SPALLANZANI's "Tracts on the Natural History of Animals and Vegetables."

† Phil. Trans. 1817, p. 252;—and 1819, p. 59.

When the ovum bursts from the *corpus luteum*, or gland in which it has been prepared, it enters the *Oviduct*, or Fallopian Tube, through which it is conveyed to the uterus. The extremity of the oviduct, next the ovarium, is somewhat funnel-shaped, and its margin is irregularly divided into leaves, or fimbriæ. This structure is considered as useful in aiding the expulsion of the germ from the *corpus luteum*, and in receiving it more readily into the tube. The oviduct itself has walls of a cellular substance, with abundance of bloodvessels, and traces of muscular threads. It is more or less tortuous in its course, and terminates by an opening in the upper part of the cavity of the uterus. Both oviducts open, in some cases, very near each other, while, in other instances, they are more remote.

The *Uterus*, or matrix, exhibits a great variety of shape in different quadrupeds. In general, however, it is lengthened in the direction of the body, receiving, at its summit, the oviducts, and terminating below in the vagina. Its walls are remarkably dense, and they are liberally supplied with bloodvessels, absorbents, and nerves. The proportion between the size of the cavity and the thickness of the walls, is likewise very various. The cavity itself exhibits several remarkable peculiarities.

In the most simple form of the uterus, the cavity is somewhat pear-shaped, and the oviducts enter it at its largest end, by very small openings. This shape prevails in the human race, the apes and sloths. It is called *Uterus simplex*. The next shape, in point of simplicity, is where the cavity divides at its summit into two processes, or horns, as they have been called. These are either straight, or variously convoluted, and terminate in the oviducts. This kind is termed *Uterus bicornis*, and may be seen in the dog, hedgehog, and the ruminating quadrupeds. In the third kind, termed *Uterus duplex*, the horns, instead of opening into a common cavity, enlarge each into a separate uterus, opening by a distinct aperture into the vagina. This kind prevails in hares and rabbits. In the American opossum and wombat, the uterus is double, with an oviduct entering the summit of each cavity. Instead of the uteri terminating separately in the vagina, as in the hare, they unite to form a common neck or opening. From each side of this neck a canal arises, and, after a semicircular course, terminates obliquely in the vagina. In the kangaroo, the uterus is simple, with a trace of a division, by means of a longitudinal ridge on each side. The oviducts terminate in the summit, one on each side of the ridge. Near their termination on each side, a lateral canal arises, which terminates in the vagina. The uterus, with these lateral semicircular canals, has been termed *Uterus anfractuosus*.

The *Vagina* is a short dilatable canal, the walls of which consist of cellular substance, supplied with bloodvessels, and lined, internally, with a soft mucous covering. At its origin it embraces the neck of the uterus, which usually projects a little, and whose opening is denominated *os tincæ*, from a supposed resemblance to the mouth of a tench. It is considered as terminating at the orifice of the urthera, in the *vulva*, the common cavity opening externally, and exhibiting, in the different species, very remarkable peculiarities.

The manner in which the uterine system is excited to action, so as to produce impregnation, appears to be involved in the deepest obscurity. Various conjectures have been proposed, and hypotheses advanced, in the absence of accurate observation or decisive experiment. That the impregnating fluid is conveyed through the uterus and oviducts to the ovarium, is rendered probable, by the circumstance, that many of the eggs or germs of those animals which produce many young at a birth, never reach the uterus, but are retained in the remote extremities of the horns, where they are brought to maturity. But the following experiment, which has been repeated by many observers, puts the subject beyond a doubt. When one of the oviducts is divided, as has been tried with rabbits, the corresponding ovarium, though capable of producing *corpora lutea*, cannot generate a *fœtus* or impregnated germ *. How the sperm penetrates the small mouths of the oviducts to be conveyed to the ovarium, will not likely be soon determined. A small portion probably suffices to produce the effect, as is known to be the case with oviparous animals and plants. It is not ascertained in the case of those animals which produce but one at a birth, whether the ovaria contribute the germs irregularly or in succession. But it is probable that, in the case of twins †, and where more males than one are admitted, as in the dog, both ovaria contribute their share.

---

* See some interesting experiments in reference to this subject by Dr Haighton, Phil. Trans. 1797, p. 159.; and by Mr Cruickshanks, ib. p. 197.

† There is a very curious instance of a white woman producing twins, one white, and the other a negro, in Dr Elliotson's valuable Translation of Blumenbach's "Institutions of Physiology," p. 331.—"A white woman of very gay character left her husband, and some time afterwards returned pregnant to her parish, and was delivered in the work-house of twins,"—"one of which," says Mr Blackaller (of Weybridge), in an account which he very handsomely sent me, "was born of a darker colour than I have usually observed the infants of negroes in the West Indies; the hair quite black, with the woolly appearance usual to them, with nose flat and lips thick; the second child had all the common appearances of white children."

Where fœtuses occur with additional parts, such as a double head, or where two bodies are united at a particular part, as in the celebrated Hungarian sisters, has the monstrosity been produced by the evolution of a double ovum, or by the union of two ova in the uterus? Many circumstances countenance the latter supposition.

The particular period of life at which the organs of reproduction become qualified for the exercise of their peculiar functions, appears to be influenced by several circumstances. When the body arrives at maturity, in respect to size and strength, by slow degrees, as is the case with man, the term of puberty is proportionally distant from birth. On the other hand, when maturity of size and strength is quickly attained, as is the case with rabbits, the term of puberty as quickly arrives. In all these cases, however, puberty precedes by months or years the maturity of the body. In the human race, puberty takes place at an earlier period in the female than in the male. Even among individuals of the same species, very remarkable differences prevail. In general, however, it is observed, that all those circumstances, whether food or shelter, which accelerate the growth of the body, exercise a proportional influence on the reproductive system, so that the period of puberty is uniformly earlier in domesticated than in wild animals, and in those which are fed plentifully with food than in those which are scantily supplied. The same circumstances seem to operate in the production of ova, domesticated animals being more fertile than wild ones of the same species *.

* There is a very remarkable difference in this respect between animals and vegetables; the reproductive system of the former exerting itself prematurely when the body is liberally supplied with nourishment, while in the latter a scanty supply of nourishment promotes, and a liberal supply retards, early flowering. A tree planted in a poor soil, or one unfavourable to its growth, will announce to the owner the unsuitableness of its situation by the production of flowers and seed, while the same species growing in a better, will not produce flowers, but merely increase in size.

Many productive operations of horticulture depend upon this singular law. Dwarfing, laying bare some of the roots of trees in winter, transplanting, paring off portions of the bark, and tying wires round the stems, are each calculated to diminish the supply of food sent to the branches, and are well known to accelerate the production of fruit, Many seeds,

In every species, except man, there is a particular period of the year in which the reproductive system exercises its energies. Domestication brings on this period earlier than ordinary, and in those animals which have several births in a season, their frequency is considerably increased. In all cases, however, as I have formerly mentioned, the season of love and the period of gestation are so admirably arranged, that the young ones are produced at the time wherein the conditions of temperature and food are most suitable to the commencement of life.

In some animals only one ovum is impregnated, and only one fœtus is produced each time. In others, many ova are impregnated at once, and a corresponding number of fœtuses are produced at a birth. The circumstances which limit the number of ova each individual is capable of producing, during life, have not been determined. Much depends on the healthy state of the ovaria, and the quantity of nourishment with which they are supplied. When one ovarium was extracted in the sow, HUNTER observed that the number of pigs produced at a farrow, was not diminished, nor the periods between the farrows lengthened; while she ceased to become pregnant much sooner, and did not produce one-half as many as a perfect sow, with which she was compared *.

When the sexual union has taken place, and the impregnated germ has been detached from the ovarium, it is deposited in the uterus, for the purpose of being brought to maturity, or at least for being prepared for birth. This

---

upon being kept two or more years, according to the same law, are less disposed to run to *straw*, or are more productive of flowers and seeds than those which are sown in the first year.

* "An experiment to determine the effect of extirpating one ovarium upon the number of young produced." Phil. Trans. 1787, p. 233.

germination or evolution appears to take place in quadrupeds according to two different plans. In the first, there is no adhesion of the germ to the walls of the uterus, while, in the second, placentation or adhesion takes place.

In the Marsupial genera, which we have seen are furnished with a complicated uterus, there is no trace, in the young, after birth, of any umbilical cord; at least Sir E. HOME could not detect any in the fœtus of the kangaroo after exclusion *. He has given a representation of a substance found in an impregnated uterus †, which he conceives to be a fœtus, in an early stage of formation; but which does not bear the remotest resemblance to the subject. From the communications made to him by Mr CONSIDEN, it appears that the impregnated uterus is filled with a gelatinous substance, of a bluish-white colour, in consistence like half melted glue, and so extremely adhesive as to be with difficulty washed off from the fingers. From the observations of Mr BELL, which were likewise communicated to Sir E. HOME ‡, it appears that the double uterus of the wombat is filled, in its impregnated state, with the same kind of gelatinous substance which closes up the *ora tincæ*. "I made a longitudinal incision into the largest of the uteri, and found its coats lined with the same jelly met with in its os tincæ. Continuing the incision through this jelly, and at the same time using gentle pressure, there issued a quantity of a thin pellucid fluid, accompanied by an embryo wrapped up in very fine membranes, which contained some of the same transparent fluid. The membranes did not appear to be at all connected by vessels either to the uterus or gelatinous matter. I

* Phil. Trans. 1795, p. 233. † Ibid. Table xx. Fig. 2.

‡ Phil. Trans. 1808, p. 309.

had no doubt of the other uterus containing a similar embryo in a less advanced state."

The origin of this gelatinous substance has not been satisfactorily ascertained. As the coats of the uterus are thin, it is scarcely to be considered as a secretion from these, but is more likely to proceed either from the oviducts or lateral canals. But the manner in which the fœtus is nourished in this jelly, is a question which remains to be determined.

The evolution of the ovum by means of placentation, is the most common method observed by Nature, and has been longest known and studied by physiologists. The ovum is first attached to the walls of the uterus, which are lined for this purpose with a layer of lymph, termed *membrana decidua*; a vascular body termed a *placenta*, is then generated, in which the minute branches of the enlarged uterine arteries terminate, and from which the vessels destined to convey the blood for the nourishment of the fœtus take their rise. The blood is returned from the fœtus by other vessels which terminate in the placenta. These vessels, which unite the fœtus with the placenta, are collected into the umbilical cord or Navel-string, so named in consequence of the form exhibited, and the place of insertion in the fœtus. At the commencement of this evolution, the os tincæ is obstructed by a glairy matter, which, while it closes the aperture, prevents the adhesion of the margins. But, in order to illustrate this curious subject still farther, it will be necessary to return to the structure of the ovum.

The ovum, after it has sufficiently increased in size, appears to be surrounded by two membranes or involucra. The innermost one is termed *amnios*, and the external one *chorion*.

The *amnios* is destined to secrete a peculiar fluid, termed *liquor amnii*, with which the fœtus is surrounded. This fluid exhibits very different properties, according to the species from which it is obtained. The skin of the fœtus is frequently covered with a fatty matter, which some consider as a deposition from the liquor amnii, but which others, with greater propriety, regard as a cutaneous secretion. Indeed, suspicions have arisen that the liquor itself is secreted by the fœtus; and these are countenanced by the circumstance, that, in the amnios of some species, as man, no bloodvessels can be traced in its structure. Whatever may be its origin, it is obviously useful in protecting the fœtus from being injured by any sudden shock, or compression.

The *chorion*, in many animals, serves the purpose of a support to the vessels which form the umbilical cord. It joins the amnios at the umbilical cord, and is united with the uterus wherever there is a placenta. In the sow and mare it is united to the whole internal surface of the uterus by numerous tubercles. In the ruminating quadrupeds these tubercles appear to collect in groups, which have been termed *cotyledons*. These consist of the *glandulæ uterinæ* or fleshy excrescence of the inner surface of the uterus, and the *carunculæ* or corresponding glands of the chorion. In the sheep and goat the glandulæ are concave, and receive the convex surface of the carunculæ, while this arrangement is reversed in the cow and deer. The third kind of placenta may be considered as arising from the more complete union of the tubercles, not into cotyledons, but into two masses, sometimes only one, either lying close together, or spread out like a belt.

In many quadrupeds there is a peculiar sac, termed *allantois*, which occurs between the amnios and chorion in some cases, as the mare, occupying the whole cavity; in others, as the sheep and cow, it is more limited in its extent *. It is destitute of bloodvessels. Its neck enters the fœtus along with the umbilical cord, and joins with the urachus of the bladder. It is the receptacle of the urine of the fœtus. In the allantois of the mare and the sow, there is a fleshy-like mass, which some have considered as a sediment from the urine, but whose nature has never been examined. It has been long known by the name of *horse-venom*, or *hippomanes* †.

In the same situation as the allantois, but unconnected with the urachus, is found, in some of the digitated quadrupeds, as the dog and cat, a peculiar organ termed *tunica erythroides*. It is connected with the mesenteric veins of the fœtus, and in the early part of pregnancy is filled with a watery fluid.

The condition of the germ, when detached from the ovarium, has not been satisfactorily determined. It is destitute of character, and too minute and delicate for accurate observation. In the uterus, however, it soon expands, its investing membranes become more apparent, and an opake spot at length appears, the rudiment of the fœtus ‡. The manner in which the whole ovum is nourished at this period, is veiled in obscurity. But after the germ has evolved, and the connection with the uterus been established by the circulating system, the means of growth are more obvious. The blood of the placenta is absorbed by the umbilical vein of the fœtus, and while part is convey-

---

* The allantois of the cow is frequently preserved in a dried state, and used to protect the surface of sores from the action of the air.

† This last term, as used by ARISTOTLE, *Hist. An.* viii. 24. refers to the mucus on the skin of a foal at birth, which the mother removes by licking; or, as in vi. 18. to the humor ex equarum equientium naturalibus distillans.

‡ HALLER found this opake spot in the sheep on the nineteenth day after impregnation. HAIGHTON observed it in the rabbit on the tenth day.

ed directly to the liver, the other portion goes to the pulmonic auricle by the inferior vena cava. The septum between the pulmonic and systemic auricle is at this time incomplete, there being a valvular aperture termed *foramen ovale*. Through this opening part of the blood in the pulmonic auricle escapes into the systemic auricle, and the remaining portion, passing into the pulmonary artery, instead of going directly to the lungs, is conveyed to the aorta, by a passage termed *ductus arteriosus*. The blood thus expelled by the systemic ventricle into the aorta, is conveyed to the different parts of the body, but a great part is conveyed back to the placenta by the umbilical arteries, which take their rise from the iliacs.

Although the fœtus is thus nourished by blood derived from the mother, the communication is so indirect, that no injection into the uterine vessels passes into the fœtus, and no injection of the fœtus reaches the vessels of the mother. The two sets of vessels, therefore, do not anastomose. The imperfect connection of the vessels is still farther demonstrated by the superior rapidity of circulation in the fœtus, when compared with the mother, the heart of the former, in the human subject, beating about one hundred and twenty times in a minute. The circulation of the fœtus even continues for some time after the death of the mother. But however indirect the communication between the two circulating systems, the condition of the mother still exercises a considerable influence over the growth of the fœtus. A derangement in the flow of her blood, by diminishing the quantity sent to the placenta, must consequently reduce the supply to the umbilical veins *. A diseased state of the mother is likewise frequently communicated to the fœtus.

* Where more fœtuses than usual are generated in the uterus, the supply of nourishment being thus divided, they seldom reach the ordinary size, and

From the very imperfect organical connection which thus prevails between the mother and the fœtus, it must appear surprising, that the latter should ever be influenced in its form or markings by the mental emotions of the former. Yet, in spite of the difficulty of accounting for the manner in which the effects are produced, the instances are too numerous and well authenticated to be disregarded, in which the imagination of the mother, in the human species, during pregnancy, has impressed upon the fœtus the marks of its high excitements. There are a few well-authenticated instances of the same kind among the inferior animals *.

---

not unfrequently one or more of them are permanently imperfect. When the cow produces twins, their sexual organs are frequently imperfect, and they are incapable of procreation, particularly when the one happens to be a male, and the other a female. Such examples are termed *Free Martins.*—See HUNTER's Account of the Free Martin, Phil. Trans. vol. lxix. p. 279, or " Observations on certain parts of the Animal Economy," p. 45.

* In the Extracts from the Minute-book of the Linnean Society of London, there is given " the following account from Mr GEORGE MILNE, F. L. S. respecting the effect of the imagination of a female cat on the fœtus in the womb: One afternoon in the month of May (1806) last, while myself and family were at tea, a young female cat, which, on account of extreme playfulness, had become a great favourite, was lying on the hearth. She was pregnant for the second time, and had arrived, as nearly as I can recollect, at the middle period of gestation. A servant handing the tea-kettle, or doing some office which led her to pass between the fire and the table, trod very heavily on the creature's tail. She screamed most frightfully, and ran out of the room; and from the nature of the noise which she emitted, it was evident that a considerable degree of terror mingled with the sense of injury. But from a circumstance so extremely common, no extraordinary result was expected, and the poor cat's tail was no more thought of, until the final period of gestation, when we were surprised with the phenomenon which has given occasion to this communication. She dropped five kittens; one, which exactly resembled herself, was apparently perfect; but the other four had the tail most remarkably distorted. About one-third of the length, reckoning from the base, there was a *nodus* equal in size to a very large pea, or about twice as thick as the tail itself; the remaining portion

These results appear still more surprising, when we consider that both the absorbing vessels of the fœtus, and likewise its nervous system, appear to act but feebly; as virulent poisons injected into the pleura, peritoneum, or cellular tissue, do not appear to produce any decidedly deleterious effects *.

The circumstances on which the sex of the fœtus depends, though removed beyond the reach of observation, have nevertheless been the subject of conjecture. Many years ago, Sir F. H. Eyles Stiles, in reference to monœcious and diœcious plants, advanced the opinion, that "in all cases where the male and female organs are found separate, the defect is not in the flower, which I suppose to be originally instructed with the rudiments of the organs of both sexes, but that it arises from some circumstances in the plant, that determines it to blow the one organ and not the other †." An opinion somewhat similar is adopted by Sir Everard Home, with regard to quadrupeds. He supposes "the ovum, previous to impregnation, to have no distinction of sex, but to be so formed as to be equally fitted to become a male or female fœtus; and that it is the process of impregnation which marks the distinction, and conduces to produce either testicles or ova-

---

being turned on one side, at an angle nearly approaching to a right angle: and what may deserve notice, all of them turned the same way, towards the left side. I was urged to rear one of them as a curiosity; but, conceiving that it might grow up rather a disgusting object, I had the whole destroyed, preserving only the one which appeared to be perfect. That one I kept about a month, when it was seized, as well as the mother, with a disorder which greatly enfeebled it; and to save the parent, I destroyed the offspring. But it was previously discovered, that this also had the tail distorted, and turned aside at a considerable angle, although free from the knot which distinguished the other four."—Lin. Trans. vol. ix. p. 323.

* M. Magendie, Precis El. ii. p. 448.

† Phil. Trans. vol. lv. p. 259.

ria, out of the same materials*." The circumstances which seem to countenance this conjecture, and which are enumerated in the note below, appear, when carefully

---

* Phil. Trans. vol. lxxxix. p. 175. where he adds: "The following circumstances are in favour of this opinion. The testicles and ovaria are formed originally in the same situation, although the testicles, even before the fœtus has advanced to the eighth month, are to change their situation to a part at a considerable distance.

"The clitoris, in fœtuses under four months, is so large as to be often mistaken for a penis. Preparations to shew the size of the clitoris at this age, are preserved in Mr HUNTER's collection; and M. FERRIEN mentions it, with a view to explain an erroneous opinion that prevailed in France, that the greater number of miscarriages between three and four months have been remarked to be males; which mistake arose from the above circumstance.

"The clitoris, originally, appears therefore equally fitted to be a clitoris or penis, as it may be influenced by the ovarium or testicle.

"In considering this subject, it is curious to observe the number of secondary parts, which appear so contrived that they may be equally adapted to the organs of the male or female.

"In those quadrupeds whose females have mammæ inguinales, the males have also teats in the same situation; so that the same bag which contains the testicles of the male, is adapted to the mammæ of the female. In the human species, which have the mammæ pectorales, the scrotum of the male serves the purpose of forming the labia pudendi of the female, and the preputium makes the nymphæ. The male has pectoral nipples, as well as the female; and, in many infants, milk, or a fluid analogous to it, is secreted, which proves the existence of a glandular structure under the nipple.

"This circumstance, when added to the instances already related, of an hermaphrodite bull, and of wethers giving suck, affords a strong presumption that the rudiments of the mammæ exist in the male, and, in some few instances, have been brought to perfection, either by an original mixture of organs, early emasculation, or other changes with which we are unacquainted.

"If it is allowed that the sex is impressed upon the ovum at the time of impregnation, it may, in some measure, account for the free martin, when two young are to be impressed with different sexes of impregnation; which must be a less simple operation, and therefore more liable to a partial failure, than when two or any greater number of ova are impressed with the

examined, rather of a doubtful character, or leading to a conclusion somewhat different from the one which they are brought forward to support.

If the determination of the sex takes place at the period of impregnation, the common origin of the testicles and ovaria only indicate that in the fœtus there is a peculiar situation appointed where the production of the essential parts of the reproductive system is to take place; since previous to the organisation of any part of that system, it is fixed whether a testicle or an ovarium is to be produced. Again, if the sex is thus determined in the impregnated ovum, we might expect the peculiar characters of the sex more strongly impressed upon the fœtus than they are known to be; and instead of finding the clitoris, (upon the supposition that it is capable of being changed into a penis,) of such singular dimensions, its total absence in the female might rather be expected. Both these circumstances rather intimate, that the ovum is neuter, that the sexual organs are in part developed previous to the determination of the particular sex, or, in other words, after the egg has germinated, and the organs are in some measure evolved. There are some facts,

---

same sex. It may also account for twins being most commonly of the same sex; and, when they are of different sexes, it leads us to inquire whether the female, when grown up, has not in some instances less of the true female character than other women, and is not incapable of having children. It is curious, and in some measure to the purpose, that in some countries, nurses and midwives have a prejudice that such twins seldom breed.

"This view of the subject throws some light on those cases where the testicles are substituted for the ovaria; since whenever the impregnation fails in stamping the ovum with a perfect impression of either sex, the part formed will neither be an ovarium nor a testicle, sometimes bearing a greater resemblance to the one, sometimes to the other; and may, according to circumstances, either remain in the natural situation of the ovaria, or pass into the situation proper to the testicle, whether it is the scrotum of the male, or the labia pudendi of the female."

however, which militate equally against the opinions, that the determination of the sex takes place in the ovum, or in the fœtus. Mr KNIGHT, whose experiments on plants have greatly enlarged the bounds of vegetable physiology, has communicated the following observations in reference to the subject under consideration. "In several species of domesticated or cultivated animals (I believe in all,) particular females are found to produce a very large majority, and sometimes all their offspring of the same sex; and I have proved repeatedly, that, by dividing a herd of thirty cows into three equal parts, I could calculate, with confidence, upon a large majority of females from one part, of males from another, and upon nearly an equal number of males and females from the remainder. I frequently endeavoured to change their habits, by changing the male; but always without success; and I have in some instances observed the offspring of one sex, though obtained from different males, to exceed those of the other, in the proportion of five or six, and even seven to one. When, on the contrary, I have attended to the numerous offspring of a single bull, or ram, or horse, I have never seen any considerable difference in the number of offspring of either sex *." Such observations rather support the conclusions, that the sex of the ovum is determined previous to im-

* "On the Comparative Influence of Male and Female Parents on their Offspring," Phil. Trans. 1809, p. 397. There are some facts, however, which daily occur in the human species, which appear to indicate that the influence of the male in determining the sex of the fœtus is considerable. The following very singular one is from p. 21. of the appendix to Dr GARTHSHORE's paper, on "A remarkable case of numerous births, with observations," Phil. Trans. 1787, p. 344. "Mr KIRWAN, a very respectable philosopher, and Fellow of the Royal Society, has often told me, that he had conversed with a gentleman, who affirmed to him, that he himself was the youngest of forty sons, all produced in succession from three different wives,

pregnation. In whatever manner the determination of the sexes is effected, we perceive, that, among the number actually produced, the males and females bear such a relative proportion, as to secure a continuance of the race.

The time which elapses between the sexual union, and consequent impregnation of the ovum, to the perfection of the fœtus and its expulsion by birth, differs greatly according to the species, and does not appear to be regulated by any peculiar systems of organization. The term of gestation in the lionness and sheep is the same, or five months; the horse and the ass, animals of the same genus, have similar terms, eleven months; while the cow and the buffalo, likewise belonging to one genus, have dissimilar terms, the former going with young nine months, while the gestation of the latter occupies twelve. Even different individuals of the same species vary in this respect a few days, and even the same individual at different times.

At birth, the fœtus is expelled the uterus, the connection with the placenta ceases by the division of the umbilical cord, the enveloping membranes are torn asunder, and the young animal leaves its watery dwelling to enjoy a more independent existence. The blood from the heart, now interrupted in its exit from the system through the umbilical arteries, is transmitted in quantity to the lungs, the thorax is thereby excited to action, and the important function of respiration commences. The bloodvessels which were necessary to carry on the modified circulation with the placenta, together with the ductus arteriosus, and urachus, change their nature, and are converted into ligaments. The foramen ovale likewise closes, unless in some of those qua-

---

by one father, in Ireland, and who all arrived at the age of manhood; and Mr KIRWAN often declared, he had no reason to doubt the truth of this relation."

drupeds which live and dive much in water. The liver, the renal glands, and the thyroid gland, now diminish considerably in size. The thymous gland, which is seated under the sternum, and ascends on each side, as far as the neck, now likewise gradually diminishes in size; and, in old age, not unfrequently disappears.

The situation of young animals at birth, in reference to the locomotive powers, points out three very remarkable modifications. In the first, the young at birth are completely formed, and capable at once of moving about, and following the footsteps of the mother. These require merely a regular supply of food, and protection from danger. In the second, the young are so imperfect, that they are incapable of following the mother, and are therefore carried about by her. Among these, some, as the human female, carry about the young in their arms; while, among the marsupial animals, there is a ventral cavity into which the young are dropped at birth, and where they are nourished for some time. The young of these last are very imperfect at birth. In the third class, the young are incapable of following the mother, and she is equally unfit for carrying them. In this case, they are deposited in a nest concealed from the light, and nourished by the mother at stated intervals. So far as I know, the young, in these cases, are born blind; and, in some species, the external orifice of the ear is likewise closed. The maternal duties imposed on animals of the first class are few, when compared with those which the species of the second and third classes have to fulfil in reference to cleanliness and temperature.

The nourishment of young viviparous animals, consists of milk secreted in the teats. These organs consist of numerous glands united into a mass by cellular substance and fat. The ducts gradually unite, and at last open, in the nipple, in women, by numerous, in the lower animals, by one or two, apertures. The teats are named, from their situation, either Pectoral, Abdominal, or Inguinal. They, in general, exceed the number of young produced at a birth, but exhibit very remarkable variations, according to the species, and even among individuals of the same species.

The milk is usually of a white colour, with various shades of yellow, differing remarkably according to the species. It is somewhat heavier than water, but boils and freezes nearly at the same temperature as that fluid. It slightly reddens vegetable blues.

When allowed to rest, it separates into two portions, one of which, termed *Cream*, is the lightest, and forms a layer on the surface of the *Skimmed-milk* below.

The cream consists of three ingredients, butter, cheese, and whey, in the following proportion in a hundred parts, of cow's milk, butter, 4.5, cheese 3.5, whey 92.0 *. It was of the specific gravity of 1.0244. The skimmed milk yield- in a thousand parts, water 928.75, cheese, with a trace of butter, 28, sugar of milk 35, muriat of potash 1.70, phosphat of potash 25, lactic acid, acetat of potash, with a trace of lactat of iron, 6, earthy phosphats 0.30. Its specific gravity was 1.033 †. The salts of milk have been still more minutely investigated by C. F. SCHWARZ, who obtained from a thousand parts of cow's milk, phosphat of lime 1.805, phosphat of magnesia 0.170, phosphat of iron 0.032, phosphat of soda 0.225, muriat of potash 1.350, lactate of soda 0.115 = 3.697. A thousand parts of woman's milk yielded, phosphat of lime 2.500, phosphat of magnesia 0.500, phosphat of iron 0.007, phosphat of soda 0.400, muriat of potash 0.700, lactat of soda 0.300 = 4.407 ‡.

* BERZELIUS's Annals of Phil. vol. ii. p. 424. † Ibid.

‡ Annals of Phil. v. p. 41.

Woman's milk contains more cream than the milk of the cow; and the cheese and butter are so intimately united that it is very difficult to effect their separation. The milk of all the other animals which has been examined, contains similar ingredients, but in variable proportions.

The most nutritious portion of the milk is usually considered as the cheese, which approaches in its nature to albumen. The claims, however, of all the other ingredients to be considered as nutritious, are equally strong. They are constantly present, and consist of ingredients needed by the young system.

During the first periods of infancy, milk is the only diet administered to the young animal. All the parts of the mouth are at the same time soft, and adapted exclusively for sucking, without injuring the nipple. By the time the teeth make their appearance, the young animal is beginning to imitate the actions of its mother, and attempting to eat of the food which she employs. Weaning now takes place, and independent existence may be said to commence. Between this period and old age, there are several important stages which may be briefly noticed. The *casting of the teeth* is the mark of increased activity in the system. The old ones drop out, or are worn away, and new ones, better adapted for acting upon the food now made use of, supply their place. *Puberty* speedily takes place, and the development and influence of the organs of the reproductive system, communicate to the individual impressive marks of sexual distinction. *Maturity* succeeds, with its accompanying strength and energy. Old age at length approaches, with its feebleness and inactivity, and a diminished power of generating heat or cold. The hair approaches to whiteness. The teeth fall out, and the means of obtaining food thus failing, the body sinks to rest. Man is the only animal that can counteract the fatal consequences attending the loss of teeth, by adapting his food to the altered state of the masticating organs, and thus outlive the period at which the life of the other species is doomed to cease.

Having thus given a general view of the sexual organs of viviparous animals, and the manner in which these exercise their functions, we are next led to consider a second class of animals with distinct sexes, in which, however, the ovum is not nourished by the circulating system of the mother.

II. Oviparous Animals.

Among the viviparous animals, the reproductive organs present many points of resemblance, and appear to be constructed according to a common model. It is otherwise with the sexual organs of the oviparous tribes, which we are now to consider. They exhibit such remarkable differences in the form and structure of all their organs, and occupy so many different situations, that it is impossible to collect them into natural groups, or assign to them characters which they have in common. It will be expedient, therefore, in this place, to avoid all minuteness of detail, and to take notice of the peculiar modifications of the sexual organs exhibited by particular tribes, when we come to treat of these in the general classification.

1. *Male organs.*—In the essential parts of the male organs of oviparous animals, few modifications of any consequence present themselves. The testes are always concealed in the abdomen. They are in general two in number, and obviously distinct from each other; but in some cases, among the mollusca, and annulosa, they appear to be collected in a single group. The spermatic duct is either double or single, according to the structure of the organ from which it proceeds; and at its external termination in birds, for example, either opens into a tubular penis,

as in the drake, and many species of the inferior classes; at the base of a solid penis, along which there is a groove, as in the ostrich, or in one or two tubercles, situated in the cloaca, the common opening of the ureters and rectum.

The influence of the season of love is much more considerable on the male organs of oviparous than viviparous animals. They increase to an extraordinary size, appear full of sperm, and occupy a large portion of the cavity of the abdomen. At other seasons they diminish so much, as in some cases to be detected with difficulty.

1. *Female organs.*—Among the oviparous animals, the female organs exhibit very remarkable differences in their structure and functions. In birds the ovarium is single, while in fishes it is double. The oviduct either leads to the cloaca directly, or suffers a previous uterine enlargement, or it terminates in a tubular elongation, to which the name of *ovipositor* has been applied.

Before considering the manner in which impregnation takes place, it is necessary to give a general view of the structure of the egg, employing that of the hen as an example. The *shell* or external covering, consists, according to VAUQUELIN, of 89.6 of carbonat of lime, 5.7 of phosphat of lime, and 4.7 of animal matter. It is penetrated by numerous pores, through which air passes during incubation. On its inner surface, and as an integument to the remaining contents, is found the firm, white membrane termed *membrana albuminis.* At the larger end of the egg, this membrane includes a cavity, filled with atmospheric air, termed *folliculus aëris*, which, by degrees, enlarges during incubation. The membrane includes the glaire or white, divisible into two layers, each surrounded by a delicate membrane. The external layer is the most fluid and transparent. The centre of the egg is occupied with the *yolk*, enveloped by its peculiar membrane, termed the *yolk-bag*. From each end of the yolk proceeds a white knotty body, which terminates in the glaire by a white flocculent extremity. These knotty bodies are called the *chalazæ* or *grandines*. What is termed the *cicatricula*, or tread of the cock, is a round milk-white spot, formed after impregnation, on the surface of the yolk-bag, and surrounded with whitish concentric circles, termed *halones*. The glaire contains chiefly albumen and water, and the yolk has, in addition, a portion of oil, to which it owes its yellow colour.

The glaire and the yolk are common to all eggs, differing greatly, however, in many of their characters. Where the egg is destined to be hatched in water, the glaire is a substance intermediate between gelatine and albumen, and capable of resisting the macerating effects of the surrounding fluid. The yolk is sometimes single, as in the eggs of insects, or compound, producing in one egg many young, as in some species of leeches. The membrana albuminis is present in many eggs, while, in several cases, the glaire of many yolks is united to form a connected mass.

In birds, the ovarium appears like a bunch of grapes, consisting of the bags containing the yolks, with their stalks of attachment. When the yolk has attained its full size, the bag or calyx in which it is contained, analogous to a corpus luteum, exhibits a white shining line, marking the intended opening, which at length takes place, and suffers the yolk to escape into the expanded extremity of the oviduct. The calyx is by degrees absorbed, and in old age nearly disappears. The yolk, in passing along the oviduct, acquires the glaire, and likewise the shell, and becomes fit for exclusion.

The manner in which the eggs of birds are impregnated by the male has not been satisfactorily determined. With the exception of the cicatricula, a bird, in the absence of a male, can produce an egg. The conjunction of the sexes, however, is necessary for the impregnation of the egg, and the effect is produced previous to the exclusion.

In many kinds of fishes and reptiles, the yolks, after being furnished with their glaire, are ejected from the body of the female, and the impregnating fluid of the male is afterwards poured over them. Impregnation can be effected readily in such cases, by the artificial application of the spermatic fluid.

Impregnation in insects appears to take place while the eggs pass a reservoir containing the sperm, situated near the termination of the oviduct in the vulva. "In dissecting," says JOHN HUNTER, to whom we owe the discovery, "the female parts in the silk moth, I discovered a bag lying on what may be called the vagina, or common oviduct, whose mouth or opening was external, but it had a canal of communication between it and the common oviduct. In dissecting these parts before copulation, I found this bag empty; and when I dissected them after, I found it full *." By the most decisive experiments, such as covering the ova of the unimpregnated moth, after exclusion, with the liquor taken from this bag in those which had sexual intercourse, and rendering them fertile, he demonstrated that this bag was a reservoir for the spermatic fluid, to impregnate the eggs as they were ready for exclusion, and that coition and impregnation were not simultaneous. It has not been determined whether the same arrangement prevails in all insects. This is a very near approach to the external impregnation of the ova, as it takes place in many fishes and reptiles.

After the ovum has been impregnated and ejected, it re-

* Phil. Trans. 1792, p. 186

quires a determinate temperature to excite the germ to action; the nourishment is obtained from the glaire and the yolk by means of umbilical vessels, and the process of aëration is effected through the pores of the shell, and the walls of the folliculus aëris, or by the intervention of the common membrane or glaire *. The hatching is accomplished at different periods, and by the use of different means.

---

* The following description of the appearances of the incubated egg of the common hen, may not be unacceptable to the reader: "A small shining spot of an elongated form, with rounded extremities, but narrowest in the middle, is perceived at the end of the first day, not in nor upon the cicatricula, but very near that part on the yolk-bag. This may be said to appear beforehand, as the abode of the chick which is to follow:"—"No trace of the latter can be discerned before the beginning of the second day; and then it has an incurvated form, resembling a gelatinous filament with large extremities, very closely surrounded by the amnion, which at first can scarcely be distinguished from it.

"About this time the halones enlarge their circles; but they soon after disappear entirely, as well as the cicatricula.

"The first appearance of red blood is discerned on the surface of the yolk-bag, towards the end of the second day. A series of points is observed, which form grooves, and these closing, constitute vessels, the trunks of which become connected to the chick. The vascular surface itself is called *figura venosa*, or *area vasculosa*; and the vessel by which its margin is defined, *vena terminalis*. The trunk of all the veins joins the venæ portæ; while the arteries which ramify on the yolk-bag arise from the mesenteric artery of the chick.

"On the commencement of the third day, the newly formed heart is discerned by means of its triple pulsation, and constitutes a threefold *punctum saliens*. Some parts of the incubated chicken are destined to undergo successive alterations in their form, and this holds good of the heart in particular. In its first formation, it resembles a tortuous canal, and consists of three dilatations lying close together, and arranged in a triangle. One of these, which is properly the right, is then the common auricle, the other is the common ventricle, but afterwards the left; and the third is the dilated part of the aorta.

The most simple mode of hatching is effected by the situation in which the eggs are placed by the mother, after

---

" About the same time, the spine, which was originally extended in a straight line, becomes incurvated; and the distinction of the vertebræ is very plain. The eyes may be distinguished by their black pigment, and comparatively immense size; and they are afterwards remarkable in consequence of a peculiar slit in the lower part of the iris.

" From the fourth day, when the chicken has attained the length of four lines, and its most important abdominal viscera, as the stomach, intestines, and liver, are visible, (the gall-bladder, however, does not appear till the sixth day), a vascular membrane (*chorion*, or *membrana umbilicalis*), begins to form about the navel, and encreases in the following days with such rapidity, that it covers nearly the whole inner surface of the shell within the membrana albuminis, during the latter half of incubation. This seems to supply the place of the lungs, and to carry on the respiratory process instead of those organs. The lungs themselves begin, indeed, to be formed on the fifth day; but, as in the fetus of the mammalia, they must be quite incapable of performing their functions while the chick is contained in the amnion.

" Voluntary motion is first observed on the sixth day, when the chick is about seven lines in length.

" Ossification commences on the ninth day, when the ossific juice is first secreted, and hardened into bony points. These form the rudiments of the bony ring of the sclerotica, which resembles at that time a circular row of the most delicate pearls.

" At the same period, the marks of the elegant yellow vessels on the yolk-bag begin to be visible.

" On the fourteenth day, the feathers appear; and the animal is now able to open its mouth for air, if taken out of the egg.

" On the nineteenth day, it is able to utter sounds; and on the twenty-first to break through its prison, and commence a second life.

" The chorion, that most simple yet most perfect temporary substitute for the lungs, if examined in the latter half of incubation in an egg very cautiously opened, presents, without any artificial injection, one of the most splendid spectacles that occurs in the whole organic creation. It exhibits a surface covered with numberless ramifications of venous and arterial vessels. The latter are of the bright scarlet colour, as they are carrying oxygenated blood to the chick; the veins, on the contrary, are of the deep or livid red, and bring the carbonated blood from the body of the animal. Their or during their exclusion. In this mode, a place is usually selected where the egg will be exposed to a suitable and uniform temperature, and where a convenient supply of food may be easily obtained for the young animals. Such arrangements prevail in insects.

In the second, the mother (aided in some cases by the sire) forms a nest, in which she deposits her eggs, and, sitting upon them, aids their hatching by the heat of her body. Birds in general hatch their young in this manner.

In the last, the eggs are retained in the uterus, without any connection, however, by circulating vessels, until the period when they are ready to be hatched, when egg and young are expelled at the same time. This takes place in some sharks and mollusca. The animals which exercise this last kind of incubation are termed *Ovoviviparous*. In

---

trunks are connected with the iliac vessels, and, on account of the thinness of their coats, they afford the best microscopical object for demonstrating the circulation in a warm-blooded animal.

"The other membrane, the *membrana vitelli*, is also connected to the body of the chick, but by a twofold union, and in a very different manner from the former. It is joined to the small intestine by means of the *ductus vitello-intestinales*: and also, by the bloodvessels which have been already mentioned, with the mesenteric artery and vena portæ.

"In the course of the incubation, the yolk becomes constantly thinner and paler by the admixture of the inner white. At the same time, innumerable fringe-like vessels with flocculent extremities, of a most singular and unexampled structure, form on the inner surface of the yolk-bag, opposite to the yellow ramified marks above mentioned, and hang into the yolk. There can be no doubt that they have the office of absorbing the yolk, and conveying it into the veins of the yolk-bag; where it is assimilated to the blood, and applied to the nutrition of the chick. Thus, in the chicken which has just quitted the egg, there is only a remainder of the yolk and its bag to be discovered in the abdomen. These are completely removed in the following weeks, so that the only remaining trace is a kind of cicatrix on the surface of the intestine."—BLUMENBACH's Comparative Anatomy, London, 1807, p. 479-484.

the Rana pipa, the eggs are deposited in a bag on the back, where they are hatched, and where the young animals reside for some time after birth. Some animals, as the Aphis, are oviparous at one season, and ovoviviparous at another.

In what manner the blood of the embryo of ovoviviparous animals is aerated, has not been satisfactorily determined.

The young, after being hatched, are, in many cases, independent of their parent, and do not stand in need of any assistance: they are born in the midst of plenty, and have organs adapted to the supply of their wants. Thus, many insects are hatched on, or within the very leaves which they are afterwards to devour. In other cases, the young are able to follow their parents, and receive from them a supply of appropriate food; or if unable to follow, their parents bring their food to the nests.

The changes which the young of oviparous animals undergo in passing from infancy to maturity, have long attracted the notice of the inquisitive observer. The egg of the frog is hatched in the water, and the young animal spends in that element a part of its youth. While there, it is furnished with a tail and external branchiæ; both of which are absorbed, and disappear, when it becomes an inhabitant of the land. The infancy of the butterfly is spent in the caterpillar-state, with organs of motion and mastication which are peculiar to that period. It is destined to endure a second hatching, by becoming enveloped in a covering, and suffering a transformation of parts previous to appearing in its state of maturity. These metamorphoses of oviparous animals present an almost infinite variety of degrees of change, differing in character according to the tribes or genera.

In birds, it is well known that one sexual union suffices for the production of impregnated eggs during the period of laying. This is a case somewhat analogous to those quadrupeds which produce several young at a birth with one impregnation, differing, however, in the circumstance that the eggs are not all produced at the same time, although they are afterwards hatched by the same incubation. In the *aphides*, or plant-lice, as they are called, one impregnation not only renders fertile the eggs of the individual, but the animals produced from these, and the eggs of those again, unto the ninth generation*.

III. Androgynous Animals.

The structure of the reproductive system in those animals which have the sexes united in the same individual, exhibits two distinct modifications. In the first, impregnation can only be effected by the union of two individuals. In the second, the hermaphroditism is complete, and impregnation takes place without any assistance.

1. In nearly all those androgynous animals, where there is sexual union, the testicle is single,—there is an external penis,—and the opening for the escape of the spermatic fluid is situated at its base. The ovarium is also single. The external openings of both organs are uniformly situated on one side of the body. In some cases, they terminate in a common cavity, while, in others, the openings of the male and female organs are removed to some distance from each other.

Impregnation, in the animals now under consideration, takes place internally, by the mutual application of the sexual organs of two individuals. The eggs, in a few species, are retained until hatched; but, in general, they are

* "Observations on the Aphides of Linnæus," by Dr Wm. Richardson, Phil. Trans. 1771, p. 182—194.

excluded a short time after impregnation. Even without being impregnated, the female organs can produce eggs similar in size and appearance to the perfect ones, but which, as possessing no vitality, soon go into decay. This we have often witnessed, when a single individual of one of the Lymnæi has been confined in a vessel of water.

Examples of this condition of the reproductive system do not occur in any of the tribes of vertebral animals. They are, however, very common among the mollusca, particularly in the pulmoniferous gasteropoda, as the snail and slug.

2. In those androgynous animals where the hermaphroditism is complete, the male organs have not been satisfactorily ascertained. During the season of conception, the ovarium is replete with a milky fluid, which is probably the sperm, and which has been conjectured to proceed from a testicle concealed by being incorporated with the ovarium.

The eggs, in some cases, are ejected from the body previously to their being hatched, while, in others, they are hatched internally. In these last, as among the Mollusca conchifera, the young are sometimes found in the gills, into which they have escaped from the ovarium.

Examples of this structure of the reproductive system occur in the whole of the molluscous animals belonging to the classes Conchifera and Tunicata.

IV. Gemmiparous Animals.

In the reproductive systems, which we have hitherto been considering, sexual organs could be distinctly perceived. In those to which our attention is now to be directed, neither male nor female organs can be detected. No separate act of impregnation is required; but the young are produced by buds forming on the surface of the body, and falling off, upon maturity, to the enjoyment of independent existence. The *gemmiparous* mode of reproduction is very strikingly exemplified in the genus Hydra, or Fresh-water Polypus. The rudiment of the future young polypus appears as a small tubercle, produced by an elevation of the skin of the parent; afterwards it projects still farther, and acquires an opening for its mouth and the tentacula surrounding its margin; a natural separation then takes place between the lower part of the young animal and the skin of the parent, when it becomes detached and independent. Two or more such buds may be observed expanding on the same parent at once; and, previous to the young dropping off, other buds may be observed evolving on their surface.

This mode of reproduction appears to be confined to the class of Zoophytes. It is not, however, the only method of generation exhibited by these animals. The Sertulariæ not only increase by the lateral evolution of their young, but by the production of vesicles containing ova.

As nearly related to the gemmiparous mode of reproduction, we may here take notice of the multiplication of animals by the *Spontaneous Division* of their bodies. The celebrated Ellis, in consequence of a hint he received from M. De Saussure, examined several species of the *animalcula infusoria*, in which he observed the body begin to contract in the middle, and at length to become divided by a transverse separation. Each of the parts assumed an independent existence *. The celebrated Muller likewise observed this singular multiplication of individuals taking place both transversely and longitudinally †. A si-

* Phil. Trans. 1769, p. 138.

† Animalcula infusoria, fluviatilia et marina, quæ detexit, systematice descripsit et ad vivum delineari curavit, O. F. Muller. Havniæ 1786, præf. p. ix.

milar mode of propagation has been observed among the Nereidæ and Planariæ.

Doubts have been reasonably entertained whether this is a natural mode of propagating individuals. ELLIS delivers his opinion on the subject with little hesitation. "The proportion, (says he), of the number of the animals which I have observed to divide in this manner, to the rest, is scarce 1 to 50; so that it appears rather to arise from hurts received by some few animalcula among the many, than to be the natural manner in which these kinds of animals multiply; especially if we consider the infinite number of young ones which are visible to us through the transparent skins of their bodies, and even the young ones that are visible in those young ones, while in the bodies of the old ones *." The multiplication of individuals by artificial division, countenances such an opinion. The common Hydra, which is naturally gemmiparous, may be cut into two or more parts; and each detached portion, by repairing the wound, and supplying what is defective, will, in a short time, become a perfect animal. Such divisions are similar to slips or grafts in the vegetable kingdom; and although they obviously are merely extensions of an individual, yet, by acquiring new organs, and becoming capable of exercising distinct volitions, they acquire an identity peculiar to themselves. They are multiplications of individuals at least, and, consequently, serve the same purposes as the products of those other modes of generation which have been regarded as more perfect †. They afford very

* Phil. Trans. 1769, p. 143.

† Many respectable botanists and horticulturists of the present day appear to regard all plants produced from cuttings, layers, roots or buds, as *extensions* merely of those plants to which they originally belonged, and as being influenced, in reference to their duration, by circumstances different

striking illustrations of that *repairing* attribute of the vital principle, to which our attention has been already (p. 19.) directed.

---

from those which regulate the continuance of plants obtained immediately from the germination of the seed. MARSHALL, in his Rural Economy of Glocestershire, published 1789, vol. ii. p. 239., remarks, "Engrafted fruits are not permanent, they continue but for a time." KNIGHT, in his Treatise on the Culture of the Apple and Pear, p. 6., has followed up the same idea, when he says, "The continuance of every variety appears to be confined to a certain period, during the early part of which only it can be propagated with advantage to the planter." BUCKNALL carries these views still farther: Trans. Soc. En. Arts vol. xvii. p. 268. "When the *first stock* shall, by mere dint of old age, fall into actual decay, a nihility of vegetation, —the *descendents*, however young, or in whatever situation they may be, will gradually decline; and from that time it would become imprudent, in point of profit, to attempt propagating that variety from any of them." From these statements, Sir JAMES E. SMITH, Introd. Bot. p. 138. and 139. seems to consider it as established, that "propagation by seeds is the only true reproduction of plants."

The sympathy which is here considered as prevailing between the parent stock and its extensions or descendents, or the dependence of the life of the latter on the duration of that of the former, the basis of this opinion, is not only unsupported by proof, but is directly at variance with a multitude of common occurrences.

The wall-flower and sweet-william plants, whose natural term of life rarely extends beyond two years, or until all the branches flower once, may be continued for many years, by being propagated by means of cuttings of the slips. Even the annual stem of the Scarlet Lychnis, may be converted into separate plants of many years duration. If the existence of this dependence of the plants derived from cuttings, on the life of the parent plant from which they were taken, can thus be disproved in those species on which we can most easily make accurate observations, it must appear unphilosophical to believe that it exists in those which outlive us by many centuries, and the laws of whose duration, therefore, have not yet been determined.

The distinction between propagation by seed, and extension by cuttings, if restricted to the manner of multiplying plants, may be harmless in science, and in horticulture useful; but when it includes an expression of a law of vegetable life, of difference in the products, as if the plants obtained by the

V. Hybridous Animals.

In the accomplishment of the important purpose of generation, it is observed, that, in the season of love, indivi-

---

former method enjoyed an individuality distinct in its nature from the results of the latter, we are disposed to conclude, that it is a distinction which has been incautiously adopted and which is apt to mislead.

That many of the valuable cider and perry fruits of the seventeenth century have already disappeared,—parents and extensions, and, that some of our present fruits are gradually wearing into decay, are facts which have been satisfactorily established. But in order to account for these extinctions, it is not necessary to admit, that all cuttings are limited in their duration to the term of life of the parent from which they have been taken. The whole phenomena seem simply to intimate, that extensions from a *diseased parent*, are, in many cases, diseased likewise, and that the skill and industry of the horticulturist cannot restore such to a healthy state.

To have combated the assertion, that *propagation by seeds is the only true reproduction of plants*, would have been unnecessary, had it not been made by a deservedly celebrated botanist, whose authority, however, is much greater in systematical than in physiological botany. That method of reproduction in plants must surely be regarded as genuine, which is employed by Nature uniformly and extensively.

In many cases, the multiplication of individuals takes place in the same plant, both by extensions and seed. Among the herbaceous plants, many of which are, in fact, annuals, the species of Orchis and Tulipa may be noticed. Seeds are produced in these by the ordinary reproductive organs. At the same time each individual, before closing its life of a year, prepares a decendent or bulb, which the following year supplies its place, when, perhaps, not a single seed which it has produced has germinated. The continuance of many species whose seeds germinate with difficulty, depends on these radical extensions or bulbs. Such natural extensions are not confined to the monocotyledonous groups; they occur in decotyledonous plants also; and among many of the acotyledonous tribes, there is no other natural method of reproduction known.

In other cases, where the continuance of the species could not be effected by the ordinary methods of impregnation and the production of seeds, the reproductive system furnishes bulbs, or extensions, to supply their place. In the case of the *viviparous grasses*, the germs which are to form the duals of a particular species are drawn together by mutual sympathy, and excited to action by a common propensity. The produce of a conjunction between individuals of the same species, partakes of the characters common to the species, and exhibits in due time the characteristic marks of puberty and fertility. In a natural state, the *selective attribute* of the procreative instinct unerringly guides the individuals of a species towards each other, and a *preventive aversion* turns them with disgust from those of another kind.

In a domesticated state, where numerous instincts are suppressed, and where others are fostered to excess, individuals belonging to different species are sometimes known to lay aside their natural aversion, and unite in the business of propagation. Instances of this kind occur among quadrupeds, birds and fishes,—among viviparous and oviparous animals,—where impregnation takes place within as well as when it is effected without the body. The product of such an unnatural union is termed a Hybridous animal. The following circumstances appear to be connected with hybridous productions.

1. *The parents must belong to the same natural genus or family.*—There are no exceptions to this law. Where the species differ greatly in manners and structure, no constraint or habits of domestication will force the unnatural union. On the other hand, sexual union sometimes takes

---

young plants, are prepared in the absence of sexual organs, and, consequently, without impregnation. Upon falling off from the parent, they strike root and enjoy an individuality of character, as decided as if the reproductive organs had gone through the whole of the ordinary process. By means of such extensions, and without ever producing seeds, the Festuca vivipara is found in every pasture field of the northern islands of Scotland. When extensions are obtained from healthy parents, or generated by such, their duration will probably extend to the ordinary term assigned to the species.

place among individuals of nearly related species. Thus, among quadrupeds, the *mule* is the produce of the union of the horse and the ass. The jackall and the wolf both breed with the dog. Among birds, the canary and goldfinch breed together, the Muscovy and common duck, and the pheasant and hen. Among fishes, the carp has been known to breed with the tench, the crusian, and even the trout *.

2. *The Parents must be in a confined or domesticated state.*—In all those hybridous productions which have yet been obtained, there is no example of individuals of one species giving a sexual preference to those of another. Among quadrupeds and birds, those individuals of different species which have united, have been confined, and excluded from all intercourse with those of their own kind. In the case of hybridous fishes, the ponds in which they have been produced have been small and overstocked, and no natural proportion observed between the males and females of the different kinds. As the impregnating fluid, in such situations, is spread over the eggs after exclusion, a portion of it belonging to one species may have come in contact with the unimpregnated eggs of another species, by the accidental movements of the water, and not in consequence of any unnatural effort.

In all cases of this unnatural union among birds or quadrupeds, a considerable degree of aversion is always exhibited, a circumstance which never occurs among individuals of the same species †.

---

* Phil. Trans. 1771, p. 318.

‡ JOHN HUNTER having succeeded in producing a breed between the dog and the wolf and jackall, hastily concluded that they all belonged to the species, by overlooking the aversion to the intercourse which was exhibited. Phil. Trans. 1787, p. 253. and 1789, p. 160.

3. *The Hybridous Products are barren.*—The peculiar circumstances which are required to bring about a sexual union between individuals of different species, sufficiently account for the total absence of hybridous productions in a wild state. And, as if to preserve even in a domesticated state the introduction and extension of spurious breeds, such hybridous animals, though in many cases disposed to sexual union, are incapable of breeding. There are, indeed, some statements which render it probable that hybrid animals have procreated with perfect ones; at the same time there are few which are above suspicion. Where such occurrences have taken place, and they are unquestionably very rare, the species have been very closely allied in structure and instincts. "If it be true," (says J. HUNTER, with great plausibility,) "that the mule has been known to breed, which must be allowed to be an extraordinary fact, it will by no means be sufficient to determine the Horse and Ass to be of the same species: indeed, from the copulation of mules being very frequent, and the circumstance of their breeding very rare, I should rather attribute it to a degree of monstrosity in the organs of the mule which conceived; not being those of a mixed animal, but those of the mare or female ass. This is not so far fetched an idea, when we consider that some true species produce monsters, which are a mixture of both sexes, and that many animals of distinct sex are incapable of breeding at all." "If, then, we find Nature in its greatest perfection deviating from general principles, why may not it happen likewise in the production of mules, so that sometimes a mule shall breed, from the circumstance of its being a monster respecting mules * ?"

Whenever we observe animals in a free state, and where there is room for choice, engaging *willingly* in the business

* Phil. Trans. 1787, p. 253.

of procreation, and giving birth to a *fertile* progeny, we may with confidence conclude that the individuals so acting belong to one species. When, on the contrary, we observe individuals in a domesticated or confined state, and without the liberty of choice, engaging *reluctantly* in the gratification of the procreative instinct, and giving birth to a *barren* progeny, we may be certain that such belong to different species, and that the union has been unnatural. This circumstance of procreating willingly, and producing a fertile progeny, is the only infallible criterion of species,—the basis of all the methods of arrangement in zoology, and all the truths of anatomy and physiology.

END OF VOLUME FIRST.

PLATE I.

Fig. 1.
Fig. 5.
a
b
a
Fig. 2.
b
a
Fig. 3.
Fig. 4.
Fig. 6.
W H. Lizars Sc
M. F del.t
Published by A. Constable & C.o Edin.r

PLATE II.

M. F del.

Engraved by W. H. Lizars

Published by A Constable & Co Edin.

PLATE III.

Fig. 1.

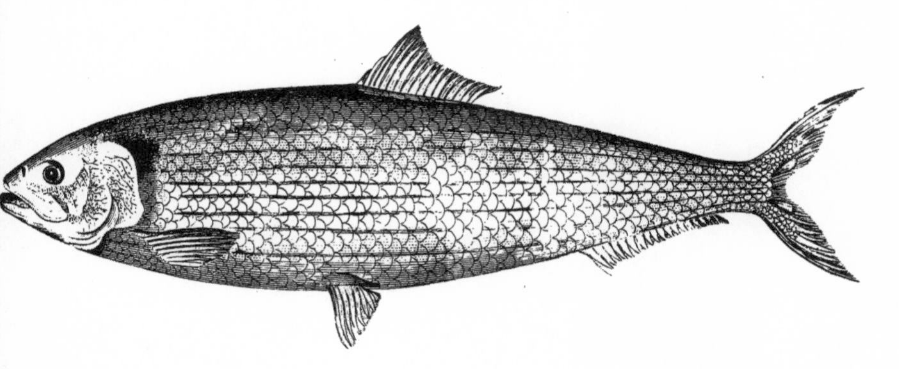

Fig. 2

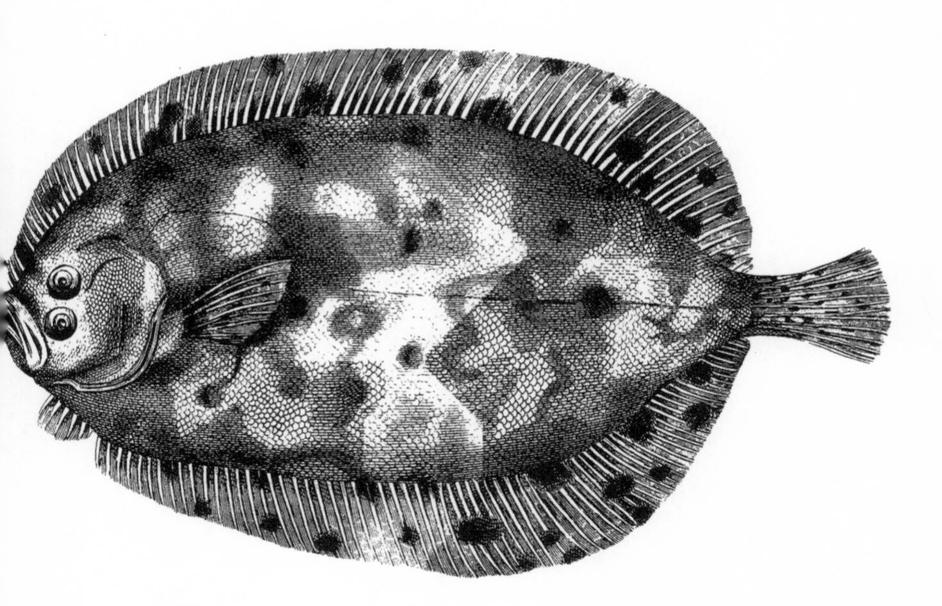

M. F. del.t Engraved by W.H.Lizars

Published by A. Constable & C.o Edin.r

PLATE IV.

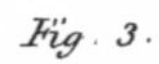

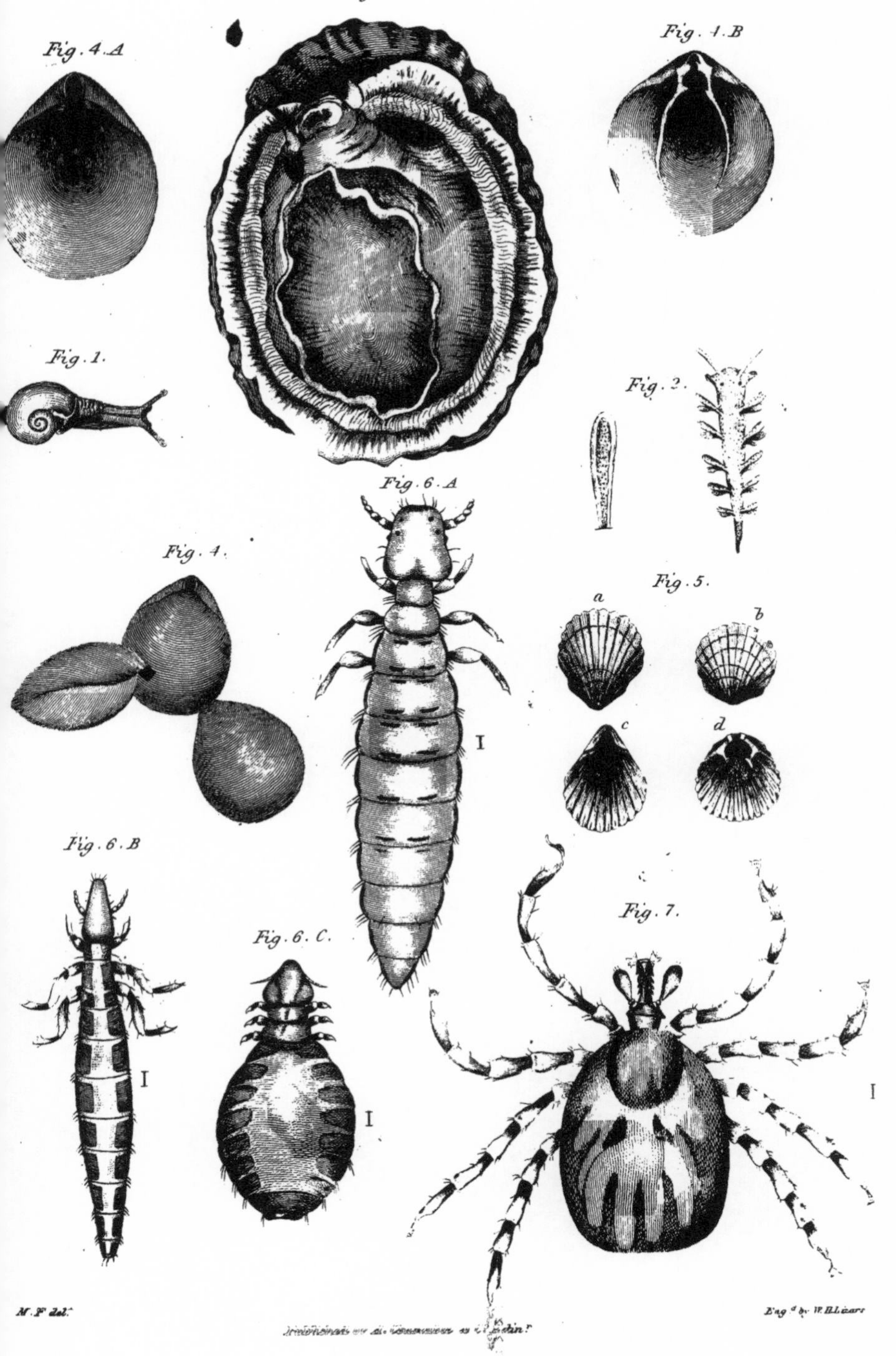
Fig. 3.
Fig. 4. A
Fig. 4. B
Fig. 1.
Fig. 2.
Fig. 6. A
Fig. 4.
Fig. 5.
a
b
c
d
I
Fig. 6. B
Fig. 6. C.
Fig. 7.
I
I
I
M. F. del.
Eng.d by W. H. Lizars

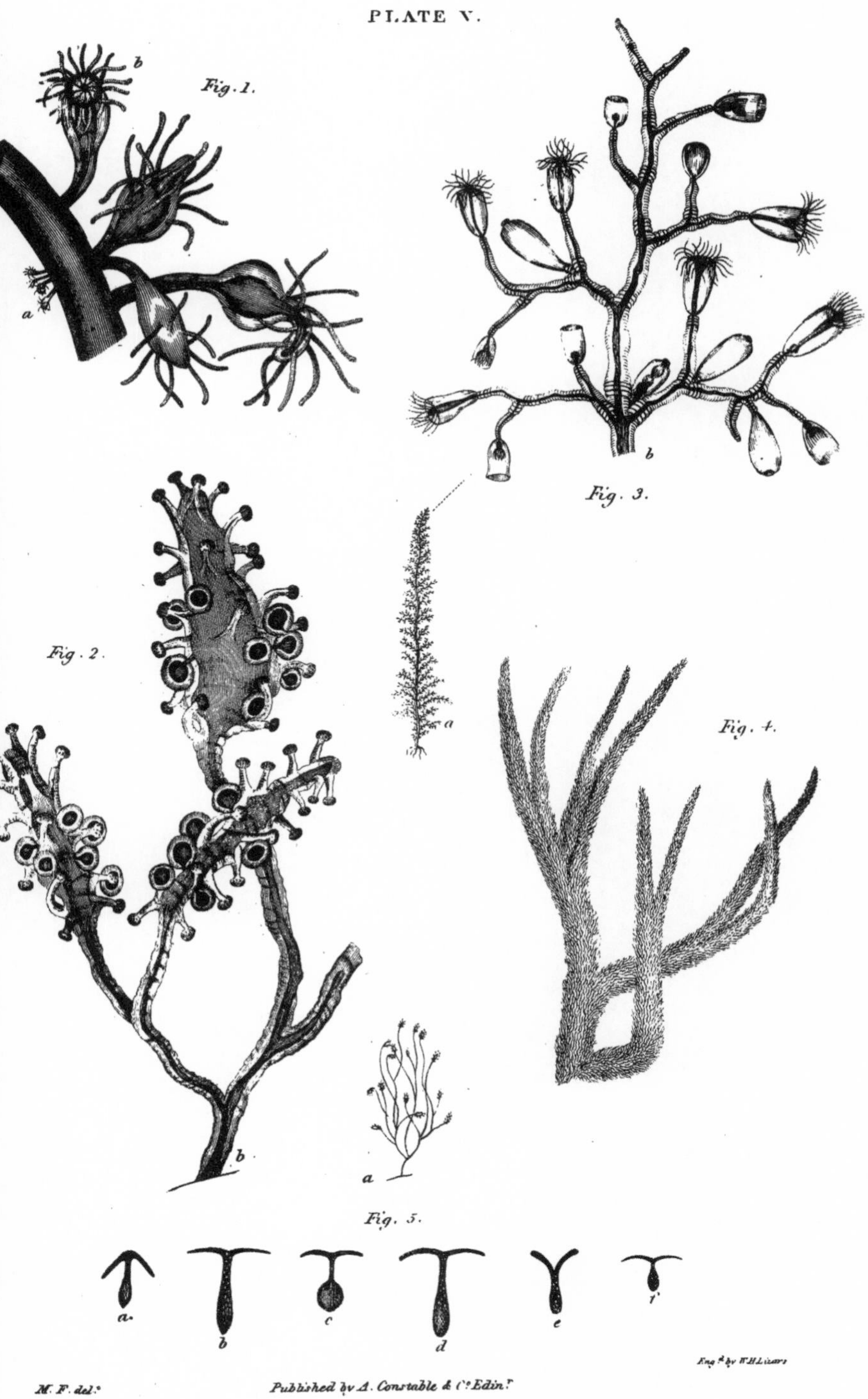

Eng.d by W.H.Lizars

M. F. del.t Published by A. Constable & C.o Edin.r

www.ingramcontent.com/pod-product-compliance
Ingram Content Group UK Ltd.
Pitfield, Milton Keynes, MK11 3LW, UK
UKHW040659180125
453697UK00010B/296